ACCOUNTS RENDERED

OF

WORK DONE AND THINGS SEEN

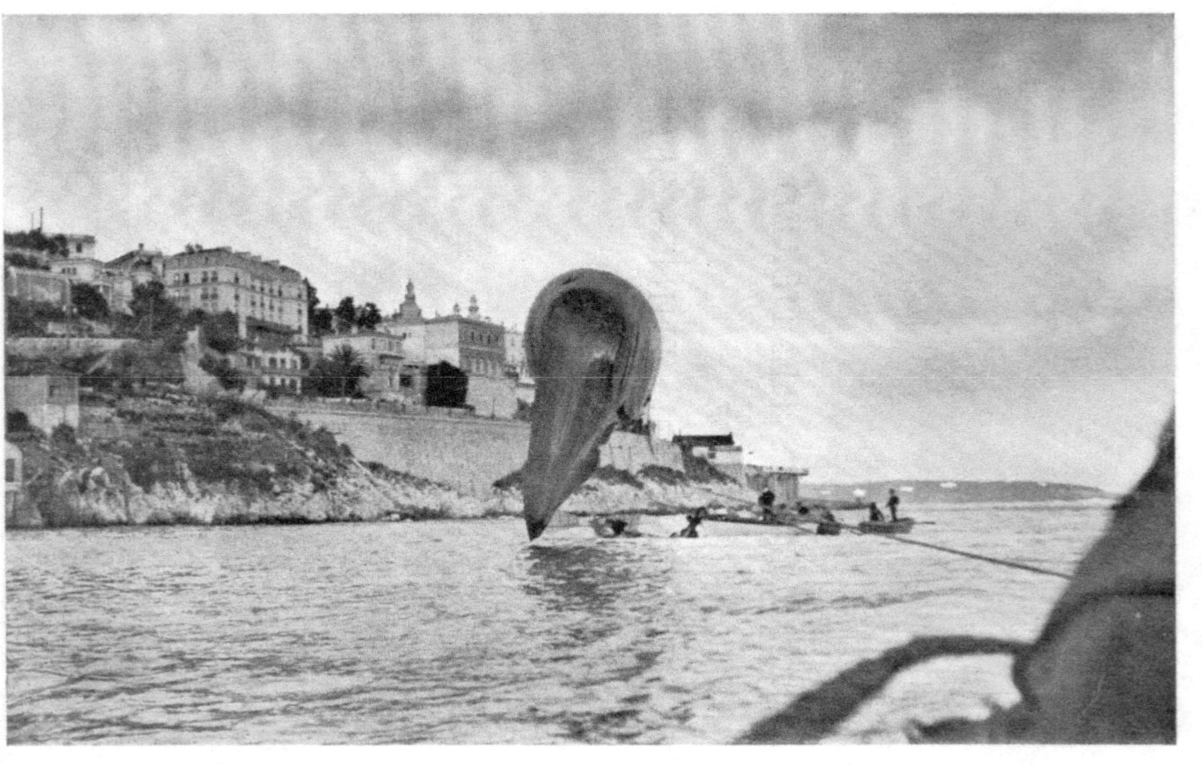

The last moments of Santos Dumont, No. 6, Monaco, 14th February 1902. The ellipsoid has reverted to the fundamental sphere.

ACCOUNTS RENDERED
OF
WORK DONE AND THINGS SEEN

BY

J. Y. BUCHANAN, M.A., F.R.S.

Commandeur de l'Ordre de Saint Charles de Monaco

Vice-President du Comité de Perfectionnement de
l'Institut Océanographique

(Fondation Albert Ier Prince de Monaco)

CAMBRIDGE:
AT THE UNIVERSITY PRESS
1919

CAMBRIDGE
UNIVERSITY PRESS

University Printing House, Cambridge CB2 8BS, United Kingdom

Published in the United States of America by Cambridge University Press, New York

Cambridge University Press is part of the University of Cambridge.

It furthers the University's mission by disseminating knowledge in the pursuit of education, learning and research at the highest international levels of excellence.

www.cambridge.org
Information on this title: www.cambridge.org/9781107683433

First published 1919
First paperback edition 2013

A catalogue record for this publication is available from the British Library

ISBN 978-1-107-68343-3 Paperback

PREFACE

LIKE my last volume, *Comptes Rendus of Observation and Reasoning*, this book contains further "accounts rendered" of work done and things seen at different times, in different places and on different subjects.

The papers are reproduced in their original form; and all notes or comments on the papers themselves, as well as reminiscences regarding the circumstances in which they appeared, are embodied in the contents, and occasionally in postscripts to the papers. It is hoped that this arrangement will be of service to the reader.

J. Y. BUCHANAN.

July 24th, 1919.

CONTENTS

Introduction. The principal theme of this lecture is the influence exercised by the application of steam to land transport in facilitating travel and stimulating trade in old countries, and in opening up new countries. By the colonisation of these the food-supply of the world is increased, with the natural consequence that the population of the world is correspondingly increased. The selection of this theme was due to the fact that the date of my birth was nearly synchronous with that of railways.

I was born in the year 1844, when there existed one or two separate local lines, but no railway system in Great Britain, and on the continent of Europe they were represented only by short lines, which connected one or two of the capitals with the summer residences of their sovereigns, as, for instance, the lines between Paris and Versailles, Berlin and Potsdam, Hanover and Herrenhausen. They had no more importance than local tramways, and did not form part of a railway system. Since then the railways of the world have increased at an ever-increasing rate, and their development has always been for me an interesting study.

The immediate reason for choosing this as the subject of my Inaugural Lecture at Cambridge was that I had just visited the Argentine Republic twice with an interval of two years and a half between the visits, and I had been able personally to witness the extraordinary effect of railway development in peopling and enriching a new country, which, before, had been only sparsely populated.

PAGE

In the end of the year 1885 I went to Buenos Aires and remained there for some months. It was at that date an old Spanish city, consisting of houses of one storey, enclosing patios and gardens exactly as they had been built since the earliest days of the Spanish colonisation.

A few local railways connected the capital with towns in the Province of Buenos Aires. In the early part of 1886 the first railway across the Pampa was completed to the city of Mendoza, at the foot of the Cordillera, and I travelled over it to Mendoza a few days after it was opened.

Having remained some weeks in that city, I crossed the Andes to Chili, as it had been crossed from time immemorial, on mule-back. The ride occupied five days; but no other five days of my life have been so well filled. From start to finish they were crowded with revelations of new experience, new incident, new accident; fatiguing they were, it is true, but worth it all and many times more.

From Valparaiso I returned to England.

In the year 1888 I returned to Buenos Aires and again spent some months there. All was changed. Buenos Aires had ceased to be a Spanish town and was rapidly becoming a European city. Railways were being built in all directions, and an idea of the rate at which this development was proceeding is given in the text.

It was impossible for me not to be struck with the complete metamorphosis of the whole country, or to be deceived in attributing it to the influence exercised by the spread of railways.

It was with this impression fresh in my mind that I wrote the lecture and it is printed here exactly as it was delivered.

Almost all sciences have a geographical significance. This is especially the case with the descriptive sciences, such as Geology, Botany and Zoology. History is nothing if not treated geographically. Geographical environment has an important share in the develop-

PAGE

No. 2. A RETROSPECT OF OCEANOGRAPHY IN THE TWENTY YEARS BEFORE 1895. Address to the Oceanographical Section of the Sixth International Geographical Congress, held in London, 1895. [From the Report of the Sixth International Geographical Congress, held in London, 1895] 28

The meeting of the Geographical Congress in London coincides with the completion of the Reports of the "Challenger" Expedition. Reminiscences of the inception of that expedition and of the men who were most closely identified with it.

The view generally held before the "Challenger" sailed was that the chalk formation was still going on at the bottom of the ocean; whence the doctrine of "the Continuity of the Chalk" 29

Discovery of the ochres of iron and manganese as a submarine formation in February, 1873 . . . 31

The Birth-day of Oceanography. When the "Challenger" sailed from Portsmouth in December, 1872, there was no word in the Dictionary for the department of Geography in which she was to work, and when she returned to Portsmouth in May, 1876, there was a heavy amount of work at the credit of the account of this department, and it had to have a name. It received the name Oceanography. It follows that the science of of Oceanography owes its birth to the "Challenger" Expedition.

On the way from Portsmouth to Teneriffe the work on board consisted in getting everything and everybody into working order. The work of the expedition proper began when the ship sailed from Teneriffe, and the first official station of the expedition was made to the westward of that island on 15th February, 1873. It was not only the first official station of the expedition, but it was the most remarkable.

Everything that came up in the dredge was new; the relation between the result of the preliminary sounding and that of the following dredging was new; and, further, from the picturesque point of view, it was the most striking haul of the dredge or trawl which was made during the whole voyage.

The distinctive feature which determines this homology is that the particular coast shall be, in the general system of the winds of the globe, a *windward coast*; that is, a coast *from* which the normal world-wind blows. Local winds, such as land-breezes and sea-breezes, do not fall within this specification.

Amongst the world-winds the trade-winds take first rank. The resultant mechanical effect of the north-east trade-wind is to drive the surface water in a south-westerly direction, and away from the western coast of the continent situated to windward of it; that of the south-east trade-wind is to drive the surface water in a

PAGE

north-westerly direction, and away from the western coast of the continent which is situated to windward of it.

The translation of water which takes place under the influence of the trade-wind is the mechanical equivalent of the translation of air which has been extinguished. The mass of water moved is very great, but its superficial extension is also very great, consequently the local velocity produced in any part of it is small, but it is continuous.

It has been shown that the water on these coasts has a temperature lower than that corresponding to the local climate, and it was this fact which arrested Humboldt's attention in his travels in Peru; therefore it must be drawn from a source which furnishes water colder than that corresponding to the local climate. Of these there are two, namely, the surface water in high latitudes situated at a great distance, and the sub-surface or abyssmal water contiguous with the coast and with the water actually being removed.

The steamer in which I travelled from Valparaiso as far as Panama, anchored at sixteen different places on the South American Coast between Valparaiso and Cape Blanco and communicated by means of boats with the shore. She always anchored in the cold fringe of water. With the ship thus lying at anchor for an hour or two at each place, I could not detect any current whatever.

If the water were supplied from the immediately adjacent and enormous store of abyssmal water of low temperature it could be furnished by an upward current of practically insensible velocity . . . 107

As a matter of mere probability the likelihood of the cold water having this abyssmal source is so great as to amount to a certainty, and all the experimental evidence confirms it.

The cold inshore water found on the Atlantic Coast of North America is a similar phenomenon in a homologous region. Like that on the Pacific coast of South America, it was attributed to a surface current from Arctic seas and was called the Labrador Current. . 108

Had there been an in-shore surface current along the coast from Newfoundland strong enough to place

PAGE

temperature of ice due to the nature of the medium in which it melts 114

General description of the voyage. It began in mid-winter in the northern hemisphere and finished in mid-summer in the southern hemisphere 115

In equatorial latitudes great uniformity of the temperature both of sea and air 116

Particulars respecting the method of determining the true temperature of the air. Observations begun before sunrise and concluded after sunset, in both cases while there was daylight enough to read the thermometer.

On the many occasions when I have enjoyed the privilege of making meteorological observations at sea I have always made my most important terms the time between dawn and sunrise and that between sunset and darkness. The temperature observed at the former is influenced principally by the integral effect of the nocturnal cooling, and that observed at the latter is influenced principally by the integral effect of the solar heating.

Impossibility, on board ship, of giving a thermometer-box a fixed position which shall secure an exposure such as to justify the assumption that the temperature of the thermometer in it coincides at any moment with the true temperature of the atmosphere outside of it.

Method of observing the wet-bulb thermometer with a film of sea-water supported by capillarity on the chemically clean bulb 117

Effect of replacing the sea-water by a film of fresh water on the bulb of the thermometer 118

When the wet-bulb thermometer is used in this way the continuous film of water resembles the surface of the sea which is exposed to the same atmospheric influences and the effect of these on the bulb of the thermometer resembles that produced on the water immediately below the surface.

Observations were made only during day-light. Description of the table of observations. With two exceptions the temperature of the sea was found to be higher than that of the air. Great heat in the northerly monsoon off the Brazilian coast. The least difference between the readings of the wet and dry thermometers

My frequent visits to the River Plate have afforded me the opportunity of remarking the astonishing development of the countries bordering on it and especially of the city of Buenos Aires in the course of less than forty years.

On my arrival in the estuary in 1885 the s.s. "Leibnitz" in which I travelled, anchored in the roads about twelve miles below the city of Buenos Aires, as had been the custom of sailing ships, and disembarked her passengers into a steam launch which proceeded to a landing in front of the Custom House; but, the draft of the launch being too great, the passengers were taken off in carts with very high wheels and landed on the beach. When I revisited Buenos Aires in the year 1888 the steamer still anchored in the roads, but the steam launch which took the passengers landed them on a jetty at the Custom House. On my visit in the year 1899 the steamer went into the port of the city of La Plata and the passengers proceeded to Buenos Aires by rail. Finally, in my last visit, in the year 1907, the steamer entered the Madero Port and landed its passengers in the city of Buenos Aires itself.

These chronological data have a geographical interest as marking stages in the rapid advance of Buenos Aires from the status of an old Spanish Colonial Residence to that of a first-class American city of the twentieth century.

There are two typical general statements, both of them negative, which are made by travellers by sea, and very often by professional seamen, namely, (*a*) the apparent blue colour of the sea is not the true colour of the sea; and (*b*) the flight of the flying-fish is not due to the use of its wings.

PAGE

When heated they give off alkaline water with an empyreumatic odour.

The number of different elements contained in them is large.

Table of the results of the analyses of six different samples 159

Nodules from different localities differ in composition but those from the same locality have similar composition.

The Manganese is chiefly present as MnO_2.

Of the rarer metals Cobalt is the principal representative.

No. 8. MANGANESE NODULES IN LOCH FYNE. [From *Nature*, Oct. 10, 1878, Vol. XVIII. p. 628] . . . 160

It was believed by every member of the "Challenger" expedition to have been established that the Manganese nodules are to be found only in the greatest depths of the ocean and to be absent in shallow water.

When I proceeded in the summer of 1878 to explore the waters round the coast of Scotland in the steam-yacht "Mallard," of 95 tons, it was without any expectation of finding evidence to upset this conviction; yet it so happened that in the very first station which I made, in the deepest part of Loch Fyne, the anchor which I used brought up a mass of mud from a depth of 104 fathoms in which Manganese nodules were present in greater relative abundance than in any mud brought from 2000 or 3000 fathoms in the open ocean. Moreover, the anchor with this valuable load was brought up at the end of a fine steel wire cord. Consequently on 21st September, 1878, the "Mallard" scored a double record, namely, the discovery of Manganese nodules in shallow water and the first use of metallic cables in oceanographical work on the eastern side of the Atlantic. Agassiz had used a metal cable in 1877 in American waters.

My cable was a cord of one-tenth inch diameter consisting of thirty-five fine steel wires in strands. It was manufactured for me by Messrs Siemens Brothers, and I used it for four seasons with uniform success. At the end of the last season it was called upon to lift a heavily loaded anchor-dredge from deep water and it parted. The cable and the dredge, both of which had served me

PAGE

well, lie in the channel north-east of the island of Rum, where there is a sounding of 147 fathoms marked on the Admiralty chart of that date.

Physical analysis of a sample of the mud showed 30 per cent. of nodules 160

Average specific gravity of the nodules 2·46 . . 161

Like the oceanic nodules those from Loch Fyne are agglomerations of the local mud united by an ochreous cement in which Manganese peroxide predominates.

The Loch Fyne nodules differ from those found in the ocean depths inasmuch as their kernels are softer than their outer shells, the reverse being generally found in the oceanic nodules.

The abundance of nodules obtained from the depth of 104 fathoms in Loch Fyne proved to be local . . 162

Later experience in the "Mallard" showed that Manganese ochre is very widely distributed among the littoral deposits of Scotland, but it was found nowhere in such abundance as in the deepest basin of Loch Fyne.

No. 9. On the Composition of Oceanic and Littoral Nodules. [From *Trans. Roy. Soc. Edin.* Vol. xxxvi. Part 2, 1891, p. 459] 163

In this paper the results of analytical work on Manganese nodules made at various dates are collected and discussed.

Samples I, II and III were from a very successful dredging in the Antarctic Ocean, about 400 miles south-west of the coast of Australia.

Data regarding the temperature, the density and the gaseous contents of the water in which they were found.

Physical description of the nodules 164

Amongst them were two ear-bones of whales and one shark's tooth, both covered with the black incrustation; also some fragments of granite, probably conveyed by icebergs from the Antarctic continent.

Samples IV and V were parts of the same nodule brought from a depth of 2740 fathoms in the middle of the North Pacific Ocean 165

Physical properties of the water in which they were found. It was much poorer in dissolved oxygen than was the bottom water of the Antarctic Ocean.

On the charts of the British Admiralty the decimal system of subdivision has always been used[1]. The fundamental unit is the nautical mile and it is subdivided into ten cable-lengths, and each "cable" has a length of 100 fathoms, whence the mile equals 1000 fathoms. Marine mensuration is thus made as simple as possible, whether it be a question of lengths, areas or volumes. This is put in evidence by the discussion in the text of the dimensions of Loch Fyne where the discovery of the nodules in shallow water was made.

[1] For the full discussion of this subject see the author's work, *Comptes Rendus of Observation and Reasoning*, page 424.

PAGE

No. 10. On a Method of Determining the Specific Gravity of Soluble Salts by Displacement in their own Mother-Liquor; and its Application in the case of the Alkaline Halides. [From *The American Journal of Science*, Vol. xxi. January, 1906] . . . 197

This research was suggested by the meteorological conditions of the summer of the year 1904. The great anticyclone which persisted over northern Europe during the months of July and August of that year produced tropical conditions which were evidenced alike by the high temperature of the air and by its insignificant diurnal variation. My hydrometric observations which demand a constant temperature of 19·5° C. were put a stop to by these conditions.

The very narrow limits of the diurnal variation of temperature suggested the realization of a long contemplated method of determination of specific gravity, namely, that of soluble salts by displacement in their own mother-liquor. I took it up at first merely as a *tour de force* in experimentation with which to occupy myself during the hot weather, but it turned out to be a valuable method of research, and the duration of the spell of hot weather enabled me to prove and to use it.

Fundamental principle of the method.

There is one liquid in which every soluble salt is quite insoluble, and that is its own mother-liquor at the temperature at which the one parted from the other . 198

By immersing the salt in its own mother-liquor at that temperature and by making the maintenance of this temperature a *conditio sine quâ non* of every manipulation during which the two are brought together again, errors due to uncertain solubility are eliminated, and contamination of valuable preparations is avoided.

It is, therefore, by the immersion of each salt in its own mother-liquor that its displacement is determined, and this, combined with the weight of the salt and the specific gravity of the mother-liquor, gives the specific gravity of the salt.

The chemist who proposes to use this method must clearly realize the experimental difficulties which he will encounter and must take such measures as his experience dictates to overcome them. Precepts will not help him.

The reason why it is not permissible for the temperature of the mixture, during adjustment, to fall ever so little below the temperature of equilibrium, while a rise of $0{\cdot}1°$ above it can be tolerated is the following. When the salt is immersed in its mother-liquor at the temperature of the system when crystallisation was complete, T, the equilibrium between crystals and solution is stable. If the temperature falls by, say, $0{\cdot}1°$ below that of equilibrium the solution, in the absence of crystals, would remain unaltered; but, in presence of crystals in abundance, the state of supersaturation is impossible and equilibrium at the temperature $T - 0{\cdot}1°$ is quickly established. If the temperature of the mixture be now raised to T the solution, though not quite saturated, is very nearly so, and the crystals of the salt are so nearly insoluble in it that the re-establishment of equilibrium at T could not be assumed to have taken place except after an exceedingly long interval of time.

A distinction must be made between the insolubility of a salt in, for instance, a hydro-carbon which, when saturated, contains only a fraction of one per cent. of the salt, and that of the same salt in a solution of itself which is so nearly saturated as to be able to take up no more than the same percentage of the salt which would completely saturate the same volume of hydro-carbon.

PAGE

When the salt is brought together with the hydrocarbon the percentage which is soluble in it is dissolved very quickly: there is nothing already in the hydrocarbon to interfere with it. When the *menstruum* is a nearly saturated solution of the salt in water the dissolution of the salt is blocked by finding the available space crowded almost to capacity by molecules of itself.

Description of Table II 204

Table II 206

This Table illustrates the relation between the displacement of the salt in crystal and its apparent displacement in saturated solution.

Table III gives the molecular weights of the salts . 207

Table IV gives the specific gravity of each salt of the ennead. It is shown that the specific gravity of the salts is a periodic function of the molecular weight within the ennead 209

Table V gives similar data regarding the density of the mother-liquors 210

Discussion of these tables 211

Consideration of saturated solutions as products of substitution 213

As a result of these investigations it is shown that the crystallisation of the Potassium and Rubidium salts of the ennead must be hindered by increase of pressure while that of the Caesium salts must be helped by the same agency 214

It will be noticed that this paper though read before the Chemical Society of London was not published by it. No exception was taken to the paper at the meeting and I was surprised when I learned that it had been decided to refuse publication of it. On my asking for an explanation I received nothing positive except the conviction that the Council was determined that the paper should not be published. I gathered, however, from what passed that it was felt that the paper did not quite come up to the pitch of standardization set for researches destined for publication in the *Transactions* of the Society.

PAGE

No. 11. THE MEDITERRANEAN SEA 216

This paper appeared in the ninth edition of the *Encyclopaedia Britannica.*

When I was asked by the editor, my old friend, the late Professor Robertson Smith, to write it, he told me that the Historical Geography of the Mediterranean had been amply dealt with in other articles, inasmuch as it was synonymous with the Geographical History of Civilization. What was wanted was the Hydrographic Geography of the Mediterranean and as he was familiar with all my work in the "Challenger" and other ships he was sure that I could furnish it.

As the work of preparing the paper promised to be congenial and the compliment of being asked to do so was great, I accepted his proposal at once.

Location and Delimitation of the Mediterranean Sea 216

It separates Europe from Africa.

It has in round numbers a length of 2100 nautical miles and a superficial area of one million square miles.

The word mile, in this paper, always means nautical mile, of which sixty go to a degree of the great circle of the terrestrial sphere.

It is divided into two principal basins which are separated by the island Sicily and the submarine plateau which connects that island with the continent of Africa. The greatest depth of water on this plateau is under 200 fathoms.

The maximum depth of the Mediterranean Sea is little over 2000 fathoms and is nearly alike in the two basins 217

Distribution of Shoals and Islands 218

Graham's Island is a singular case. It appeared in July, 1831, and reached its maximum size at the end of August; then gradually dwindled and by the end of the year had disappeared below the surface of the sea.

It affords a terrestrial parallel with the *Novae* in the firmament 219

Description of the Adriatic 220

The northern part shallow; in the middle a deep basin with maximum depth 765 fathoms is shut off from the outer Mediterranean by a ridge covered by water having a depth less than 400 fathoms.

PAGE

No. 13. On the Oxidation of Ferrous Salts. [From *Proc. Roy. Soc. Edin.* 1881, Vol. xi. p. 191] . . . 243

This paper is a study in chemical dynamics and it was suggested by my earlier work in that field, namely, the investigation of the rate of decomposition of Chloracetic Acid at 100° C. in the systems—$C_2H_3ClO_2$ + 164 H_2O—$C_2H_2ClO_2Na$ + 159 H_2O and $C_2H_2ClO_2Na$ + NaHO + 159 H_2O. These investigations were made in the Edinburgh University Laboratory in the year 1871 and were published in the *Berichte* of the German Chemical Society (1871) iv. pp. 340, 863.

It is of great importance, in the study of deep sea muds, to determine both the ferrous and the ferric iron present in them. The most convenient solvent of substances of this kind is moderately strong hydrochloric acid, and the most convenient reagent for the estimation of ferrous salt in a solution is Potassium Permanganate. But, according to the orthodox canon, potassium permanganate must not be used in hydrochloric acid solution. I was quite aware of the existence of the canon, but I was not aware of any experimental evidence which supported it. Accordingly, my first task was, not to look up the history of the dogma, but to find out for myself the circumstances under which the presence of hydrochloric acid interferes with the quantitative oxidising action of permanganate, and how this disability is to be removed. The solution of this problem is given on page 244.

Details of Ferrous solutions used 243

Other oxidising agents used besides Permanganate, namely, Chlorate, Perchlorate and Nitrate of Potassium 244

Table I shows that, using proper precautions, Hydrochloric Acid does not affect adversely the oxidation by Permanganate 245

Table II. Action of Chlorate and Perchlorate of Potassium at the ordinary room-temperature . . 246

Table III. Rate of oxidation by Chlorate of Potassium at room-temperature. In the same conditions the Perchlorate and the Nitrate of Potassium are inert . 247

Table IV. Rate of reduction of Potassium Chlorate by Ferrous solution at ordinary temperatures in durations up to 200 minutes 248

At the invitation of my old friend, the late Robert Kaye Gray, then the manager of the Cable Department of the India Rubber, Gutta Percha and Telegraph Works Co. (Limited) of Silvertown, I joined him on the "Dacia" which was commissioned to survey the route and to lay the cable to connect the Canary Islands with Spain.

It was while the cable staff were accomplishing the difficult operation of laying the shore-end to connect up the island of La Palma with the deep-sea cable that I was able to make the observations on the changing visibility of the Peak of Teneriffe, seventy miles distant, before and after sunrise. This was also an admirable station from which to study the sunrises, and it will be seen in the paper that they exhibited no unusual feature on the ten consecutive mornings on which I observed them from that station.

The observation of the alternating visibility and eclipse of the Peak of Teneriffe every morning furnishes an excellent example on a large scale of what is well known to every pilot. When he is steering a vessel through a fog after dark, nothing handicaps his efforts so much as the lights which his ship is obliged to carry, and the more powerful they are the more difficult becomes his task.

In a weak diffused light the fog particles are nearly invisible and affect the visibility of objects behind it only by their mass; when, however, these particles are illuminated by a strong direct light, the light which they reflect may be relatively so intense, even in the case of the very thin haze of a depth of seventy miles above cited, as to dazzle the eye and prevent it seeing

PAGE

through it objects which would be clearly perceptible under less powerful illumination.

Sunset of 27 October, 1883, behind the Peak of Teneriffe, distant 115 miles 250

Remarkable effect of reflection by waves.

Sunsets less brilliant on land than at sea . . 251

The sunrises presented nò unusual feature.

Variable visibility of the Peak of Teneriffe at a distance of seventy miles before and after sunrise.

Analogous phenomenon well known to Pilots.

Difficulty in accepting the usual explanation of these remarkable sunsets 252

No. 15. SUR LA DENSITÉ ET L'ALCALINITÉ DES EAUX DE L'ATLANTIQUE ET DE LA MÉDITERRANÉE. [From *Comptes Rendus*, 1893, Vol. CXVI. p. 1321] . . 254

This is an account of a series of simultaneous determinations of the density and the alkalinity of samples of sea-water taken on a voyage from Dartmouth to Genoa, in the months of August and September, 1892, in the steam-yacht "Princesse Alice," which had just been built on the Thames for His Serene Highness Prince Albert I of Monaco.

The yacht was fitted with all the engines and apparatus for exploring and investigating even the deepest seas of the ocean.

The Prince's long experience in this class of work, carried on in the "Hirondelle," a schooner of only 200 tons, made him, without question, the highest authority on the practical details of deep sea investigation. The more primitive fittings of the "Hirondelle," on which everything was done by hand, were replaced by steam winches and wire lines and cables, and every novelty and invention which his experience suggested.

His Highness did me the great honour to invite me to accompany him and to work with him on board during the opening cruise to Genoa where I had the privilege of assisting as a guest of His Highness at the magnificent fêtes given by that city in honour of the four-hundredth anniversary of the Discovery of America by the most famous of her sons, Christopher Columbus.

Description of Apparatus and of experimental Method 254

PAGE

Simony's comparison of the bottom temperature of the G'munder See with the winter temperature of the air 296

My observations in the Lochs of the Caledonian Canal in Scotland 297

The bottom temperature of even the deepest of them varies from year to year.

His observations covered the two very cold winters 1878–79 and 1880–81.

Up to this date it was universally believed and taught that water below the ice of a frozen lake must necessarily have a uniform temperature of 4° C. or 39·2° Fahr. This was held to follow as a corollary of the demonstration that the density of pure water has its maximum value at that temperature. In the first of these winters, 1878–79, and on the first day that I sank my thermometer below the ice of Loch Lomond in the region between Balloch and Luss, I found all the water from the ice to the bottom to have a temperature below 34° Fahr. and thus proved the corollary to be a fallacy 300

Effect of wind in circulating the water of a Lake.
Observations in other frozen Lakes 302
Changes of level and seiches 303
Remarkable seiche in Loch Tay 305
Changes of level due to meteorological causes . 307
Lake Tanganyika a remarkable example.
Great interest of Central African lakes 308
Lakes without outlet 309
Table of data regarding some of these 310

No. 21. On the Distribution of Temperature in Loch Lomond during the Autumn of 1885. [From *Proc. Roy. Soc. Edin.*, 1885, Vol. XIII. p. 403] . . 312

The observations reported in this paper were made during a summer residence on the banks of Loch Lomond, and the object of them was to obtain information regarding the relation which exists between the gain and loss of heat of the water of the lake and the change of season.

Loch Lomond is naturally divided into three basins,

Introduction. The regularity of the form assumed by the beds of all streams, and especially by those of large rivers flowing through plains consisting of alluvial or other matter easily displaced by any current of water, even the most insignificant, had always appeared to me to be a fact in nature which expressed a law which had not been stated.

It was not until the meeting-place of the British Association in 1892 was fixed in Edinburgh that I began to put together my observations and reflections on this subject, with a view to laying them before my scientific brethren who would be there assembled.

But on such an occasion there is rarely time for the reading of extended papers; therefore I devoted myself principally to preparing maps and diagrams to fully illustrate the facts observed and the conclusions arrived at. By doing so I hoped to be able at the meeting to use to the best advantage whatever time might be available for the exposition of their meaning.

By taking this precaution I was able to make clear to the very representative audience there assembled the interpretation of the writing on the soil of the earth

PAGE

Appearance of brass tube after the experiment . 351

The conditions of the experiment with the copper sphere were different and the destructive effect of the collapse of the internal glass sphere was less serious 352

Appearance of the copper sphere on returning from a depth of 6000 metres 353

Discussion of the experiment in the tube . . 354

Discussion of the experiment in the sphere . . 355

Analogy with earthquakes 356

Resemblance to the case of a gun bursting when fired with even the lightest plug of snow in the muzzle 357

Importance of time in physical considerations.

No. 25. PRELIMINARY NOTE ON THE COMPRESSIBILITY OF GLASS. [From *Trans. Roy. Soc. Edin.* 1880, Vol. XXIX. p. 589] 358

The instrument described in this paper with which the absolute compressibility of glass was determined was designed by myself and constructed by Messrs Milne of Milton House, Edinburgh. It was an elaboration of the principle of the instrument which I constructed and used for the same purpose on board the "Challenger," using a column of sea-water both as the means and the measure of the pressure.

Description of the instrument 358

The substance of which the compressibility is sought must be in the form of rod or wire.

It lies in a metal tube with glass ends through which the two extremities of the rod or wire can be observed and exactly fixed by micrometers.

When pressure is raised in the envelope it reduces the length of the rod or wire by an amount which depends on its compressibility.

The displacements of the ends caused by the pressure are measured with the micrometers and the shortening of the rod is thus directly determined. The pressure being known the linear compressibility of the material is obtained and from it the cubical compressibility, in the ordinary way, both compressibilities being absolute.

Method of using the instrument 359

Analysis of glass rod the compressibility of which was determined.

PAGE

the rate of 14·74 feet per minute. Consequently she has the power to prevent her sinking at all.

The "Vanguard," which was rammed by the "Iron Duke" took 72 minutes to sink.

By making the main deck of a ship, to which the water-tight bulkheads are carried up, air-tight, she would be proof against sinking from damage to her bottom 407

Dealing with leaks in this way is equivalent to transferring the leak from the ship's bottom to her deck 408

A ship fitted with air-tight subdivision is thereby also protected against destruction by fire 409

The principle of air-tight subdivision is that of the very old nautical device of filling the damaged hold of the vessel with empty casks 410

Compressed air every day occupying a wider field as a means of transmitting power.

No. 30. THE NORTHALLERTON ACCIDENT, 1894. [From *The Engineer*, 1894, Vol. LXXVIII. p. 382] 412

This accident, happening as it did at the season when the south-going trains from Scotland are most crowded, and almost on the anniversary of a more serious accident to the same train on the same district in the previous year, naturally attracted considerable public attention.

Since the adoption of the continuous air-brake by the British railways I had noticed, with surprise and some anxiety, that, though the power of this brake was produced on the engine, its usefulness was confined to the coaches.

In the time which necessarily elapsed before the passengers could be taken up and sent on by relief trains, I was able to study the accident carefully, and I arrived definitely at the conclusion that, had the two engines which hauled the train been supplied with brake power equivalent, mass for mass, to that supplied to the coaches hauled, the accident would have been prevented.

This was the more certain because, in the same instant that I heard and recognised the danger blast, the brakes on the whole train were put hard down by

PAGE

the unfortunate engineers, who were absolutely on the alert, and used all the means with which they were supplied for avoiding collision.

My experience in a car in the rear of the train . . 412
The condition of the train after the accident . . 413
The rear half of the train undamaged being adequately braked.
Ideal distribution of brake-power in a train.
The accident was due to the absence of sufficient brake-power on the engines 414
The postscript contains matter of less technical kind which formed part of my letter to *The Times* of 5 October, 1894 . . . , ... 415

No. 31. THE WRECK OF "SANTOS DUMONT NO. 6" AT MONACO, 14 FEBRUARY, 1902. [From *The Times*, February 21, 1902] 416
M. Santos Dumont's ascent and accident on 14th February, 1902, marks an epoch in the history of the development of the air-ship 416
Owing to the kindness of H.S.H. the Prince of Monaco, I was able to observe and follow the flight of the balloon from start to finish, in the steam launch of the yacht "Princesse Alice" and to get an instructive series of photographs of it.
Previous excursion of the balloon over the waters round Monaco 417
Ease and certainty with which the air-ship was taken into and out of its hangar.
Speed attained by the air-ship over 15 knots per hour.
On February 14th M. Dumont started about 2.30 P.M. 418
Plate I, Fig. 1. The Balloon leaving the hangar.
Fig. 2. The Balloon passing overhead, the guide-rope trailing in the sea.
Dangerous pitching began immediately . . . 419
Plate II, Fig. 3. The Balloon pitching: critical situation.
When threatening to turn over, gas was released and Balloon buckled.
Fig. 4. Balloon buckling on the escape of gas.
Means used by the crew of the "Princesse Alice" to render assistance 420

No. 1.

GEOGRAPHY

IN ITS PHYSICAL AND ECONOMIC RELATIONS

INAUGURAL LECTURE[1]

AS UNIVERSITY READER IN GEOGRAPHY
DELIVERED AT CAMBRIDGE IN OCTOBER, 1889

THE word Geography means literally the description of the Earth—not only the delineation of the form of its continents and seas, its rivers and mountains, but their physical conditions, climates and products, and their appropriation by plants and animals, and by communities of men. In fact it implies knowledge of the World as well as of the Earth. The subject is therefore one of the widest extent, and cannot possibly be compassed in one series of lectures. Even the smallest portion of the earth such as a country, a town, a parish, or a street, presents so many aspects, that if a description of any one of them were asked, it would be necessary to carefully specify the point of view from which it was to be regarded. One person might wish to have his district described orographically with reference to its hills and valleys, streams and lakes, roads and bridges, canals and railways. Another with regard to its agricultural, manufacturing, mining or trading capacities; a third with regard to its population and its social, political and religious tendency; a fourth with regard to the history of its population, buildings and institutions; a fifth with regard to its geological structure; a sixth with regard to its botany and so forth. With so many different ways of treating a single district, it is evident that, when we have the whole world to describe, it will be necessary to limit our view by some definite and circumscribed horizon.

Most sciences have reference to phenomena taking place on the surface of the globe and have necessarily a geographical significance, and are in the natural course of things treated

[1] For Introduction see Contents, p. vii.

more or less geographically. This is especially the case with regard to the descriptive sciences or rather the descriptive departments of these sciences, such as Geology, Botany and Zoology. It would be absurd to attempt to repeat in a necessarily perfunctory manner in a course of lectures on geography what has already been thoroughly done in the special lectures on these subjects. It will, however, be necessary to make occasional use of the teachings of these sciences in elucidating the form and development of our globe and the vital activity upon it.

Again, we hear of Historical Geography, but history is nothing if not treated geographically, though its treatment will differ somewhat, according as it is approached from the moral or the physical side. The historian treats the earth as the platform on which he parades his men and he has a tendency to ascribe to their physical and mental superiority all their successes whether in the barbarous contests of man against man, clan against clan and nation against nation, or in the more elevated, but not less keen, contests in the domains of science and art, of industry and commerce. The geographer has to endeavour to restore the balance, by reclaiming for Mother Earth her share in the shaping of the destinies, not of the human race only but of all her offspring whether vegetable, animal or indeed mineral.

The portion of the earth which we inhabit is the solid crust, about three-fourths of which is covered by water, and the whole surface, land and water, is overlaid by a gaseous envelope. Roughly speaking therefore the globe as we see it consists of air, earth and water. If we include the further element of *fire*, and understand by it the energy which is continually supplied by the radiation of the sun, and by virtue of which all the changes and vital phenomena are produced and sustained, we have, under the ancient headings, *fire*, *air*. *earth* and *water*, a rough classification of the elements of nature

. Owing to the fostering influence of the sun's heat there is a constant interchange of matter going on between the sea and the land through the medium of the atmosphere. Water is removed from the sea and is absorbed as vapour by the atmosphere, the equilibrium of which is thereby disturbed, producing winds. The air charged with moisture is liable to lose it again when its temperature is lowered, much of it falling back into the sea, but much of it also falling on the land, especially the mountainous regions. This again destroys equilibrium, producing other motions of the atmosphere. None of the mineral constituents of the earth are proof against the action of rain, especially when exposed at the same time to a hot sun; the result is chemical decomposition of the mineral matter, the transference of some portion of it to the water by solution and the removal of some of it by the mechanical action of the water during its return to the sea, and much of it remains *in situ* in an altered and disintegrated state.

All the elements of physical geography are continually acting and re-acting on one another, each altering and modifying the other, and producing, even in the inorganic world, a series of changes which may fairly be called a kind of life. The Land and Water of the globe possess besides, each for itself, a highly developed organic life which in its turn acts and re-acts on the inorganic world.

The atmosphere, the real external shell of our planet, has from the earliest times been the object of observation and speculation by all peoples who by experience and practice have learned to connect atmospheric phenomena with the production of the thing called weather, which exercises so important an influence on the physical condition of the inhabitants. But until quite recently it was only in its lowest layers, those immediately lying upon the land or water surface of the globe, that it had been in any way thoroughly investigated. It is only within the last few years that the upper strata of the atmosphere have been taken more notice of, especially in regard to the upper air currents and their importance in weather forecasting and generally in the climatology of the earth. High level observatories also are being multiplied in all countries and are doing for the heights of the atmosphere what recent ocean exploring expeditions have done for the depths of the sea.

The general laws governing the weather at the surface of the globe depend chiefly on the actions and reactions between the atmosphere and the surface of the ocean. Hence the investigation of the ocean has a double value both on account of the interest which attaches to itself and on account of its influence on the land. The sea has, so far as its surface is concerned, and in parts contiguous to the coast, been at all times an object of observation by the inhabitants who happened to dwell in its vicinity. What, however, lay beneath the surface of the ocean was until the last twenty or thirty years entirely unknown.

Oceanic Exploration. The foundation of our knowledge of the conditions affecting the ocean and its surface and of the department of maritime meteorology is due to the late Captain Maury who united remarkable powers of observation and generalisation with the imagination and enthusiasm without which great things cannot be accomplished. Through Maury's efforts the U.S. Government issued an invitation for a maritime conference which was held in Brussels in 1853 and attended by representatives of the governments of most of the European States. The main object of the conference—to devise a uniform system of meteorological observations and records—was accomplished. According to the agreement, ship's logs were to have columns for recording observations of the following subjects: latitude, longitude, magnetic variation, direction and velocity of currents, direction and force of wind, serenity of the sky, fog, rain, snow and hail, state of the sea, sp. gravity and temperature of the water at the surface and at different depths. It was also proposed that deep sea soundings should be taken on all favourable occasions and that all other phenomena such as hurricanes, water spouts, dust showers, etc., should be carefully described and tidal observations made when practicable.

The practical results of this conference were great. The systematic and uniform collection of data by men of all nations is going on uninterruptedly to-day and is furnishing the means for the solution of many of the problems relating to the geography of the sea.

Maury's enthusiasm in the cause is shown by the terms in which he refers to the results of this conference which was assembled by his exertions. He says: "Rarely has there been such a sublime spectacle presented to the scientific world: all nations agreeing to unite and co-operate in carrying out according to the same plan one system of philosophical research with regard to the sea. Though they may be enemies in all else, here they are to be friends. Every ship that navigates the high seas with these charts and blank abstract logs on board may henceforth be regarded as a floating observatory, a temple of science."

An epoch in the progress of this science is marked by the appearance of Maury's wind and current charts, his physical geography of the sea, and his sailing directions, which contain the record of the first deep soundings taken by the ships of the United States; and to the United States, through Maury's efforts, belongs the honour of having inaugurated the first regular cruise for the purpose of sounding in great depths. Under his instructions the U.S. brig "Dolphin," commanded by Lieut. Lee and subsequently by Lieut. Berryman, was detailed in 1851–53 to search for reported dangers in the Atlantic and *to sound regularly* at intervals of 200 miles going and returning. The "Dolphin" was provided with Brooke's sounding apparatus, and with it succeeded in obtaining specimens of the bottom from depths of 2000 fathoms. The practicability of the work was thus fully demonstrated and although some of the earlier results through defective appliances and lack of experience were not entirely trustworthy, it is impossible to deny to the Americans as a nation and to Maury as an individual the credit of having been the first to undertake systematic deep sea exploration.

The civil war naturally put a stop to these operations by United States' ships, but the work was quickly taken up by other governments, and we find from that time to the present the records of a large number of expeditions for diverse scientific observations in all parts of the world.

With the extension of telegraphic enterprise the investigation of the depth and configuration of the ocean bed became of vital importance; and the work of sounding for that purpose was taken up with activity, and continues to be prosecuted with diligence, revealing much which could neither have been suspected nor looked for, but for the necessity on commercial grounds of having as complete a survey as possible of the bed on which a cable is to lie before placing it there. Continued improvements in the appliances and instruments have made the results of sounding more precise than was possible in earlier times, and as the data accumulate the bathymetric charts of oceans are becoming more accurate. Submarine eminences are discovered by the act of measuring their height; terrestrial eminences are discovered first and their height measured afterwards: hence it will be a long time before we can map the bottom of the ocean in the same detail as we can the surface of the land.

The civil war in America was doubly disadvantageous to the cause of deep sea investigation. It of course prevented their ships being employed in this service; and, as Maury took the side of the South, he was never employed again and died in retirement in the year 1873. This was one of the many deplorable but unavoidable accidents of the war; but even if it had been otherwise, the exhaustion following on the long war would have allowed little scope for the active prosecution of such investigations. He died, however, with the satisfaction of having established the meteorology of the ocean at its surface on a solid basis and of having given the starting impetus to deep sea investigation, the far-reaching results of which were already becoming every day more clear.

Submarine research labours under a certain disadvantage as compared with subterranean research. In the latter wherever the exploring instrument penetrates there the explorer himself can accompany it. This is not the case in submarine exploration. The utmost depth at which a man has been able to work is 25 fathoms and was accomplished a few years ago by a diver employed to recover specie from a steamer wrecked on the steep volcanic coast of the Grand Canary. The pure blue ocean water which washes the shores of the island possesses great transparency, so that there was no obstacle to the salvage operations except the physical conditions under which the explorer had to live. It is difficult to realise the great change of condition which this diver's body had to endure every time he descended to his work from an atmosphere equivalent to a column of 30 inches of mercury to one equivalent to 15 feet of the same metal. Every inhalation in his diving dress supplied oxygen for the combustion of his tissues, five times greater than that taken in at the surface. No wonder then, that though he recovered the treasure, half of which was to be his, he died of his exertions. The extent to which submarine regions can be personally explored is thus so limited as to be negligible, and we have to trust for our information to the indications of instruments and the collection of specimens. It need hardly be added that our knowledge of the solid land underlying the ocean is still more restricted, being limited to the few inches of thickness which can be penetrated by the sounding tube or scooped up by the dredge, and it is unlikely that it can ever be directly extended to any greater depth.

But while the explorer of the ocean labours under the disadvantage that he cannot penetrate into its interior, he has this compensating advantage that he can freely traverse its surface in every region and place himself vertically over every point of the three-fourths of the surface of the solid globe, which is covered by water, and explore it in detail from the bottom to the top to an extent limited only by the instrumental means at his hand.

This method of, and field for, research was unknown or neglected until about thirty years ago. On looking back it seems almost incredible that the men of all nations burning with scientific and exploring zeal should have entirely overlooked and apparently despised this large portion of the world, so accessible and so easy of exploration, and have incurred incalculable hardships and dangers in the exploration of inhospitable deserts, fever-stricken jungles and inaccessible mountain tops. If they *did* turn their eyes towards the sea, they did not allow them to rest on what was before the door, on the seas and oceans traversed in the ordinary course of trade, which might be presumed to be of interest to the nations separated by them, but carried them away to the dreary polar regions and sent their best men entrusted with no higher aim than to thrust their ships against ice as far as it was possible. Every one must admire, and it would be impossible to overestimate, the moral qualities exhibited by the courageous army of Arctic explorers which has been sent out from time to time by our own and other countries; but the results obtained and obtainable are quite incommensurate with the labour and money expended and the dangers incurred. It is well that Arctic regions should be explored, but it is astonishing that they should have had the preference over more accessible and more interesting regions. No doubt the reason for it is a very human one. An Arctic expedition appeals to the imagination of every one, it *is* a service in which hardships and discomfort are continuous and in which dangers are daily and require the continual exercise of courage, perseverance and endurance in order to meet and overcome them. Further, the Arctic and in a greater degree the Antarctic regions are the only parts of the globe remaining where discoveries in their earlier sense can still be made. Formerly, that term was used to indicate discoveries of new lands, the delineation of fresh coast lines. It is possible that there may still be an undiscovered islet in the vast expanse of the Pacific Ocean, where passing ships even on the principal routes are a rarity, but it is on the whole unlikely; so that we may safely say that, outside of polar regions, there are no new countries to be discovered. A well-appointed expedition to the Antarctic regions would afford plenty of opportunity for the exhibition of those higher qualities which enable men to overcome difficulties and despise dangers, and at the same time could not fail to supply *data* of the highest scientific interest. But for purely marine investigation there are fields nearer home as yet barely scratched, but ready to yield rich harvests for the mere reaping. The warm and pleasant waters of tropical oceans are still almost untouched and teem with objects of interest and in their exploration and investigation there are no difficulties to be overcome, discomforts to be endured or dangers to be faced. Oceanic investigation in low latitudes is, so far as natural causes are concerned, always carried on under the most favourable and most comfortable conditions.

Communication and Population. In the history, especially the quite recent history, of the world the cultivation of science and art has resulted in such an enormous development of industry and commerce that geography can no longer rest satisfied with the mere objective contemplation of lands and seas as statical items; it has to do with, if not the countries themselves, certainly with a large amount of their population and productions in a state of motion from place to place. The communications therefore of nations, and the economical circumstances which attend them, have a distinct and important geographical significance.

Until within the last fifty years the population of the world had increased and spread over the land slowly and gradually, and the method and rate of spreading had probably undergone no serious change from the earliest times. Especially favoured districts like the shores of the Mediterranean became first comparatively crowded, and the people collected in cities, producing the artificial conditions of a population living on an area which produces nothing for its support. In order to maintain a multitude so situated, food had to be raised in and removed from other districts where, as a necessary condition, the population was smaller than could be supported by the land on which it dwelt and which it cultivated. The industry of the peasantry in the environs of villages and small towns is usually sufficient to provide food for the inhabitants of the town. But where the only means of transport is by road there is a limit to the radius of the area from which the daily food of the town can be drawn, and consequently to the number of people that can be kept alive in it. But for the continuous victualling of a city it is necessary not only to have a large area of supply, but also easy means of communication, by which to transport the produce. Hence while villages and small towns in fertile countries could easily subsist on the agricultural industry of the neighbourhood, large cities, whose wants were much greater than could be supplied by any district within a moderate distance, or than could be collected from a greater distance by ordinary methods of traction, were forcibly dependent for their growth and existence on water communication, and a position on or very close to the seaboard became a necessity. Without these maritime advantages, large cities were an impossibility. Inland towns could grow to a certain size and then they were obliged to stop. Provincial towns in all old countries, as for instance the county towns of England, were before the spread of railways almost stationary as regards population from one century to another. A study of the size and population of such towns (if reliable data could be obtained in the absence of an official census) compared with the situation and area of its food supply would be a useful and interesting geographical exercise, and it might thus be possible to arrive at an expression for the limiting size to which a town could so attain. It would probably depend principally on the fertility of the soil and the industry of the agricultural inhabitants as well as on the condition of the roads. Depending on local resources such towns were exposed also to local accidents; and any excessive or immoderate growth of their dimensions would be surely checked by an occasional failure of crops or local famine. The town on the seaboard and having command of the sea could tide over a calamity of this kind and would go on growing while its inland neighbour was being killed.

The inhabitants of ancient Rome were fed only to a very insignificant extent by the products of the fields of the Campagna, and with all her power a limit would have been set to her size had she not occupied a secure position on the margin of an extensive and much indented inland sea, a sea which in all directions cuts up and penetrates into the heart of some of the most productive countries of the world.

Further, just as it was necessary for the existence of a large city to have free communication with the sea in order to receive food from foreign countries, so it was necessary for those supplying countries to be furnished with free water communication in order that when they had grown and harvested their crops they might be able to get them easily and quickly away. A fertile country is of little use unless it either carries a sufficient population of its own, or by means of its communications is in a position to support a similar population elsewhere.

The length of coast and navigable river line was, before the introduction of railways, the most determining physical feature of a country's development and importance. The greater the extent of this line the greater was the available "working face" if I may use a mining term. By means of it the more advanced inhabitants of crowded nations and great cities introduced the developments of science and art and removed the natural products of the countries—whether food for the population or raw material for its industries: and the contact of the two classes of people was promotive of civilisation. The great wealth and importance of ancient Egypt was given by its unique position in what must be considered the centre of the old world, with the shortest water communication from every part of the East and West to its door and the noblest of rivers carrying it into its very heart. The decadence of Egypt and the rise of Athens and Rome with their subsequent decay show that geographical position alone neither makes nor preserves a great nation.

With the fall of the Roman Empire and the disappearance of a recognised centre, life in Europe lost much of its artificiality and every part of the continent pursued its own local interests, and population spread slowly over the parts capable of permanently supporting the settlers. In the same way, after the discovery of America, the settling of it by Europeans took place very gradually and was confined to the sea coast and the margins of the great rivers such as the St Lawrence, the Mississippi and the Parana.

In North America the Atlantic coast was colonised principally from England, while the margins of the St Lawrence and the Mississippi were peopled principally by the French immigration. In South America the colonisation, when it left the sea coast, followed the streams of the Parana and the Uruguay. The State of Paraguay which is more than 1200 miles from the estuary of the La Plata owes its settlement and colonisation to the Jesuit missionaries in the 16th century;

and just as the country towns and districts of the New England States of North America remind one of similar districts in Old England, so do the cathedrals and monasteries (generally much dilapidated) of Paraguay and the province of Cordova remind one of the towns of Old Spain.

But the occupation did not wander far from the waterways, and, with existing means of communication, it would have required many hundreds of years before by mere increase of population and the consequent necessity to extend the area of occupation the interiors of the great continents would have been opened up.

Transportation and Civilisation. Up to the commencement of the present century the means of transport and communication were essentially the same as had served the world for thousands of years. Between distant countries the only means of communication was by sea in ships which depended entirely for locomotion on the strength and direction of the winds. For communication and distribution of traffic on land dependence was placed on the beast of burden, the waggon and the barge.

With the introduction of steam everything was changed. It was then that the great rivers of the continents asserted their value and a town on a navigable river, however far inland, suddenly acquired all the advantages of the oldest established seaport. The bed of the river became for commercial purposes a useful indentation or extension of the coast line. Indeed the introduction of steam navigation in rivers virtually carried the sea inland as far as their waters were navigable; and as both banks of the river were available for distribution of traffic to the landward districts, *the actual effect, expressed numerically, was to increase the existing coast line by double the length of the navigable portion of the rivers.* This sudden extension or multiplication of length of the effective coast line created an enormous demand for the means of locomotion so as to utilise the advantages conferred, and the consequence was that the steamship builder devoted himself at first almost exclusively to the construction of steamers for rivers and inland waters.

It is just a hundred years since the first experiments in steam propulsion were being made simultaneously in America and in this country. In 1787 Rumsey made some short voyages on the Potomac with a boat about fifty feet long propelled by the reaction of a stream of water drawn in at the bow and forced out at the stern, by means of a pump worked by a steam engine. In 1788 Mr Millar of Dalswinton in Dumfriesshire fitted his pleasure boat with a boiler and engine and the first steamer in British waters was a steam launch on Dalswinton lake. Of him Carlyle says: "he spent his life and his estate in that venture, and is not now to be heard of in these parts, having had to sell Dalswinton and die *quasi-bankrupt*, and I should think broken-hearted." Steam navigation, however, assumed no commercial importance until Fulton, who had followed the experiments on both sides of the Atlantic, returned to America, and in 1807 launched the steamer "Clermont" of 160 tons burden which made several trips on the Hudson river and in 1808 took its position as a regular trading vessel running between New York and Albany[1].

It was not till 1811 that Henry Bell, of Helensburgh on the Clyde, built the "Comet," nor till January 1812 that she began to ply regularly on the river. From that time to this numerous experiments and an uninterrupted succession of improvements have brought steam navigation to its present high state of perfection.

Steamship construction thus originated simultaneously and to some extent independently in North America and in Great Britain, and in both countries the adaptation of the steam engine to the propulsion of river craft has been prosecuted with energy and success. It might have been expected that the two nations working, not out of communication with each other, on the same material and for the same end would have turned out much the same article; yet how different, not only in outward appearance but in almost every detail of construction, are the American and the European river steamer.

In America steamers were built so as to suit American rivers, which though of great volume are in many places

[1] The ship "Clermont" was built on the East River, New York, but her engines were imported from England. They were built to Mr Fulton's order by Messrs Boulton and Watt of Birmingham. See *Life of Robert Fulton* by C. D. Colden, New York, 1817, pp. 120, 165, 168.

very shallow. The size of the rivers was too great for the idea to be entertained of dealing with the shoals so as to allow vessels of the sea-going type to pass over them, consequently the vessels were built to pass the shoals. Hence the form developed was of the broad flat-bottomed type, suitable for the navigation of shallow waters. Also the engine used for propelling them was, and to a great extent continues to be, of the type of the early pumping engine with the ponderous so-called "walking beam."

In Europe steam navigation began on the Clyde, and the form of vessel was given by its local peculiarities. Although originally a shallow river, the commerce of Glasgow had already assumed such importance that the channel had been artificially deepened so as to admit the passage of vessels of sea-going build of a size far beyond what even the most sanguine expected to see fitted with engines. Consequently the ordinary type of sea-going sailing vessel was not departed from in drawing the lines of the early steamers; more especially as what is known as the river steamer traffic of the Clyde is to a great extent on the deep fiord-like inlets or sea lochs, where a ship's draft of water rarely comes in question.

Owing to a number of conspiring favourable circumstances the Clyde at once became the chief centre of steamship building, while at the same time its picturesque and mountainous surroundings, being deeply indented by the sea, became available for settlements, especially for summer quarters for the citizens of the towns. With this double advantage the development of the Clyde river steamer advanced with great rapidity, and by the year 1850 it had almost reached its present model. Few of the thousands of tourists who every summer pass through the Caledonian Canal and along the west coast of Scotland are aware that the steamers in which they travel have nearly all been running for upwards of forty years.

The engines which up to 1850 were almost exclusively of the upright single cylinder type, so-called "steeple engine," were changed to the double cylinder oscillating, and have again to a great extent reverted to the single cylinder, but diagonal type.

The American type of steamer prevails exclusively on North American rivers, but on South American and Eastern rivers, as for instance the Parana and the Irrawady, the European type has been introduced. Inhabitants of savage and uncivilised countries who are accustomed to ocean ships and steamers have no difficulty in recognising a Clyde or even a Rhine river steamer, but they would be at a loss to know, at first sight, what purpose one of the magnificent Mississippi or Fall river steamers was intended to serve.

At first no immediate need was felt for steam communication across the ocean. All the traffic was adequately met by the sailing tonnage, which was capable of almost indefinite extension. Moreover steamships mean coal depots, and it was a long time before they were to be found in distant regions.

Steamers have been running now for eighty years. During the first forty they were almost exclusively used for inland navigation, the second forty years has been the period of the development of the ocean steamer, and every day we witness fresh triumphs of the shipbuilder and the engineer in providing for the ever-growing traffic between the new and the old worlds.

The paramount importance of rivers as a means of distributing inland traffic was only short-lived. The application of steam to propulsion on land broke the monopoly, and with the invention of railways the existence and development of towns and cities far from the sea or any navigable waterway became possible. At first railways were made only in populous districts and between places where there was known to be large traffic; and of course, both in Europe and America, the ground where traffic was assured and financial success, under reasonable direction, a certainty was so large that outlying districts obtained no attention from the railway concessionist or contractor of the earlier period. No one cared to run the risk of making railways on the chance of developing a traffic as long as populous districts remained unprovided. Attention was however called to the existence of enormous tracts of country, presumably inhabitable, fertile and rich, but either uninhabited or peopled by wandering tribes. It would be difficult to say how long this might have lasted, but for accidental cicrumstances.

Soon after the discovery of America the Spanish conquistadores thoroughly explored Mexico, Central America, Peru and the whole west coast of South America, and the determining motive was gold. All these countries were peopled by a civilised and industrious race who were well acquainted with the artistic and industrial uses of the mineral wealth of their country, and the gold and silver treasure was enormous and lay ready to the unscrupulous hand of the conquerors. It was the discovery of gold in possession of the inhabitants that led to the exploration and opening of these populous countries by Europeans; and it was the discovery of gold in 1848 in the soil of California and a few years later in that of Australia that led to the opening up of these vast countries, till then with hardly an inhabitant, but capable when in the hands of civilised people of supporting a population which might rival that of China.

The immigration, which began with the gold rush, quickly peopled California and the adjacent territories, which proved to have wealth much beyond their hidden metal. But the immigrants were almost all conveyed by sea and landed on the Pacific coast as their forefathers had done on the Atlantic coast. At the date of the Civil War the Eastern and Western States were separated by more than a thousand miles of Prairie and Rocky Mountains, forming a barrier and separation much more complete than an equal width of ocean. So separated were they that at no time during the Civil War or afterwards did the depreciated paper money pass current in California. The bridge was made by the construction of the first transcontinental railway. The inducement to Contractors to make this Railway which passed through such an extent of at that time almost uninhabited country was that of the free grants of a portion of the land for a certain depth along the line, a plan which has since been followed with success in opening other new countries.

In Australia the want of railways did not make itself so quickly felt because it was long before any considerable settlements were made at a distance from the sea which gave facilities for exploring the whole of the valuable coast land before proceeding to the less hospitable and more arid inner regions. Latterly, however, railways have extended in Australia as elsewhere though not at a rate comparable with what is observed either in North or South America.

In the Argentine Republic, which corresponds geographically with Australia and with the United States of North America, the fertile pampa is accessible for commercial purposes only from the Atlantic and up to the present almost exclusively from the River Plate. Hence as soon as the country became settled and people began to prosper, railways became a necessity. A few miles of the present Western Railway were laid down in 1857 and a year or two afterwards the Northern of Buenos Ayres now one of the most valuable lines in the country was begun with the old material of the Balaclava military railway. Unsettled political circumstances retarded the construction of railways during the next twenty years; but in 1880 there were 2300 kilometres of trunk line open which by branches and extensions was doubled by 1885 and tripled by 1887; so rapid is the progress after ground has been well broken and it must go on at an ever-increasing ratio. For in such countries railways take the place of or at least precede roads, the roads being an evidence of greater advance in prosperity and civilisation.

The function of railways is different according as the country is an old or a new one. In old countries they have been put down to facilitate the communications of existing populations, in new countries they are put down in order to introduce a population which then makes roads and more railways to enable it to move about.

Another similar region in the East, the Transcaspian countries of Asia, is being rapidly opened up by the Russians, and even without any view to conquest the Transcaspian Railway is a great work from a purely commercial point of view. It is destined before long to form one of the most important trunk lines of communication of the world.

It is remarkable and must at once strike the geographer

that the new countries to which almost the whole of European emigration has been directed occupy similar or, better said, homologous positions on the earth's surface. Their principal extension is in zones between 30° and 40° Latitude (N. and S.), as will be seen from the position of the cities giving access to the countries and which therefore are physically their capitals—thus San Francisco lies in 37½° N., New York 40½°, Buenos Ayres in 35° S., and Sydney and Cape Town in 34° S. The prairies and pampas and the corresponding part of Australia and South Africa which have no distinctive name all lie in latitudes which, if covered by sea, would be subjected to the constant influences of the trade winds. This is somewhat modified by the land but the distinctive character of the trade wind climate prevails; it is a dry and on the whole salubrious one. It is chiefly owing to the droughts that these districts in their natural state have always been inhabited by nomadic populations, as is shown by the inhabitants of the Russian Steppes, the African Bedouin, the North and South American Indians and the cross-breed Gauchos. The soils also, which are as similar as the climates, are always fertile under cultivation and require but little labour to produce heavy crops of boundless extent.

But the population is at first very small and a single family can raise grain sufficient for many times its own number. All the neighbours are in the same position, and except close to the shore of the sea or the banks of the river, there is no means of getting rid of the crops. Ten years ago the pampas of the Argentine Republic lying between the Atlantic and the Andes were like a coal seam in the bowels of the earth. The proprietor of the mine wishes to utilise his possession and to do so he must open it out. First he has to sink his shaft to the seam, and when he arrives at it and finds it as good as he expected, he is still only at the threshold. Although he may have abundance of willing labour he cannot utilise it until he has opened out workings so as to admit of the labourers at his command getting free access to the mineral and working their best at its extraction. Having secured the rapid mining of the seam, he must have arranged to have sufficient haulage to move what has been mined or the works will be checked by the activity of the workers themselves. And when the coal has been brought to the surface he must have means of transport at his command to convey it to the market at least as rapidly as it is put out. Similarly the fertility of the pampa is of little use while it lies between the ocean and the mountains. It must be made accessible to labourers and facilities must be offered for the conveyance of the fruits of their toil to the market. Until within the present decade railways were few, and as has been shown, until a certain amount has been built to start on, extension is necessarily slow. But when a few trunk lines had been driven in different directions into the country, secondary lines began to branch off from them in all directions; and after the first slow process of breaking ground had been got over, the railway mileage increased in rapidly rising proportion. Now the hardy North Italian immigrants who form the working elements of the country are absorbed and draughted away the moment they land, instead of wandering about the streets of Buenos Ayres as they did a few years ago, until they had to be assisted to return home. Very soon after they have settled on their land they provide return traffic for the railways in the shape of agricultural produce raised by their labour. Indeed their value is so great and at certain times of the year is so sought after that many Italian peasants, after gathering their own crops at home, go out to the River Plate for the harvest there which is in the middle of the Northern winter and return with a handsome sum in hand.

The introduction of railways into America and Australia has in a very few years increased to about double the extent of producing country in the world. In the same time the population of the world has increased comparatively slightly, consequently the supply of food and other commodities for the existing population has been enormously increased, with the natural economical effect, that, being more easily obtained, they can be more cheaply sold and their price has fallen. Apart from local or artificial circumstances such as wars for instance, prices are likely to remain at a low though not at a constant or stationary level, until in the course of years the countries, which we now look on as merely growing grain and raising stock, shall have also raised a population on the ground to eat the bread and meat and to wear the wool and leather which it produces. In proportion, as this condition approaches fulfilment, there will be less food to spare for old countries whose populations have outrun their means of subsistence, and the present abnormal and indeed monstrous over population of countries like our own will be reduced by natural causes.

The same effects have been produced in other branches of production. Thus in mining industries, the means which have so largely increased the area of food production have increased the area of metal production in perhaps even a greater ratio. The consequence is that the prices of all metals have fallen to a fraction of what they were a quarter of a century ago. Here again what is wanted to raise the prices is population to start industries to use the products of the mines.

In whatever direction we look the introduction of steam carriage on sea and especially on the land, along with the spread of the electric telegraph for the instantaneous communication of information, has produced a revolution in the affairs of the world such as those who have not seen it could not have conceived. The freedom and facility of communication over every part of the globe have afforded an opportunity of expansion to the population which a hundred years ago could not have been afforded even by the annexation of another planet.

SUPPLEMENT ADDED 1918.[1]

Civilisation and Extermination. It will naturally occur to the critical reader that only one side of the account between civilisation and the world has been stated; and it is not quite certain whether he will call it the credit or the debit side. In the text attention is directed almost exclusively to the advantage of improved methods of transportation in assisting and for-

[1] See Contents, p. x.

warding the aims and objects, if not in fostering the cupidity, of the "population"; and the word population, without being defined, is tacitly taken to mean the White Man. No consideration is given to the rights or claims of the black or red man, and still less to those of the brute population.

The example of the improvement of facilities of transportation between California and the eastern states of North America by the construction of the first transcontinental railway, and the advantage which it conferred on the "population" by enabling it to reach easily the great gold deposits of California and to remove the metal quickly to the world's markets, is stated shortly on page 16. The reader will naturally conclude that the construction of this railway was like that of any other railway connecting neighbouring towns;—when it was finished there was just a railway as well as a road between the two places. But there was no road across the continent. The White Man crossed it as an adventurer with his life in his hand. There was an antecedent and legitimate population, both man and beast, by which his trespass was likely to be disputed. After the opening of the railway this population had practically ceased to exist.

How many of the millions of people who in Europe or America, have seen "Buffalo Bill's Wild West Show" have attached to the name "Buffalo Bill" any greater significance than that he was a successful hunter of big game? Although it happened less than fifty years ago, it will be news to most people, that Colonel Cody, but acting for others, was in fact the exterminator of the grandest example of the animal creation on the American continent. In order to feed the navvies and other workmen who made the railway through regions where provisions could not be purchased, Colonel Cody was commissioned to furnish food from the wild animals of the country. The workmen were fed, the railway was built, people went west, gold went east, but the Buffalo disappeared.

The same thing but on a much greater scale has happened in Africa.

Africa may be said to be the tropical and equatorial part of the Old World. It is in this climatic region that the fauna and flora of the land take their greatest extension and their highest development. In the New World the sea cuts far too deeply into the land on both sides of the American continent in intertropical latitudes to permit its having the extension required for multiplication and development on such a grand scale as we see in the Old World and especially in Africa.

Big Game, in the meaning applied to it in Africa, was represented in America practically only by the Buffalo. In Africa it is, or was, represented by at least a score of different animals, such as the Elephant, the Rhinoceros, the Hippopotamus, the Giraffe, many Buffaloes and, above all, the Lion, King of Beasts. All of these names have been familiar to us since our childhood, not as matters of history, but as the actual and rightful owners and occupants of a large part of the habitable world.

But besides the magnificent brute inhabitants of the continent there is the human species, the Black Man. It may be that he is the only representative of the fauna of Africa that will successfully resist extermination by civilisation. If he does succeed in this it will be because, on his own ground, he is stronger than the White Man with all his perfected instruments of death.

However, the black men are still the human population of the greater part of the continent of Africa. They are very numerous and perfectly adapted for work in an equatorial climate. It may therefore suit the white man, who shall dominate their country, to preserve them for economic and military purposes. But even though he persisted only in a state of slavery, his survival would not be artificial. It would be as natural as that of his master, because he is the fitter of the two in the conditions obtaining, and they are permanent. Therefore the retention of the black man, however accomplished, preserves the possibility of a reaction, which is out of the question when the race, whether of man or beast, has been exterminated.

In order that the reader may be able to form an idea of the conditions of life in the equatorial part of the African continent, before civilisation had made much progress in its murderous work, his attention is directed to Gordon Cumming's well known book, entitled *Five Years of a Hunter's Life in the Far*

Interior of South Africa. He spent the whole of the years 1844 to 1848 as a hunter and a naturalist among the great mammalia of Africa, which abounded in these years and is now all but extinct.

His book is unique as a first-hand document regarding the natural life and the habits of these animals in what was their own home. It is, of course, concerned mostly with his success as a sportsman; but, in the absence of the sporting instinct with the courage and the enterprise required to gratify it, he could not have made the intimate acquaintance with these animals, which is exhibited in the notes descriptive of what may be called their personal habits in a state of freedom, which frequently interrupt, and, to a certain extent, relieve and embellish the narrative of the hunter.

Remarkable examples of these are the descriptions of:

—the Gemsbok, in Chapter V;—the Wild Dogs, in Chapter VIII;—the Lion, in Chapter IX;—the Rhinoceros, in Chapter XI;—the Giraffe, in Chapter XII;—and the Elephant, *passim*.

The short extracts, which follow, may be taken as samples, the first, from Chapter IX, of his descriptions as a naturalist, and the second from Chapter XI, of his narrative as a hunter. They should encourage the reader to study the book, because the thing has passed forever[1].

THE HABITS OF THE LION

Extract from Chap. IX

One of the most striking things connected with the lion is his voice which is extremely grand and peculiarly striking. At times, and not unfrequently, a troop may be heard roaring in concert, one assuming the lead, and two, three, or four more, regularly taking up their parts. They roar loudest in cold, frosty nights; but on no occasions are their voices to be heard in such perfection, or so intensely powerful, as when two or three strange troops of lions approach a fountain to drink at the same time. When this occurs, every member of each troop sounds a bold roar of defiance at the opposite parties; and when one roars, all roar together, and each

[1] It was my good fortune, as a boy, to be taken to see his collection of heads and other trophies when it was exhibited in Glasgow in the year 1850. It was the most remarkable collection ever made by one man; and, in making it, he used nothing but the muzzle loading gun and rifle, black powder and hard bullets.

seems to vie with his comrades in the intensity and power of his voice. The power and grandeur of these nocturnal concerts is inconceivably striking and pleasing to the hunter's ear. The effect is greatly enhanced when the hearer happens to be situated in the depths of the forest at the dead hour of midnight, unaccompanied by any attendant, and esconced within twenty yards of the fountain which the surrounding troops of lions are approaching. Such has been my situation many scores of times; and, though I am allowed to have a tolerably good taste for music, I consider the catches with which I was then regaled as the sweetest and most natural I ever heard.

FIRST EXPERIENCE IN HUNTING THE RHINOCEROS

Extract from Chap. XII

It was on the 4th of June, 1845, that I beheld for the first time the rhinoceros. Having taken some coffee, I rode out unattended, with my rifle, and before proceeding far I fell in with a huge white rhinoceros with a large calf, standing in a thorny grove. Getting my wind, she set off at top speed through thick thorny bushes, the calf, as is invariably the case, taking the lead, and the mother guiding its course by placing her horn, generally about three feet in length, against its ribs. My horse shied very much at first, but, the ground improving, I got alongside, and, firing at the gallop, sent a bullet through her shoulder. She continued her pace with blood streaming from the wound, and very soon reached an impracticable thorny jungle, where I could not follow, and instantly lost her. In half an hour I fell in with a second rhinoceros, being an old bull of the white variety. Dismounting, I crept within twenty yards, and saluted him with both barrels in the shoulder, upon which he made off, uttering a loud blowing noise, and upsetting everything that obstructed his progress.

Shortly after this I found myself on the banks of the stream beside which my wagons were outspanned. Following along its margin, I presently beheld a bull of the borèlé, or black rhinoceros, standing within a hundred yards of me. After standing a short time eyeing me through the bush, he got a whiff of my wind, which at once alarmed him. Uttering a blowing noise, he wheeled about, leaving me master of the field, when I sent a bullet through his ribs.

One hundred years ago Africa with its great fauna was practically intact. Seventy years ago it was in the state of exuberant life pictured by Gordon Cumming. Sixty years ago communication between Europe and South African ports was still maintained by occasional sailing vessels; and persons landing at these ports had only one means of penetrating into the interior, namely, the ox-waggon. Behind such means of travel and transportation the big game of Africa could maintain itself indefinitely.

Fifty years ago steamships were already replacing sailing vessels on all ocean lines, including those to South Africa; and ground had already been broken for a railway in the immediate vicinity of Capetown. By the year 1870, the discovery of diamonds about five days' drive, by Cape-cart, from Capetown produced a rush of miners and speculators to the place afterwards called Kimberley; and very soon there was a large population, which called for a railway and naturally it was constructed with all speed.

About the year 1873 gold was discovered, in quantity, on the Rand and it attracted a still greater rush than did the diamonds, and to a distance still further inland.

Perhaps the great African fauna might have withstood the desire of civilisation for diamonds, but its greed for gold was quite another thing. Railways spread into the interior and in the course of forty years the process of invasion, extermination and re-settlement by immigration has become nearly complete.

The great fauna is almost extinct and the indigenous human races may follow the same course.

If we wish to know by what they have been replaced a page taken at random from the *Business Directory* of the city of Johannesburg will furnish an average sample.

It is probable that I and my shipmates on the "Challenger," when she visited New Guinea and the Admiralty Islands in the beginning of the year 1875 were about the very last of our race who came into personal contact with the primitive uncultured man, proud of being undisguised by clothing, who met his visitors adorned with shells and flowers, and armed, the New Guinean, with the bow and arrows, more than a cloth-yard long, tipped with the hollow bones of birds, and the Admiralty-Islander, armed with spears fitted with massive obsidian heads, deftly fashioned from the volcanic glass of his mountains.

To young people the process of rapid destruction and extermination which has altered the whole world in so short a time may appear to be quite natural, but to older people and to those who reflect that the world is not there for them alone, the injustice, the cruelty, and the ferocity of the whole thing are appalling and humiliating.

Even without the high-pressure development, in the last four years, of instruments of death the final stage cannot be far distant and the nature of it is no longer in doubt. Everything must go down before the White Man excepting what he desires for selfish purposes to preserve. This is no longer a speculation; it is a necessity. In a short time, it may be in the lifetime of an infant born to-day, the wealth of beautiful living things that has lived and developed during countless ages will be as strange to it as the fossils that are dug out of the earth are to us to-day.

The world will be dominated, if not exclusively inhabited, by the White Man and the animals which he has domesticated, their parasites and their vermin.

In order to reckon the time which it takes to exterminate a creation by the instruments and methods of civilisation, our ordinary units of days, months and years are sufficient. When, however, we attempt to estimate the time covered by the development and persistence of a creation before it came in contact with civilisation, we find that we are in want of an adequate unit.

If we make the unit one thousand years, great as this may seem to us, we cannot say with certainty that a hundred or a thousand or a hundred thousand or perhaps even a million of these units would be an excessive estimate of the time which would cover the development and persistence of the great African fauna on its own ground. But we know that one hundred of our years ago its extermination had hardly begun and we see that it is now nearly complete.

If we demand a reason for this sudden annihilation of almost all that is beautiful on the earth, we are told that the advance of civilisation demands it and has the power to effect it. But, as a reason, this can satisfy no one. If, however, we are asked: What, if anything, could have prevented it? the answer is ready. If the steam-engine had not been invented and developed as it has been, the organic life of the world would still be continuing its progress up the gentle incline of development, and creation would have lost nothing of its beauty.

Had the steam-engine been invented and developed a

thousand years ago, we may be sure that we would now be inhabitants of a desert of hideous uniformity. It is this that we are leaving as a legacy to posterity; and it will not be retarded by Man rivalling both the Eagle and the Whale in their own elements.

But the further question may be asked:—What would have prevented the invention and development of the steam-engine? Evidently nothing but some fundamental difference in the nature of man.

No positive cause suggests itself; but, in the absence of the faculty of speech, it is hardly conceivable that an invention such as that of the steam-engine, with its developments, could have taken place.

There is evidence in very ancient history that interference with freedom of communication by speech can produce a corresponding interference with the realisation of great enterprises, undertaken in the service of the spread of civilisation.

The story of the Tower of Babel is the record of an enterprise of this kind. It is told in very few words in the eleventh chapter of the book of Genesis. Beginning at the fourth verse, we there read:—

> "And they," the people of Shimar, "said Go to, let us build us a city, and a tower, whose top may reach unto heaven; and let us make us a name, lest we be scattered abroad upon the face of the whole earth.
>
> And the Lord came down to see the city and the tower which the children of men builded.
>
> And the Lord said, Behold, the people is one, and they have all one language; and this they *begin to do*; and *now nothing will be restrained from them, which they have imagined to do.*
>
> Go to, let us go down, and there confound their language, that they may not understand one another's speech.
>
> So the Lord scattered them abroad from thence upon the face of all the earth; and they left off to build the city."

Thus were all inventions indefinitely delayed; and the world got a respite.

It must not be forgotten that the Fall of Man was directly attributable to his possessing the faculty of speech; and the consequence of it was that the last and best of his creations was the first that the Creator had to punish.

No. 2. [*From Report of the Sixth International Geographical Congress, held in London,* 1895.]

A RETROSPECT OF OCEANOGRAPHY IN THE TWENTY YEARS BEFORE 1895

ADDRESS TO THE OCEANOGRAPHICAL SECTION OF THE SIXTH INTERNATIONAL GEOGRAPHICAL CONGRESS, HELD IN LONDON, 1895; REVISED AND AMPLIFIED

I.

THE Geographical Congress meets in London at an opportune time, when the publication of the *Challenger Reports* has just been completed by the appearance of the summary volumes. This great work marks an epoch in the science of geography, and deserves some words of notice here. The history of the "Challenger" expedition is well known to all students of oceanography, which, as a special science, dates its birth from that expedition. It must be remembered that when the "Challenger" expedition was planned and fitted out, the science of oceanography did not exist. The chief men to whose influence the expedition was due—Carpenter, Huxley, Wyville Thomson—are dead. Those who at present are most active in the furtherance of the science did not take any interest in oceanography then, and, notwithstanding the voluminous reports of the voyage, it is impossible for the student of to-day to realise what were the views and expectations twenty years ago which determined the procedure of the "Challenger" in breaking ground in the vast dominion of the sea. In the few remarks which I have to offer here, I propose to illustrate this in one or two cases. Isolated observations of a physical and biological nature had been made by many observers previous to the summer expeditions of the "Lightning" and "Porcupine" belonging to the British navy, and the "Pomerania" belonging to Germany. The most remarkable of these were the researches of the "Bulldog," when surveying the route for the first Atlantic cable. The route followed by the "Bulldog" did not lead across any water which would now be called very deep. Out in mid-ocean, and away from the influence of land, the mud brought up was calcareous, and consisted mainly of the dead and comminuted shells of foraminifera. Deeper soundings which had been reported by earlier observers were generally discredited at this time. Many of them were obviously faulty in execution and untrustworthy in result, but there were several soundings which showed that depths of over 2500 fathoms were to be found in the ocean. Thus Sir James Clark Ross, in the course of his memorable Antarctic voyage, found 2677 fathoms in lat. 32° 21′ S. and long. 9° E. He also made a remarkable sounding in lat. 68° 34′ S. and long. 12° 49′ W., when, under very favourable circumstances, he ran out all his sounding-line without having touched bottom in 4000 fathoms. Opinions differ with regard to the importance to be attached to this sounding, but if his lead had touched bottom in anything under 3500 fathoms, Sir J. Ross would certainly have known it.

The effect of the "Bulldog's" work, and of Dr Wallich's report on it, was to produce the belief which was generally expressed by saying that at the present time *chalk* is being laid down all over the deep ocean; that, therefore, geologically speaking, the bottom of the ocean is a cretaceous formation[1]. This was undoubtedly the prevalent belief when the "Challenger" sailed from Portsmouth on December 22, 1872.

[1] Wyville Thomson's *Depths of the Sea* was published only a few weeks before the "Challenger" sailed, and the last chapter of it is entitled "Continuity of the Chalk." After pointing out that, whereas the chalk of our downs consists of almost pure carbonate of calcium, with no silica, the chalk mud of the Atlantic contains as much as twenty or thirty per cent. of silica, and that the English chalk is the very purest of its kind, he goes on to say (p. 470), "There can be no doubt whatever that we have forming at the bottom of the present ocean a vast sheet of rock which very closely resembles chalk; and there can be as little doubt that the old chalk, the Cretaceous formation which, in some parts of England, has been subjected to enormous denudation, and which is overlaid by the beds of the Tertiary series, was produced in the same manner and under closely similar circumstances; and not the chalk only, but most probably all the great limestone formations."

Further on (p. 495) he says, "I have said at the beginning of this chapter that I believe the doctrine of the continuity of the chalk, as understood by those who first suggested it, now meets with very general acceptance; and in evidence of this I will quote two passages in two consecutive anniversary addresses by presidents of the Geological Society, and we may have every confidence that the statements of men of so great

Regular work commenced only in February, when she left the Canary islands to run her first line of trans-oceanic soundings, and fundamental discoveries were made in the very first week. On February 15, 1873[1], a sounding gave 1525 fathoms in lat. 25° 45′ N., long. 20° 14′ W., and no sample of bottom. It is remarkable, as indicating the views of the time, that the absence of a sample of mud was not supposed to afford evidence that the bottom was not soft, but only that the tube had not acted. So far it was deemed to be inconceivable that any part of the basin of the ocean could avoid being covered with mud. One of the very earliest dredgings was made on this spot, and its harvest was one of the most remarkable of the cruise. It came up full of masses of jet-black branching coral attached to masses of black-banded rock, and bearing in its branches huge siliceous sponges. It was a sight which I see now before me as clearly as I did when on the deck of the "Challenger." Clearly the sounding-tube had not failed in its duty. The black colour of the coral and of the stony masses adhering to it was held to be carbonaceous, and I did not examine it till some days afterwards. As the ship proceeded westwards the depth increased, until, on February 21, a sounding was obtained in 2740 fathoms, and the dredge which was put over came up full of pure red clay, apparently without any calcareous matter, and quite amorphous. In a similar

weight, made under such circumstances, indicate the tendency of sound and judicious thought. Professor Huxley, in the address for the year 1870, says, 'Many years ago (*Saturday Review*, 1858) I ventured to speak of the Atlantic mud as "modern chalk," and I know of no fact inconsistent with the view which Professor Wyville Thomson has advocated, that the modern chalk is not only the lineal descendant of the ancient chalk, but that it remains, so to speak, in possession of the ancestral estate; and that from the Cretaceous period (if not much earlier) to the present day, the deep sea has covered a large part of what is now the area of the Atlantic. But if Globigerina and Terebratula, *Caput serpentis*, and Beryx, not to mention other forms of animals and of plants, thus bridge over the interval between the present and the Mesozoic periods, is it possible that the majority of other living things underwent a sea change into something new and strange all at once?'"

The other quotation is from Mr Prestwich, in 1871, but that from the late Professor Huxley will suffice to show the tendency of opinion at the time.

[1] 15 February 1873, the Birthday of Oceanography, see Contents, p. xii.

clay obtained a few days later, certain black nodules were found. These had no apparent connection with either plants or animals, and I onsidered that my department had to render an account of them. At the same time, I took them to the laboratory with the expectation of finding that their black colour was due to carbon. A portion heated before the blowpipe did not change colour, but in the closed tube it gave off a large quantity of water, which had a strong alkaline reaction and an empyreumatic odour. The sample was then submitted to the ordinary process of mineral analysis, and it was found to be an ochreous nodule, consisting of the hydrated higher oxides of manganese and iron, with traces of copper, nickel, and cobalt, and varying amounts of the clay in which it had apparently taken its rise. The physical structure of the nodules reminded me at once of the stony masses to which the black coral of February 15 was attached. They were examined in the same way, and were found to be identical in structure and composition.

In these few days, at the beginning of the voyage, the carbonate of lime had been observed to disappear gradually and completely, giving place to an apparently perfectly amorphous and non-organic deposit. This deposit had the character of a clay, with much ochre, both ferric and manganic, diffused through it, and locally concentrated into nodules. Proceeding westwards, the water became shallower again, and on March 4 a sounding was obtained in 1900 fathoms. As the depth became less the quantity of carbonate of lime became greater, and the mud from this station was a globigerina ooze containing 75 per cent. carbonate of lime. Going further west, the water deepened again to over 3000 fathoms before it shoaled on nearing the West Indies, and here again the same variation of the nature of the bottom with the depth of the water was observed. On this line the dredge was used almost exclusively, and it acted well, generally bringing up a full load of whatever mud was at the bottom. This afforded precious opportunities of verifying the temperature shown by the deep-sea thermometers used in the sounding. The mass of mud was so great that it preserved inside an almost ice-cold temperature until it was emptied on the bridge. An ordinary thermometer plunged into it showed a temperature almost exactly the same as that given by the registering thermometer on the sounding-line. Indeed, I think that it was in the first dredgeful of red clay that Professor Wyville Thomson cooled a bottle of champagne to celebrate the discovery at dinner. Similar observations were made on the way from St Thomas to Bermuda.

By this time it had been firmly established that the nature of the deposits in the open ocean varies in a definite way according to the depth of the water. Between 1000 and 1500 fathoms the predominant constituent was the pteropod shell. At greater depths they disappeared, and the calcareous portion of the mud consisted of the shells of foraminifera up to a depth of about 2500 fathoms. Beyond this depth the foraminifera rapidly disappeared, and at a depth of over 3000 fathoms the mud consisted almost entirely of red ochreous and argillaceous matter.

Bermuda island, separated from the nearest land by 600 miles of sea of over 2500 fathoms depth, consists entirely of the *débris* of marine calcareous shells bedded mostly under the influence of the wind. The island, however, is covered with a rich soil, and in many places between the calcareous beds there were layers of red earth. Professor Thomson was much struck with this occurrence, and it suggested to him an analogy with the distribution of the calcareous and the earthy matter at the bottom of the ocean. The disappearance of the calcareous matter with increasing depth had already been attributed to solution in passing through a greater amount of water; and it was natural that the thin and delicate pteropod shells should disappear before the smaller and stouter globigerinae.

Professor Thomson imagined that a similar action had taken place in the calcareous beds of Bermuda under the action of the rain, which removed the carbonate of lime, leaving the red clay as a species of "ash." It was here supposed to have formed an integral portion of the substance of the shells, which thus consisted partly of carbonate of lime and partly of the silicates and ochres of the red clay. When the carbonate of lime was removed, the red clay was revealed. In order to test this theory, Professor Thomson asked me to dissolve a quantity of a pteropod ooze from which the finer parts had been removed by washing, leaving what appeared to be nothing but shells. These shells, on being dissolved in dilute hydrochloric acid, left a very large residue of argillaceous matter, and it was held that the "ash theory" had been confirmed.

Now, although this theory was only half sound, in so far as it accounted for the disappearance of the carbonate of lime, and was very soon found to be untenable as regarded the source of the clay, it was of great use in giving a provisional form to the ideas at that early period of the cruise, and it was in the testing of it that the truth was very gradually worked out. It must always be honourably connected with the memory of Sir Wyville Thomson.

On the voyage between Bermuda, Halifax, and the Azores, the deposits showed modifications due to the carrying power of ice. Ice, starting from continental land bearing its *débris*, drops it on melting, and thus distributes terrigenous matter to a much greater distance from the shore than would otherwise be possible. The characteristic feature of such deposits is the presence of quartz sand at great distances, and stones and boulders in closer proximity to the ice sources. Between Bermuda and Halifax, in lat. 41° 14′ N., long. 65° 45′ W., the dredge brought up a boulder of syenite weighing over 5 cwt. The far-carrying power of ice was thus strikingly illustrated, and it was thought that it might be sufficient to account, by the disintegration and decomposition of the stones thus dropped, for all the mineral matter on that part of the floor of the ocean. The fine sand from primary rocks which was found in the muds on the way between Bermuda and the Azores furnished less striking but equally convincing proof of the distributing power of ice in the North Atlantic.

The question, which had been still agitated when the ship left England, whether the foraminifera, whose shells make up the globigerina oozes, live at the surface or the bottom, had come to be answered in favour of the former. And it was not found that the fresh animal caught with the tow-net contained the mineral matter which would be required to account for the red clay or for the argillaceous matter left on dissolving the lime of a globigerina ooze. Hence that portion of Wyville Thomson's ash theory, which ascribed an organic origin to the clay, had to be abandoned. But mineral matter had been found to be abundant all over the ocean-bed, and beyond certain limits it could not be held to have been distributed by the agency of ice. Hence it was necessary to call in some other agency, if the mineral matter was to be taken from continents or islands. It was to Mr Murray that the idea first suggested itself to account for the mineral matter on the floor of the ocean by the decomposition of pumice, which, after floating about for a long time, had finally sunk at the spot. The eruption of Krakatoa, and the fields of pumice which were met with for many months afterwards floating all over the Indian Ocean, and all of which in the end found its way to the bottom, came as a welcome support to the theory ascribing a volcanic origin to much of the mud found on all parts of the bottom of the ocean. The microscopic analysis of the muds further showed, by the abundance of glassy felspar and the absence of quartz, that their parent rock was volcanic and not primary. This happy idea of Mr Murray's, and the development which it received in his hands, formed one of the greatest achievements of the expedition.

In the Antarctic Ocean primary as well as igneous rocks were found in the droppings of icebergs; and on more than one occasion the trawl came up full of stones of all sizes, as if it had been dragging over a moraine or a river-bed. On such occasions there was a peculiar charm in going over them with the hammer. It amounted in effect to a geological excursion in the unknown regions of the Antarctic continent. But the Antarctic Ocean afforded also a typical class of purely pelagic deposit, the diatom ooze. Patches of deep olive-green water were very common amongst the floating ice, which itself was often stained by the same colour. This was due to the chlorophyll of the minute diatoms. When they died and sank to the bottom, their siliceous tests persisted as a fine infusorial earth. Later on, when passing through the Arafura sea, lying to the westward of New Guinea, diatoms were found in abundance, though the conditions, especially those of temperature, were very different; but the salinity of the tropical was nearly identical with that of the glacial water. The density of both waters was low, the one on account of the melting ice, and the other on account of the equatorial rains. While carrying out operations in the Gulf of Guinea on board the s.s. "Buccaneer," belonging to the Indiarubber, Gutta Percha, and Telegraph Works Company of Silvertown, I received abundant confirmation of the suitability of dilute sea-water for supporting diatom life. Here, however, the fresh river-water, which produces by mixture the almost brackish sea-water, is accompanied by abundant land *débris*, which prevents the formation of a pure diatom ocean ooze. The deposits here have a predominant terrigenous character, associated, however, with many siliceous organisms.

Another pelagic deposit of a siliceous character is the radiolarian ooze. Like the diatoms, the radiolarians are found in water of low salinity, and in the Pacific, far from land, they form true pelagic deposits.

Starting, therefore, with expectation of finding a more or less universal chalk formation at the bottom of the ocean, the result of the "Challenger's" work in the first two years was to open up a new geological world, and to show its dependence on the physical conditions of the oceans.

The Ship and her Equipment. The "Challenger" was a spar-decked corvette, and, when serving an ordinary commission, she carried twenty-one guns. These had been removed and the large ports enabled the ship to enjoy the most perfect ventilation. She was ship-rigged and her engines were able to drive her 11 knots at full speed. Her displacement was about 2300 tons. Like all the men-of-war of her time she was built of wood, with very solid timbers. Her screw propeller could be hoisted up out of the water. This was a great convenience because all the *passage* was made under sail. The whole amount of coal which she could carry was very little more than that required for manœuvring the ship at the sounding and dredging stations. The work at a station generally took the whole day from sunrise to sunset, and every one familiar with steamers knows how expensive in coal is the operation of keeping station.

The material collected at each station had to be examined, preserved and stored, before the ship arrived at the next one. The stations were generally about 200 miles apart, so that in the passage from one port to another a station was made every second or third day. This was easily accomplished under sail and it added enormously to the comfort and the interest of the voyage. All the advantages of having a wooden sailing ship were not fully realised at the time. It was not until I had taken part in one or two expeditions in well found iron or steel ships in tropical waters that I found out the discomfort which we escaped by being on board of an "old wooden ship." The temperature of the air in the ship was, of course, never lower than that of the air outside; but, on the other hand, it was never higher. Nothing astonished me more than the perfect uniformity of temperature of the air of the main deck of the ship in the tropics. I was able to make experiments on the effects of pressure on the deep sea thermometers in a hydraulic apparatus on the main deck, which I could not have made anywhere else. The temperature of the air did not vary by one-tenth of a degree (C.) during the whole of the day.

Iron or steel ships, even the magnificent yacht of the Prince of Monaco, get heated through by the sun in the course of the day, and at first they do not cool as much during the night. They are like a black bulb thermometer, they do not lose as much heat as they gain until their temperature has risen a good many degrees above the mean temperature of the air, and that can be pretty high.

The voyage of the "Challenger" lasted three years and five months. Of this time three years were spent between the parallels of 40° S. and 40° N., and the greater part of that time between the tropics. I have no hesitation in saying that the work could not have been carried on continuously in these tropical seas for such a length of time in any other kind of ship. The principal points of advantage were, the thick wooden walls, which completely prevented over heating and over cooling, the splendid ventilation which was provided by the twenty-gun ports on the main deck, and the practice of making the passage under sail.

A word with regard to the equipment. Throughout the voyage hemp sounding-line and hemp dredging-rope were used, and much of the success of the expedition is due to this fact. There was really no temptation to use anything else, because wire sounding had not passed the experimental stage, and all that was known for certain was that the same wire could not be expected to be used in many soundings without breaking. As our sounding line was on every occasion to carry a load of valuable instruments, this risk could not be run. Captain Nares knew that he could do all that was wanted with sounding line, and he was brilliantly justified by the result. There was no question of using wire rope for dredging. It was first used in America by Agassiz in 1877, and on this side of the Atlantic by myself in 1878.

Wire is suitable only for sounding, pure and simple; and the detailed investigation of the form of the bottom of the ocean cannot be carried out with anything else. As there is nothing at the end of the wire but a sounding tube or a sinker, it can be hove up as quickly as the engine will run, and the loss owing to a breakage of the wire is insignificant. For deep-sea research it is entirely unsuitable, because it has to carry valuable instruments. Their value increases largely with the number of times that they are employed, and when they are lost, for instance, in the Pacific, it takes the best part of a year to replace them, and then they are replaced by new instruments which have no history.

The question is not infrequently asked why hemp line was used for everything in the "Challenger" expedition, and not wire and wire rope. Although wire had already been used for sounding many years before, it was due to Lord Kelvin, then Sir Wm. Thomson, that the use of wire was made practically available. He studied it on board his yacht, the "Lalla Rookh," and designed the apparatus necessary for giving it a place on shipboard, and thus reduced it to practice. It had been brought so far in 1872 as to have emerged from the tentative experimental stage, and was ready for being tried on a large scale. Had the "Challenger" been going to run a line of soundings across the Atlantic for the purpose of selecting a bed for a telegraph cable, it would have been perfectly reasonable that she should be fitted with wire sounding-apparatus and a large supply of wire. She would, no doubt, have got to the other side of the ocean with wire to spare, and by filling up again with fresh wire, she might have continued similar work. But such was not the work that the "Challenger" was fitted out for. Determining the depth of the ocean was only a small part of her work at the 354 stations which mark her course round the world. Each of these stations marks, on an average, ten to twelve hours' work. Had wire been equally trustworthy with hemp sounding-line, then about one hour of this time would have been saved. So far from wire being equally trustworthy with hemp, it is the very emblem of treachery, and had the leaders of the expedition allowed themselves to trust thermometers and other precious instruments to wire, the store of instruments would soon have been exhausted, and the physical and chemical work would have been at a standstill. Wire rope had not been proposed for dredging until some years after the "Challenger's" return. In the first year of the cruise the sounding-line carried away nine times. After August 16, 1873, and until the "Challenger" reached home in May, 1876, it only carried away once, namely, on June 14, 1874. Consequently, after the first six months of the cruise there was no loss of instruments; and deep-sea temperatures, for instance, were, station after station, taken with the same thermometers. How different the conditions are when wire has to be depended on alone I experienced in the "Buccaneer," where, in spite of every precaution which care and practice could suggest, and an abundant supply of instruments, the stock of thermometers was almost completely exhausted before the work was done. I never attached a thermometer to the wire without feeling that I was guilty of a form of cruelty—cruelty to instruments.

In the taking of serial temperatures and water specimens, no time would have been saved by the use of wire, as with the powerful steam-winch the sounding-line came up in perfect safety from depths such as 1500 fathoms quite as quickly as wire could have been brought up with great risk and with the certainty of frequent loss.

The thermometers used on board the "Challenger" were of the Six type with protected bulb, the original Millar-Casella type. These thermometers did good service, and, like the hemp sounding-line to which they were attached, their appreciation at present is undeservedly low. In the Antarctic seas they were found wanting, as it was known that they would be wherever the temperature of the water did not fall with increasing depth. But of the three years and six months that the cruise lasted, three years and a quarter were spent in regions where this condition is fulfilled. Their use, therefore, was amply justified, and their convenience was great.

In the instruments of the ordinary pattern, the graduation was into single Fahrenheit degrees on glass slips fixed on the vulcanite to which the thermometer was attached. After the return of the "Challenger," I had them made for my own use with a millimetre scale etched on the stem as well as the Fahrenheit scale on the slips. As the length of a degree Fahrenheit was 3 millimetres, it was easy to read accurately to $\frac{1}{30}°$ Fahr. This is of great importance in the very deep water, where the extreme variations of temperature are limited to fractions of a degree. In the survey of the Gulf of Guinea, these thermometers gave very interesting results. When temperature observations came to be made in high latitudes in winter—and they are generally more instructive than those made in summer—the conditions observed by the "Challenger" in the Antarctic regions were found to be quite usual, and it was necessary to adapt the thermometer to the circumstances.

With the Millar-Casella thermometer there was furnished a "pressure correction," to be applied to the readings in proportion to the depth to which the thermometer had been sent. I always doubted the applicability of this correction, because I could not imagine that it had been determined in any other way than by observing the apparent rise of temperature as indicated by the maximum index when exposed to pressure in an hydraulic receiver. The portion of this which was due to rise of temperature by the compression in the receiver was certainly not applicable to a thermometer passing rapidly through the limitless waters of the ocean, and that part of it which was due to actual compression of the thermometer could only be applicable in the proportion of the exposed portion of the minimum limb above the index to that of the whole stem up to the maximum index, or to less than one-tenth part. But the tenth part of the pressure correction was always less than the probable error of reading an instrument. The truth of this conclusion was demonstrated by experiments made in the hydraulic apparatus on board. It was fitted on the main deck; and in the tropical regions, where the experiments were made, a mass of water equal to that in the experimental receiver did not vary by a fraction of a tenth of a degree Centigrade during the whole period of the experiments. I mention this, not only to show that the change of apparent temperature indicated by the maximum index of the thermometer under compression was produced by action in the receiver, and was in no way due to changes of temperature in the atmosphere, but also to call attention to the very favourable temperature conditions offered by a ship in tropical regions for carrying out physical experiments. The choice of such a laboratory would usually be dismissed without examination, while a little experience would reverse the decision.

While in the Antarctic regions, the impossibility of determining the temperature of the water with the thermometers available gave me cause for much thought, and it was not until we had left the icy regions that a method of adapting them to the occasion occurred to me. As similar conditions might be found to obtain in the waters near the Falkland islands, which we should pass through on the way home, I caused some special Six's unprotected thermometers to be sent out to me. My intention was, if the temperature gradient was found to be inverted, to open the extremity of one of these thermometers, and send it down open. It would then be subject to the variations of temperature and pressure obtaining at the different depths. The pressure increases regularly with the depth, and produces a sliding scale for the temperature. The effect of pressure alone would be determined afterwards in an hydraulic receiver. The opportunity for using these instruments did not occur, but the mercury piezometer, which was a development of them, was successfully used by Professor Mohn in his researches in the Norwegian sea. It is usual now to use capsizing thermometers, and to overturn them by a messenger sent down the line when the thermometer has arrived at the required depth. Both of these devices are very old, and were used by Aimé in his remarkable researches in the Mediterranean between 1843 and 1846. For the investigation of the really deep water of the ocean, we require thermometers with such a scale that tenths of a degree Fahrenheit or the corresponding portion of a Centigrade degree can be determined with certainty. In the course of my researches in the "Buccaneer," I passed over the point where three ridges meet, almost on the equator to the north of the island of Ascension. These ridges delimit three basins of the Atlantic, which are distinguished by the temperature of their bottom waters. The temperature at the bottom of the basin lying to the south and west of this point is 35·5°, in that to the north and west it is 36·0°, and to the eastward in the Gulf of Guinea is 36·4° Fahr. It is obvious that, in order to be certain of differences of this character, the thermometers must have a wide and open scale.

We have already a large number of determinations of the distribution of temperature with depth made at different localities in the ocean, but we have very few determinations of the distribution in the same locality at different seasons and in different years. In the "Buccaneer," I made a point of repeating the serial temperature observations at all of the "Challenger" stations in the neighbourhood of the Guinea coast, and very considerable differences were found, especially in the surface layer of 100 fathoms in thickness. In the Gulf of Guinea I also carried out systematically the determinations of the temperature gradient in the layer of water extending from the bottom to 250 fathoms above it (*Scottish Geographical Magazine*, 1888, p. 13). This is a branch of inquiry which has received very little attention, but it deserves to be assiduously cultivated. We have many determinations of distribution of deep-sea temperatures, but we may say no discussion of them from a calorimetric point of view. In the *Proceedings of the Royal Society of Edinburgh*, 1885–1886, p. 423, I discussed a series of observations made at different dates in Loch Lomond from this point of view, giving the heat exchanges which take place in the course of the year. The heat-unit used in this discussion was the *fathom-degree* (Fahr.)—that is, a depth of 1 fathom heated one degree (Fahr.). If the fathom has a sectional area such that the volume of water weighs 1 pound, then the fathom-degree is the same as the ordinary British heat unit, the water-pound-degree. It is rather remarkable that, if the metre be used for measuring depth, and the Centigrade degree for measuring temperature, the resulting heat-unit for depth is the same as when the fathom and the Fahrenheit degree are used, because the fathom is 1·8 times the length of the metre, and the Centigrade degree is 1·8 times that of the Fahrenheit degree.

The record of the "Challenger" in freedom from accidents is a very brilliant one. She was supplied with two qualities of sounding line designated Nos. 1 and 2, of which No. 2 was inferior in size and quality. It was used during the first two months of the cruise and was found so untrustworthy that it was condemned, and from February 1873 line No. 1 was used exclusively. Before reaching Lisbon three No. 2 lines were lost, with three thermometers. After leaving Lisbon the No. 2 line was used for the last time on February 19, 1873, when it parted, and two thermometers were lost. From this date until the end of the voyage in May 1876, No. 1 line was used exclusively, and it parted on five occasions, namely, April 28, 1873, losing one thermometer; on May 1, 1873, losing one thermometer; on August 16, 1873, losing two thermometers; and on June 14 and June 16, 1874, losing on each of these days two thermometers. After June 16, 1874, no sounding-line was lost. In all, therefore, nine sounding-lines were lost, and along with them thirteen thermometers. During the whole voyage only two temperature-lines were lost with eight thermometers. This immunity from accident was due not only to the excellence of the line, but in a far greater degree to the constant care which was taken of it by the officers and men who had charge of it.

The immunity from breakage of the sounding-line enabled the temperature at the bottom of the ocean to be determined time after time with the same thermometer, and even with the same pair of thermometers. The following is a list of the thermometers which were used more than ten times for determining the bottom temperature.

Thermometer .. No.	93	69	68	87	86	83	92	66	89
Times used	78	74	31	29	23	15	13	20	11

In the following table will be found the number of times that certain pairs of thermometers were used together for the determination of the bottom temperature at the same locality.

Thermometers .. .. Nos.	69 / 68	69 / 83	69 / 66	93 / 86	93 / 87	93 / 92
Times used together ..	28	13	12	23	22	13

The advantage of being able to use the same thermometers for determining the bottom temperature at so many different localities, does not require to be pointed out to any scientific man.

It was part of the regular routine of a station to determine the temperature of the intermediate water from the surface down to 1500 fathoms at every hundredth fathom. In addition, that of the water between the surface and a depth of 100 fathoms was generally determined at every tenth fathom. Sometimes it was determined at every twenty-fifth fathom from the surface to a depth of 300 fathoms. It was usual to use from six to eight thermometers on the line at once, so that the temperature at every hundredth fathom down to 1500 fathoms was effected in two operations. The temperatures at closer intervals, in the water near the surface, were determined also at the rate of eight per operation. Therefore in obtaining the intermediate temperatures, first 1500 fathoms had to be run out and hauled in; then 700 fathoms; then one or perhaps two shorter lengths, according to the number of temperatures near the surface which were desired. This service alone entailed the handling of something like 2500 fathoms of line, and, as we have seen, during the whole voyage only two accidents occurred to the temperature line. The following table shows the work which was done by the temperature line.

Interval of Depth	Number of separate Temperatures Observed	Mean of the Depths Corresponding to these Temperatures
Fathoms		Fathoms
From 10 to 100 incl.	1278	59·4
,, 110 ,, 200 ,,	992	160
,, 225 ,, 300 ,,	522	278
,, 400 ,, 1000 ,,	1396	700
,, 1100 ,, 1500 ,,	711	1300
From 10 to 1500 incl.	4899	465

The number of stations which furnished these 4899 temperature observations was 262, which gives an average of 18·7 observations per station. A full station in deep water included either three or four operations and furnished from 20 to 30 temperatures. At many stations the depth was less than 1500 fathoms, and it was not necessary always to take temperatures at such close intervals. The main result of the above table is to show that the enormous number of nearly five thousand deep-sea temperatures was obtained with an expenditure of only eight thermometers and one line. It was at the rate of 612 observations for every thermometer lost.

Although specimens of the earliest pattern of Negretti and Zambra's reversing thermometer were received on board the "Challenger" before the end of the voyage, all the temperatures above referred to were taken with the *Millar-Casella maximum and minimum* protected thermometer. It is owing to the use of this instrument that the temperature work of the "Challenger" during little over three years is comparable in amount with all the temperature work which has been done by other ships in the thirty years since her date. The reversing thermometer is an indispensable instrument for observations in isolated depths, and for series of temperatures in the very restricted localities where the great law of the decrease of temperature with increase of depth does not hold. The total extent of these localities is less than one-tenth of that of the whole ocean. They cover the two polar areas and the neighbouring waters which are affected by the presence of ice. In lower latitudes they include only the so-called *enclosed basins*, the largest of which is the Mediterranean, and in these the law holds rigorously down to a definite depth. The whole of the open ocean lying between the parallels of 50° N. and 50° S. can be thoroughly investigated with the protected maximum and minimum thermometer, and if hemp line be used, seven or eight of them, as experience showed, can be safely risked in each operation. The actual pattern to be used is the one with which I supplied the late Mr Casella on the return of the "Challenger"; and with it I have made all my later temperature investigations, notably the thermal survey of Loch Lomond[1] and other Scottish Lakes, as well as that of the Gulf of Guinea which I carried out on board the "Buccaneer"[2] in the early part of the year 1886. It differs from the original "Challenger" pattern in being longer and having two scales. The one scale carries either Celsius' or Fahrenheit's degrees on enamel slips fixed to the vulcanite backing of the thermometer and close alongside the stem. The other is a scale of millimetres, etched on the stem itself. This is the real scale of the instrument, and the value of its divisions is determined by careful comparison with a standard thermometer. At every observation both scales are read and the readings recorded, and the one always corrects the other in the case of a misreading. In my instruments the length of one Fahrenheit's degree was from 2·5 to 3 millimetres, which enables the temperature of very deep water to be determined with great exactness. An exploring ship should always carry

[1] *Proceedings of Royal Society, Edinburgh*, 1885, Vol. XIII. p. 403.

[2] *The Scottish Geographical Magazine*, April and May, 1888.

some thermometers reversing by *messenger* to test cases where from the indications of the maximum and minimum thermometer the law of decrease of temperature with increase of depth seems to be departed from.

Samples of intermediate water were collected at depths of 800, 400, 300, 200, 100, 50 and 25 fathoms, and a separate operation was required for each depth. At a full station this necessitated the handling of 1875 fathoms of line. This service was performed *without the loss of any material whatever.*

The mean depth corresponding to 251 soundings with temperatures, tabulated in the *Report* on Deep-sea Temperatures, is 2060 fathoms, and 403 separate observations of the temperature at the bottom were obtained. All the accidents, excepting two, happened to the sounding-line in the first sixty of these soundings, namely before, and including, the 16th August, 1873. At the beginning of the voyage only one thermometer was used at each sounding and exactly sixty temperature observations correspond to these sixty soundings. The breakages of the sounding line which occurred during this period occasioned the loss of eleven thermometers, and this was due principally to the use of the inferior sounding-line (No. 2) during the first months of the voyage.

At almost every one of the remaining soundings, 191 in number, two thermometers were used. The exact number of individual observations of temperature at the bottom was 343. We have seen that during this time, from August 1873 to the end of the voyage in May 1876, only two sounding-lines were carried away, namely, those of the 14th and 16th June, 1874, entailing a loss of four thermometers. Therefore, in two years and nine months, 343 independent observations of the temperature at the bottom, in an average depth of 2060 fathoms, were made at an expenditure of four thermometers. This is at the rate of, in round numbers, 86 determinations per thermometer lost. After leaving New Zealand, the whole of the exploration of the Pacific Ocean, occupying eighteen months, was carried out without the loss of a sounding-line or of a thermometer, or other instrument attached to it. One thermometer was lost by the parting of the temperature-line on June 18, 1875. It may be added that, during the whole voyage, eleven thermometers collapsed under the high pressure, at very great depths.

We see then that, at a full station of average depth the length of sounding line handled was:

For the sounding	2060	fathoms
For intermediate temperatures ..	2500	,,
And for intermediate waters ..	1875	,,
Total	6435	,,

When the depth was 3000 fathoms or more the length of sounding-line handled in the day amounted to from 7500 to 8000 fathoms. In addition to this there would be the 3000 fathoms of dredge-line.

In dredging and trawling the "Challenger" was equally successful. From the date of sailing from England to October 3, 1873, or nine months, the dredge or trawl-line parted six times. From October 3, 1873, till June 26, 1875, 113 stations, no line parted. In 1875 three lines, and in 1876 two lines parted. In all during three-and-a-half years, during which 354 stations were made, there were only eleven cases of the parting of the dredge or trawl-line.

In the ordinary routine on board the "Challenger" the sounding was made with a sinker weighing 336 lbs. The "Baillie" tube which carried it weighed 25 lbs., and the water bottle weighed 20 lbs., so that the total weight at the end of the line was 381 lbs. in air. Excepting the water bottle which was of bronze, all this weight was of iron, and we find its weight in water by deducting one-eight or 48 lbs. which leaves 333 lbs. as the effective sinking weight. The No. 1 line, which was in daily use, weighed in water 6 lbs. per 100 fathoms (in air it weighed 20 lbs.). Therefore, every 100 fathoms of line used in the sounding added 6 lbs. to the effective sinking weight, but at the same time by its friction it produced a retardation which depended on the velocity of descent. The retarding effect, except in very shallow water, is greater than the accelerating effect, therefore the net effect is one of retardation. With wire the opposite is the case. After the first 50 or 100 fathoms of line have run out, there is a continual and progressive retardation. The line was always allowed to run out free from coils on the deck and without any break or resistance.

Perhaps the greatest advantage which hemp line has over wire for sounding, and more particularly for dredging in deep water, is that it loses about 70 per cent. of its weight when immersed in water, whereas the wire loses only 13½ per cent. Thus, the No. 1 line weighs in air 200 lbs. per thousand fathoms and only 60 lbs. in water. The same length of sounding-wire weighs 14½ lbs. in air and 12·6 lbs. in water. The breaking strain of the line is 14 cwt. or 1568 lbs., that of the wire is 210 lbs. The length of the line which weighs 1568 lbs. in water is 26,000 fathoms, while that of the wire which weighs 210 lbs. in water is 16,700 fathoms. Therefore, granting that we can sound in 16,700 fathoms with wire, there are nearly 10,000 fathoms more that can be explored only with hemp, and beyond 26,000 fathoms, if such depths existed, we should not be able to explore them at all.

Of course these limiting depths are purely theoretical, because, each being at its breaking strain, neither the line nor the wire could be hove up from them. They serve, however, to accentuate a very real advantage which the hemp line has over the metal wire. This advantage will make itself practically felt in dredging in the great depths of 4000 and even 5000 fathoms which are now known to exist. For instance, the wire rope used by Agassiz on board the "Blake" had a circumference of 1⅛ inch. One fathom of it weighed 1·14 lbs. in air and 1 lb. in water. Its breaking strain was 8750 lbs., so that its *breaking length* was 8750 fathoms in water and at rest. It is obvious that if it were to be used for dredging in 5000 fathoms the remaining 3750 lbs. would be quite inadequate to bear the weight of the dredge with its contents, and the strain which would have to be exerted in order to bring it up in a reasonable time, to say nothing of the margin for safety, or of the resistance while being dragged over the ground.

Much misunderstanding prevails about the relative rapidity with which sounding and dredging operations can be carried on with hemp and with wire. In well-appointed cable ships, soundings in 2000 fathoms of water can be made in an hour from start to finish, but in order to do this the wire must be hove in as fast as possible and breakages frequently occur. When the wire carries deep-sea thermometers or other valuable instruments, it is impossible to work at this rate with any regard to the safety of the instruments. With hemp line the maximum rate of working can be observed whether the line carries instruments or not, because, if the line has been properly cared for, breakages do not occur.

The following short table shows the ordinary rate at which deep soundings were carried out on board the "Challenger":

Depth (*d*)		Time running out, with Sinker weighing		Time heaving in, Sinker slipped	Average Rate of heaving in per minute		Total Time of Sounding in Depth (*d*)
		3 cwt. or 150 kilos	4 cwt. or 200 kilos				
fathoms	metres	minutes	minutes	minutes	fathoms	metres	minutes
0	0	0	0	0	0	0	0
1000	1800	11·5	10·5	20	50	90	30·5
1500	2700	19·9	17·7	35	42·9	77·2	52·7
2000	3600	29·4	26·2	50	40	72	76·2
2500	4500	39·8	35·75	70	35·7	64·3	105·75
3000	5400	51·1	46·2	90	33·3	59·9	136·2
3500	6300	63·1	57·4	115	30·4	54·7	172·4
4000	7200	75·1	68·7	140	28·5	51·3	208·7
4500	8100	87·1	80·0	170	26·5	47·7	250·0

With regard to the dredge or trawl line, when clear of the bottom it was brought up from 3000 fathoms at the average rate of 1000 fathoms per hour.

In pointing out the good fortune that it was for the "Challenger" that she was fitted with hemp sounding and dredging lines, it must not be supposed that these lines can be allowed to take care of themselves, and that they guarantee the safety of the instruments attached to them. On the contrary, deep-sea sounding-line must, to begin with, be made conscientiously out of the very best long-fibred hemp, and from the time it is first used until the day it is condemned as being worn it has to be most carefully attended to, especially in warm latitudes. After every sounding it has to be thoroughly dried before being used for the next one, and it has to be constantly surveyed in case of chafes or weaknesses. It was partly to the goodness of the material, but very much more to the unremitting care and watchfulness of those who had charge of it, that after the first beginning only one sounding-line was lost. Fortunately, the novelty of the single wire for sounding purposes has worn off, and the safety of instruments is thought more of. The principal advantage possessed by the wire was that a great length of it occupied very little space, and in small vessels operating in deep waters this is of some importance. A small wire cable is now manufactured; it takes up very little more room than the single wire, and is comparatively trustworthy. H.S.H. the Prince of Monaco has a cable of this kind only 2·25 millimetres in diameter, for work on board the "Princesse Alice," and with it several water-bottles can be used at once, enabling the great bulk of the deep water of the ocean to be studied physically and chemically.

We owe, however, to the single wire and the free use which has been made of it, chiefly in the interests of the submarine cable industry, the pretty detailed knowledge which we now have of many portions of the ocean bed. Its usefulness as a means of obtaining many soundings was shown in the clearest way by the performance of the U.S. S. "Tuscarora" in the Pacific, on a line from California by the Sandwich islands to Japan, in the course of which the astonishing depths of over 4000 fathoms were first discovered; and all this work was done by hand. In the "Challenger" the distance between the soundings varied from 100 to 300 miles. In the "Tuscarora" they were taken at intervals of 30 miles, and they revealed differences of relief in the bottom of the ocean which had not been suspected. The report of the soundings of the "Tuscarora" was received with the greatest interest on board the "Challenger" when at Hong-Kong.

Although at the date of the sailing of the "Challenger" the chart of the ocean was almost a blank as regarded deep soundings, there were studded over it numerous shallow soundings, generally marked with a *D* to indicate doubt. In many places in mid-ocean shoals were indicated on which actual soundings under 100 fathoms had been obtained. Now, although at that time the art of deep sounding was not very widely distributed among mariners, every seaman could be trusted to know if he struck bottom in 40 or 50 fathoms, and it seemed to me that, if properly looked for, many of these shoals would be found to be perfectly genuine. But ships that were sent to look for them, not being fitted for rapid deep-sea sounding, generally returned satisfied of their absence because a sounding of 1500 or 2000 fathoms had been found within 10 or 12 miles of the position. But the "Challenger" and "Tuscarora" showed by abundant evidence that in the case of oceanic islands, as for instance Bermuda, a depth of 2000 fathoms may be found within 5 miles of the shore, and that such islands have often almost precipitous escarpments. When I had the good fortune to accompany Mr Robert Kaye Gray in the "Dacia" to survey the cable route between Cadiz and the Canaries, we passed over ground where one such shoal was known, and where there might be more. By following every indication of shallowing, and by rapid work, we were able to discover three new shoals rising to near the surface out of water 2000 fathoms and over, each one of which would have been overlooked by a ship sent to look for it as above. These discoveries in the "Dacia" gave body to the mass of isolated observations of shoal water in the open ocean, and instead of their being removed from new editions of the charts, their verification was undertaken seriously. Our surveying-ships, fitted with wire sounding-gear, went on their track, and, especially in the Pacific, old ones were identified and new ones were found, and all were submitted to a more or less detailed study, to the great advantage of the science of oceanography.

It is right here to point out that the search for oceanic shoals on board the "Dacia" was not undertaken on purely scientific grounds; their existence or non-existence had a commercial importance. The shoal which was known to exist when the "Dacia" began her short three weeks' cruise was the Seine bank, and it had been discovered the year before by the s.s. "Seine," belonging to the Telegraph Construction and Maintenance Company. While she was laying the second Brazilian cable between Portugal and Madeira, and in water which she had every reason to believe to be about 2000 fathoms, the cable suddenly parted, and, on sounding, the depth was found to be 110 fathoms. Instead of being laid along a plain, the ship was quite unawares laying it over the top of an isolated peak some 12,000 feet high. It is needless to say that the precautions to be observed in laying over such ground are different from those demanded by a level bottom, and the result was rupture of the cable and detention of the ship. Unwittingly and unwillingly, the ship had made an important oceanographical discovery. It was to avoid making such discoveries in a similar way that the "Dacia" did her best to find them in the preliminary survey, and was successful. The Lisbon-Madeira route had been surveyed by soundings about 30 miles apart, and when this enormous submarine mountain, the Seine bank, occurred, there was nothing in the soundings to indicate its existence but a slight shoaling from 2100 to 1800 fathoms. It was by taking this lesson to heart, and looking on a slight shoaling, even in very deep water, as the possible indication of the existence of a formidable shoal in the neighbourhood, that the "Dacia's" researches had the success that rewarded them.

It is not only in laying cables that the ships engaged in this work add to our knowledge of the bed of the ocean; it is in the repairing and recovering operations that we obtain minute and detailed information which cannot be got otherwise. The causes themselves of the rupture of a cable, which may have been laid for years and worked well, are of interest, but not always easy to ascertain. In one place the cable may have got covered up by mud, and all attempts to hook it with the grapnel are fruitless; the ship has to follow its line, dragging across it frequently with the grapnel, until she comes on it lying bare on the bottom. What are the causes which cover it in one place and leave it bare in another? In warm and shallow seas the cable gets quickly grown over by corals and other calcareous growths, providing it with a rocky pipe or tunnel for its bed, which is not always an advantage. When a break occurs, even in very deep water, it is not seldom found that, in spite of the most careful survey, the cable has been laid over a patch of bare rocky ground which is there usually found to be slightly shallower than the surrounding soft ground. Then the existence of movement in the waters, even at the bottom of deep oceans, is made manifest by the worn-through ends of the broken cable; each wire in its strand is sharpened out into needle-points.

The "Dacia's" cruise was fruitful, not only in the discovery of shoals, but also in the determination of their nature. It was in steering toward a slight shoaling in the soundings on the Lisbon-Madeira line, similar to that which was found to have indicated the Seine bank, that, when 50 miles short of the position, bottom was unexpectedly struck in about 450 fathoms.

In the detailed study of this bank, which was at once undertaken, the dredging-apparatus improvised out of grappling-gear, being very strong, succeeded in wrenching away masses of the rocky substance of the bank and bringing them to the surface. They consisted of luxuriant branches of a coral, determined by the late Professor Moseley as *Lophohelia prolifera*, the living ones rooted to and flourishing on dead branches of the same species, which were beginning to get coated with peroxide of manganese. No depth less than 400 fathoms was found on the bank, and it increased gradually to about 500 fathoms near the edge, from which it plunged very steeply to a depth of 900 to 1000 fathoms, when the slope became more moderate. In sounding round it, the lead was dropped in one place just on the edge of the precipitous slope. It struck bottom in 550 fathoms, then tumbled off, stopped again in 620 fathoms, tumbled off again, and finally brought up in 800 fathoms, the ship being motionless, the sea calm, and no wind. It is probable that this sounding indicates the general character of the escarpment of the upper 300 or 400 fathoms of the bank. It was important to observe this coral growing luxuriantly in deep water of temperature under

50° Fahr. (10° C.), and building up a veritable reef, towards which it had already contributed a pedestal some 400 fathoms high. It had also a particular interest for me, because until then there was an objection to Dr Murray's theory of coral islands to which I could never find a sufficient answer. In Murray's theory, as stated by him, the shoals on which the reef-building corals of tropical seas finally settled, and, by their vital activity, increased so as to form an atoll, were formed by the more rapid accumulation of calcareous sediment on the shallower parts of the ocean than on the deeper, the rate of accumulation increasing as the depth diminished. Now, knowing that the effect of our eminence on the bottom of the ocean is not only to restrict and to locally intensify any current that may exist in the region, but also, by interfering with the all-pervading tidal undulation, to convert it, *pro tanto*, into a tidal current; the effect of either or both of these agencies would be to arrest the growth of a shoal formed of sediment, at a depth so great and in water so cold, that the lodgement on it of tropical reef-builders would be impossible. But with the deep-sea corals stepping in as an intermediary, and taking up the building in depths of 1000 fathoms or more, where the current begins to sweep away as much sediment as falls, thereby stopping its accumulation, and by the same act producing the conditions most favourable for the growth of the deep-sea corals, the difficulty was removed, and the so-amended theory became an accurate expression of the facts.

While a certain amount of time was spent in the "Dacia" in ascertaining the slopes in the oceanic shoals, which had only a scientific interest, much more time was spent in ascertaining the slopes leading up to the shores of the islands on which the cable was to be landed. This is the most difficult and delicate part of the submarine-cable engineer's work. The Canary islands are all volcanic, and have been built up from the bottom of the ocean by successive overflows of lava, which takes the form of more or less raised streams having a more or less arched surface, like a glacier. The valleys of the islands are primarily the intervals between these streams, subsequently accentuated by the meteorological decay of the rock. The landing of the cables on the islands of Teneriffe, Gran Canaria, and especially La Palma, was no easy matter, and an enormous number of soundings were taken, during which it was conclusively shown that the valleys which form so remarkable a feature of the islands are continued with much the same features under the sea, and down to a depth of 700 or 800 fathoms. This was an important observation as a fact in the morphology of islands; it was also commercially valuable, as the cables were, in the end, laid in a bed, where they remained in good order for many years. The very detailed sounding work made in this expedition showed how great the advantage of it may be; and when surveying the route for the West African cable, to connect up places along the coast, from Conakry, near Sierra Leone, on the north, to St Paul de Loando, on the south, it was surveyed in a series of profiles, run from the deep water to within the 100-fathom line, or in the reverse direction. The rule in running these profiles was to have soundings not differing by more than 200 fathoms. The steepness of the continental escarpment varies considerably round the Gulf of Guinea, in certain positions being quite precipitous, in others very gradual. The average slope of the steep part of the escarpment was about 150 fathoms in the nautical mile; that is, the tangent of the inclination was 0·15, and the angle 8½°. Near Cape Three Points the slope was as much as 322 fathoms per nautical mile; whence the tangent of the inclination is 0·322, and the angles 18°. The approaches to the volcanic islands, St Thomas and Principe, had to be surveyed, and, as in the Canaries, they were found very steep, and with submarine valleys. Here a slope of as much as 356 fathoms in half a nautical mile was measured, giving the tangent of the slope 0·712, and the angle 35½°. If precipitous slopes on land be examined with reference to the ratio of height to horizontal distance, which gives the tangent of the slope, these slopes will be found to be steeper than the average of precipitous mountain-sides. On the moderately steep parts the soundings were taken at distances of 1 nautical mile apart; where the slope was gradual, at 2 miles apart. In work of this kind, the great convenience of the units commonly used by all seafaring people is very apparent. The units of distance still in use on board the ships of all nations is the nautical mile, which is the length of a minute of arc on the meridian at the place. The unit of depth used by all nations except the French up to the last eight or ten years was the fathom, and the British fathom, for all purposes of comparing depth with distance, is the $\frac{1}{1000}$ part of a nautical mile. More nearly, 1010 fathoms go to the nautical mile, so that a correction of one per cent. is all that is required for work of great accuracy. In the work which we have been describing, 10 fathoms is a small fraction of the ship's length, and accuracy to that extent cannot be guaranteed. The difference of depth in fathoms per nautical mile of distance gives at once the tangent of the slope. Using nautical miles for distance and metres for depth is exceedingly inconvenient, and only produces confusion. Up to the year 1880, all the oceanographical work that had been published was expressed in terms of the nautical mile and English fathom, and at the present day quite nine-tenths of what has been done has been done with the fathom. It is a very great pity that continental nations, other than France, should introduce confusion by using a new unit of depth, and one which has no simple relation with the unit of distance. As an example of the great convenience of the fathom and nautical mile, the report on the magnificent work of the U.S. S. "Albatross," between California and the Sandwich islands, published at Washington in 1892, may be cited.

The routine chemical work on the "Challenger" consisted in boiling out the atmospheric gases, and determining the carbonic acid in as many samples of water as possible. The apparatus for boiling out the gases was the same as that used and described by Professor Jacobsen, to whose visit to Leith in the "Pomerania" I was indebted for many useful hints. The apparatus for determining the carbonic acid differed from his in some details. He boiled down the water to very near dryness, collecting the distilled water and evolved carbonic acid in a receiver holding baryta water. Dr Jacobsen told me at the time that there was danger in his method of the carbonic acid of the neutral carbonates being driven out by the concentration of the chloride of magnesium, which was obvious; but he thought that what he had obtained was mainly the free and half-bound carbonic acid. He found that carbonic acid came off during the whole of the distillation, and his view was that it was loosely bound to the chloride of magnesium. In the autumn of 1872, I made some experiments on the absorption of carbonic acid by saline solutions. These were not very conclusive, but they seemed to point to the sulphates as being capable of retaining carbonic acid. Acting on this indication, I thought it could do no harm to remove the sulphates before boiling out the carbonic acid, and an essential feature in the method which I used was the addition to the volume (225 c.c.) of sea-water of 10 c.c. of saturated chloride of barium solution, which was more than sufficient to precipitate all the sulphates. The effect of this addition was to accelerate greatly the evolution of the carbonic acid, and to produce very smooth and regular boiling.

When Jacobsen's "Pomerania" results were published in 1874, and were received on board the "Challenger," I compared my results up to date with those which he had obtained in northern waters, and I was concerned to note that the carbonic acid which he found was almost exactly double what I found. It seemed unlikely that this could be due to locality, and I very thoroughly examined my method, and found nothing to find fault with. The well-known risk in such an operation was that, from decomposition of neutral carbonates by the chloride of magnesium, the carbonic acid obtained would be too high. As mine were all much lower than Jacobsen's, they were probably freer from that source of error than his. It must be remembered that what both Jacobsen and I wished to determine was the carbonic acid which is *not* bound to the base as neutral carbonate. It was only the free or half-bound carbonic acid that could be held to form part of the atmosphere offered to the inhabitants of the water, which it was our first business to analyze. As chemists, both of us were aware that the total carbonic acid could be easily and accurately determined by acidifying the water and collecting the carbonic acid. But this was not what was wanted. It subsequently turned out that by the addition of chloride of barium, I had taken exactly the best means of avoiding the action of the chloride of magnesium on the neutral carbonates which vitiated Jacobsen's results; for it had been observed by H. Rose, though at the time I was ignorant of it, that the precipitate formed by chloride of barium in a mixed solution of carbonates and sulphates is a double salt of the carbonate and sulphate of baryta, which is extremely stable even in presence of strong acids, to which it yields its carbonic acid only with great difficulty. It was, therefore, practically quite proof against the action of a weak acid like chloride of magnesium. My results, therefore, gave quite accurately what was wanted, namely, all the half-bound and free carbonic acid, and none of that engaged as neutral carbonate.

Part of my regular work was to examine the samples of bottom which came up in the sounding tube or the dredge. Amongst them the most interesting were the ochreous deposits, of which the best example is the manganese nodule which was found to be widely distributed over the bottom of the ocean.

The experience of the "Challenger" was that manganese nodules are found all over the ocean, but principally in the great depths where calcareous deposits are rare or absent. On September 23, 1878, while prosecuting oceanographical researches in the Firth of Clyde[1] on the steam yacht "Mallard," which I had built in that year expressly for such work, I found in Loch Fyne in water of about 100 fathoms a rich deposit of mud which contained over 20 per cent. of its bulk of manganese nodules, which, in outward appearance and characteristics as well as in chemical composition, were not to be distinguished from oceanic nodules. This was a very important discovery. Some years later these nodules were found in other parts of the Firth of Clyde. The submarine manganese nodules are a distinct geological formation. Their essential constituent is an ochre, that is, a higher hydrated oxide of one or more of the metals of the iron group. The hydrates of the peroxides of manganese and of iron are present in preponderating quantity, and they are always

[1] See Paper No. 8, p. 160.

accompanied by the homologous oxides of nickel and cobalt. The association of these four metals and the constancy of character observed in the nodules, suggested to me as a first idea that they were perhaps simply the products of the oxidation of meteorites. Further acquaintance with them rendered this explanation very improbable. A characteristic feature of the nodules is that when heated in the closed tube they emit a strongly empyreumatic odour and give off steam which condenses to an alkaline liquid[1]. As my attention was thus early directed to the formation of ochres, I carefully studied every occurrence of them. The organic matter revealed by heating in the closed tube was as invariably present as the ochres, and in the many instances, principally in the Pacific, where large fragments of pumice were brought up from great depths, these masses were perforated by annelids and the holes produced were almost always clothed with a black ochreous lining of the same composition as these manganese nodules, and the pumice in the neighbourhood of the holes was stained of blackish brown colour from the same cause.

This frequent occurrence of the ochreous formation in connection with the deep sea annelids and the invariable occurrence of organic matter in freshly collected nodules, suggested the connection of the formation of the ochreous deposits with the organic life on the bottom. Ochres, especially hydrated ferric oxide, are essential constituents of the oceanic "red clay." When the sounding tube brings up a sample of bottom from one of these regions, it is quite usual to find that the upper layers of the samples are of a red colour, while the mud immediately below is of a bluish-black colour. As the dredge furnished evidences of the abundance of life in the mud, as the difference of colour of the upper and lower layers of the mud was evidently due to a different state of oxidation of the iron in it, and as the water in contact with the surface of the mud always showed a deficiency of oxygen, there was little difficulty in concluding that the existence of animal life in the mud had some effect in modifying its chemical composition.

[1] They lose this property on being preserved in the air.

When a freshly collected sample of submarine mud is carefully washed with a jet of water, until the finer flocculent particles are removed, the mud which remains is in the form of elongated casts of ellipsoidal form. Pressure with the finger breaks them up into flocculent particles, which can be washed away with the jet of water, leaving still some ellipsoids. By continuing this treatment, finally all the flocculent matter can be washed away, but the ochreous deposits thus freshly collected and carefully examined are always found to be made up of these ellipsoids which are nothing more nor less than present-day coprolites. The animals, which live in abundance in the mud, live by passing it through their bodies and extracting from it what nutriment they can. The trituration of the mud in the interior of the animals and in contact with living organic matter reduces the sulphates of the sea-water to sulphides. These, in contact with the ferric oxide of the mud, reduce it to ferrous sulphide with separation of sulphur. Hence the mud not immediately in contact with the water has a bluish-black appearance. When it comes in contact with the water which contains free oxygen, the ferrous sulphide is oxidised and the surface layer becomes red. If this is the true explanation we ought to be able to find traces of free sulphur in the mud, although the finely divided sulphur which is produced in this class of reaction is easily oxidised.

Acting on this idea, and connecting it with Oscar Peschel's brilliant application of the *Relicten fauna* of lakes and rivers in the diagnosis of morphological terrestrial changes, I treated a series of oceanic muds and manganese nodules with chloroform for the extraction and determination of any sulphur that they might contain. The experiment was successful in every case, and the results are given in a paper[1] on the occurrence of sulphur in marine muds, read before the Royal Society of Edinburgh. When surveying the Gulf of Guinea in 1886 in the "Buccaneer," I found this coprolitic character of the mud near the mouth of the Congo so highly developed that in the reports of the soundings I had to introduce a new designation for this class of mud, namely *coprolitic mud*.

[1] See Paper No. 6, p. 133.

Bathybius. When the "Challenger" started on her voyage, it was not only expected that the bottom of the sea would be found everywhere covered by a calcareous deposit, but it was believed that it had been shown that the mud at the bottom of the ocean was everywhere associated with an all-pervading organism to which Huxley[1], its discoverer, had given the name of *Bathybius.*

The following extract from Wyville Thomson's *Depths of the Sea*, p. 410, gives a description of a mud in which this mysterious being was believed to be present.

"In this dredging, as in most others in the bed of the Atlantic, there was evidence of a considerable quantity of soft gelatinous organic matter, enough to give a slight viscosity to the mud of the surface layer. If the mud be shaken with weak spirit of wine, fine flakes separate like coagulated mucus; and if a little of the mud in which this viscid condition is most marked be placed in a drop of sea-water under the microscope, we can usually see, after a time, an irregular network of matter resembling white of egg, distinguishable by its maintaining its outline and not mixing with the water. This network may be seen gradually altering in form, and entangled granules and foreign bodies change their relative positions. The gelatinous matter is therefore capable of a certain amount of movement, and there can be no doubt that it manifests the phenomena of a very simple form of life.

"To this organism, if a being can be so called which shows no trace of differentiation of organs, consisting apparently of an amorphous sheet of a protein compound, irritable to a low degree and capable of assimilating food, Professor Huxley has given the name of *Bathybius haeckelii*. If this has a claim to be recognised as a distinct living entity, exhibiting its mature and final form, it must be referred to the simplest division of the shell-less rhizopoda, or if we adopt the class proposed by Professor Haeckel, to the monera. The circumstance which gives its special interest to *Bathybius* is its enormous extent; whether it be continuous in one vast sheet, or broken up into circumscribed individual particles, it appears to extend

[1] *Journal of Microscopical Science* (1868), Vol. VIII. p. 1.

over a large part of the bed of the ocean; and as no living thing, however slowly it may live, is ever perfectly at rest, but is continually acting and reacting with its surroundings, the bottom of the sea becomes like the surface of the sea and of the land,—a theatre of change, performing its part in maintaining the 'balance of organic nature.'"

Although *Bathybius* was discovered by Huxley it was Haeckel who popularised it. His paper on "*Bathybius* und das freie Protoplasma der Meerestiefen[1]," is one of the most fascinating memoirs that has ever been written.

In reviewing Huxley's article, he says that the most important fact brought out by Huxley's investigations is that the bottom of the open ocean, even in the greatest depths, is covered with enormous masses of free-living protoplasm which exists there in the simplest and most original form, that is, it has no definite shape and is hardly individualised. The fact that these enormous masses of living protoplasm cover the great depths of the ocean in preponderating quantity and under quite peculiar conditions, suggests so many reflections that a book could be written on them. Haeckel asks, "What is this *Bathybius* for an organism? How did it come into being? What becomes of it? What place are we to accord to it in the economy of nature in these abysses?"

Haeckel recognised clearly the far-reaching importance of the discovery. He concludes with the inquiry, "Have we not here the case of protoplasm coming continuously into being by creation? We stand here face to face with a series of dark enigmas, the answer to which we must hope to receive from future investigations."

It must be remembered that the material for the study of *Bathybius* was rare and valuable. Specimens of the mud from the bottom of the open ocean were then very scarce and were jealously guarded. It was quite legitimate for Haeckel to look forward for more light when material would be more plentiful.

In the early part of the cruise all the naturalists sought for *Bathybius*, but they found nothing answering to it which

[1] Haeckel, *Zeitschrift für biologische studien.*

showed motion. Apart, however, from the motion, the white gelatinous matter like coagulated albumen seemed to be present.

It was obvious that, if an organic body like albumen were present all over the bottom of the sea, the water taken from the bottom must necessarily contain enough of it to show clear evidence of organic matter when evaporated to dryness. Experiments which I made repeatedly in this direction gave a negative result.

As chemist of the expedition I looked at the matter from a different point of view from that of the naturalists. The nature of the experiments which I made and their result are best given by quoting from my report to Professor Wyville Thomson, which was published in the *Proceedings of the Royal Society*[1].

"If the jelly-like organism which had been seen by some eminent naturalists in specimens of ocean bottoms and called *Bathybius* really formed, as was believed, an all-pervading organic covering of the sea-bottom, it could hardly fail to show itself when the bottom water was evaporated to dryness and the residue heated. In the numerous samples of bottom water which I have so examined, there never was sufficient organic matter to give more than a just perceptible greyish tinge to the residue, without any other signs of carbonisation or burning. Meantime, my colleague, Mr Murray, who had been working according to the directions given by the discoverers of *Bathybius*, had actually observed a substance like 'coagulated mucus,' which answered in every particular except the want of motion to the description of the organism, and he found it in such quantity that, if it were really of the supposed organic nature, it must necessarily render the bottom water so rich in organic matter that its presence would be abundantly evident when the water was treated as above described.

"There remained then but one conclusion, namely that the body which Mr Murray had observed was not an organic body at all; and on examining it and its mode of preparation, I determined it to be sulphate of lime, which had been eliminated from the sea-water, always present in the mud, as an amorphous precipitate on the addition of spirits of wine. The substance when analysed consisted of sulphuric acid and lime; and when dissolved in water and the solution allowed to evaporate, it crystallised

[1] "Report on chemical work done on board H.M.S. 'Challenger,'" by J. Y. Buchanan, *Proceedings of Royal Society* (1876), Vol. XXIV. p. 605, and Paper No. 5 of *Collected Scientific Papers*, Cambridge, 1913.

in the well-known form of gypsum, the crystals being all alike, and there being no amorphous matter amongst them."

Haeckel relied chiefly on its faculty of being stained by carmine as evidence that the body which he was examining was organic. Sulphate of lime as prepared by the precipitation of an aqueous solution of a calcium salt by alcohol is a perfectly amorphous flocculent precipitate which is coloured intensely by carmine, and the colour is fast as against treatment with spirit. The naturalists on board had great difficulty at first in believing that this reaction was not, as Haeckel thought it was, absolutely decisive of the organic character of the body. To remove this view, however, it was only necessary to point out that the production of pigments by the staining of amorphous mineral precipitates with organic colouring matters was a very old chemical industry. The pigments so produced are called by the generic name of *Lakes*; and the mineral precipitate most commonly used is hydrate of alumina; but many other substances can be used for the purpose, and it appeared that sulphate of lime when freshly precipitated by alcohol was to be added to the list.

Huxley, the discoverer of *Bathybius*, frankly acknowledged the mistake which had been made; Haeckel, who adopted *Bathybius*, ceased to use the word.

I have dwelt at considerable length on these two doctrines relating to the conditions at the bottom of the ocean, namely that of the continuity in time of the chalk and that of the continuity in space of organic plasma, not only because they characterise the views held by leading naturalists between the years 1868 and 1873, but also because the proving of these doctrines was the immediate motive of much of the early work done on board the "Challenger." That the result showed that it was impossible to uphold either doctrine, diminishes in nothing the usefulness of their having been put forward as hypotheses, nor does it afford any reason for their being allowed to pass into oblivion.

The waters of different localities of the ocean are distinguished by the amount and nature of the saline matter dissolved in them. It has been found that the nature of the dissolved contents can, for almost all purposes, be held to be constant, and that, therefore, a water is generally characterised by the amount of its dissolved contents, by its salinity. This salinity, within the limits met with in the ocean, varies directly with the density. The density can be determined with great accuracy even at sea by means of a suitable hydrometer. It has been found, also, that the preponderance of chlorides over other salts in sea-water is such that the salinity of a sea-water varies approximately as the amount of chlorine which it contains. I myself always use the hydrometer, with which I determine directly the density to one or two units in the fifth decimal place, as against distilled water of the same temperature determined at the same time and with the same instrument. The chlorine method is quite unsuitable for use at sea; first, because the quantity of chlorine is so large that the amount of water convenient for analysis is very small, and it cannot be weighed at sea. Then at sea nothing is free from chlorine—the air and everything is impregnated with chlorides; so that, as a means of specifying and distinguishing oceanic waters, I consider the chemical method absolutely untrustworthy, except when made with all refinements in a laboratory on land. There is, of course, no comparison in the amount of time required compared with the hydrometer method.

Many writers, in passing judgment on the hydrometer as an instrument for the determination of the density of liquids, have only in their minds the hydrometer whose indications are determined by comparison with another or standard instrument; or by immersion in solutions, the densities of which have been otherwise ascertained. These instruments have no greater value than that of more or less carefully constructed copies of a standard, the method and the principle of the construction of which is not always given. Rightly, therefore, they prefer the density as determined by weighing a vessel filled with the liquid and comparing it with the weight of distilled water of the same temperature filling the same vessel.

The hydrometer which I constructed for the "Challenger" expedition, and used during the whole of it, is not an hydrometer in the above sense: it does not give comparative results; it gives absolute ones. By its means, the weights of equal volumes of the solution and of distilled water of the same temperature are determined directly. It is neither more nor less than a pyknometer, where the volume of liquid *excluded* up to a certain mark is weighed instead of that *included* up to a similar mark. In the pyknometer, the internal surface per unit of length of the stem can be made smaller than the external surface per unit of length of the stem of the hydrometer. On the other hand, the volume of the hydrometer can safely be made many times larger than that of the pyknometer, the dimensions of which must always be kept small on account of the difficulty of ascertaining its true temperature, which is always a matter of guesswork, because it is not measured directly. The temperature of another mass of liquid is measured, and the two are assumed to be identical. With the hydrometer, the liquid being in large quantity and outside of the instrument, its temperature can be immediately ascertained with every required accuracy.

Again, for every determination with the ordinary pyknometer, the weight of the liquid contained in it has to be determined by a separate operation of weighing. With the hydrometer the weight of the liquid displaced, being always equal to its own, is determined once for all by repeated series of weighings, where every refinement is used to secure the true weight of the instrument. This weight can then be increased at will by placing suitable small weights on the upper extremity of the stem. Their weight is also most carefully determined once for all, so that at any moment the total weight of the displacing instrument is accurately known. The stem of the instrument is divided over a length of 0·1 metre into millimetres, and its diameter is chosen so that 100 millimetres of it will displace 0·9 to 1 cubic centimetre; the total volume of the instrument is intended to be about 180 c.c., and the glass-blower who supplies them generally fulfils this specification very closely. He loads the instrument so that it floats at 0 millimetre in distilled water of 30° C. The small weights used are in the form of spirals of aluminium wire for fractions of a gramme, and of brass or silver wire for greater weights. The system is such that any weight up to 10 grammes, increasing by steps of 0·05 gramme, can be added. It is thus possible, by making the first reading when the instrument is loaded so as just to be immersed to the lowest division (0 mm.) of the stem, to make a series of twenty-one independent determinations of the weight of twenty-one different volumes of the same liquid in a very few minutes. If the liquid is replaced in the cylinder by distilled water of the same temperature, twenty-one determinations of the weight of the same twenty-one volumes of distilled water of the same temperature can be made in as short a time, and we have as the result twenty-one perfectly independent determinations of the specific gravity of the liquid, that is, of the ratio of its density to that of distilled water of the same temperature; and the accuracy of each determination depends on nothing but the accuracy with which the original weighings have been carried out—that is to say, it depends on the operation, which is capable of being performed with the greatest precision in the laboratory. In actual practice I use steps of 0·1 gramme, and I aim at having at least nine separate observations both in the liquid and in distilled water. It never happens that the successive readings in distilled water are identical with those in the liquid, but by repeated immersions in distilled water the stem is accurately calibrated, so that a correction can with safety be made for the difference of one or two millimetres between them. We then have a series of scale readings, and opposite each a pair of weights giving the weights of these identical volumes of the liquid and of distilled water, and the ratio of each pair gives the specific gravity of the liquid at the common temperature. There is no difficulty, when working in the circumstances which are alone suitable for determinations of the kind, in securing identity of temperatures within 0·1° C.

For all ordinary purposes it is not necessary to make a determination in distilled water along with each sample of sea-water or other liquid under examination. When a sufficient number of separate observations have been made in distilled

5—2

water at different temperatures, we may either take the series made at the temperature nearest to that of the liquid, and compare the two after making the necessary small corrections, or we may construct a table by interpolation, giving the weights required to immerse the hydrometer up to, say, every tenth division of the stem in distilled water at different temperatures. From such a table we should be able at once to find the weights required to depress the hydrometer to the same scale divisions as had been observed in the liquid, and from them obtain the specific gravities. The table may, however, take another form. The weight of distilled water displaced at every observation is known by the weight of the hydrometer and added weights. If we know the volume of a cubic centimetre of distilled water at all the temperatures covered by the experiments, we have directly the volume of the immersed portion of the hydrometer, and as such observations are made at different temperatures, we obtain the volumes of the hydrometer at different temperatures, and its rate of expansion. In constructing a table of the volumes of the hydrometer, it should always be stated what factors have been used, so that the absolute values depending on weighing alone can be recovered. For all important or normal determinations, the parallel series of observations in distilled water of the same temperature should not be omitted.

Assuming the correctness of our knowledge of the density of distilled water of different temperatures, and deducing the volume of the hydrometer from observations with it in distilled water of known temperature, we obtain directly the volume of a unit weight of the liquid, or its density; and for many purposes this is convenient.

The following table gives the results of determinations of specific gravity of samples of Mediterranean water collected in 1893 by H.S.H. the Prince of Monaco, which happened to be made at identical temperatures with different hydrometers. The specific gravities given are each the means of from nine to eleven separate observations on sea-water and distilled water at the same time and at the same temperature. The greatest difference between any pair of values is 3·3 in the fifth decimal place, and the individuals of each pair depend on perfectly distinct sets of weighings, and are therefore quite independent.

It may safely be asserted that, working in this way, the specific gravity of a sea-water or similar solution can be determined with a probable error of not more than ± 1 in the fifth

Water, No.	Hydrometer, No.	Temperature, $t°$ C.	Sp. gravity at $t°$	Difference
8	6	8·05°	1·030156	16
	1	8·10	140	
21	1	8·00	1·030211	15
	6	8·05	196	
56	6	8·00	1·030036	4
	38	8·00	032	
51	1	10·00	1·029966	15
	16	10·10	951	
28	38	8·30	1·029234	1
	29	8·30	235	
56A	1	10·00	1·029944	0
	6	10·05	944	
8A	1	10·00	1·030271	27
	16	10·05	244	
52	6	10·00	1·029927	33
	1	10·00	894	
20	38	8·00	1·030002	1
	1	8·10	001	
20A	6	8·70	1·029945	8
	38	8·75	937	

decimal place. In a water whose specific gravity is 1·03000, 1 in the fifth decimal place represents $\frac{1}{3000}$ of the whole solid contents; so that, by the careful use of the hydrometric method, the salinity to one part in 75,000 of water, or differences of 1 grain per gallon, can be determined; and it has proved itself of great use in general chemical practice, especially in cases of pollution of streams.

It has been shown that the specific gravity of a sea-water can be determined with the greatest possible accuracy with the hydrometer, and, in reporting on a water, this is the first and fundamental result to be stated. It is an absolute value depending only on the accuracy with which the weight of the hydrometer has been determined, and is free from all other errors. In order that observations on different waters may be comparable when they have not been made at the same temperature, it is necessary to effect some reduction. It may be taken that the variation of density of distilled water with change of temperature has been determined with the greatest attainable accuracy. If we multiply the specific gravity found by the density of distilled water at the temperature, the result will be the density of the sample at the same temperature; or, as it may be otherwise expressed, the specific gravity referred to that of distilled water at 4° C. as unity. This value is affected only by whatever error may be inherent in the number taken for the density of distilled water.

To make further reductions, we require a knowledge of the law of thermal expansion of the sea-water. This cannot be known with the same precision as that of distilled water, consequently its use should be kept within the narrowest possible limits. The density of a sea-water is determined for two perfectly distinct purposes—the one is chemical, in order to obtain a knowledge of its salinity; and the other is physical or mechanical, in order to know its effect on the equilibrium of the waters. In order to compare waters with regard to their salinity, their densities must be reduced to their value at a common temperature. Such reduced densities give a very accurate representation of the relative salinity of different waters. In order to obtain absolute values, resort must be had to one of the tables connecting these two variables. For mechanical purposes, it is requisite to know the density at the temperature which the water had *in situ*. A further reduction might be made to allow for the effect of pressure in the case of water collected below the surface, but it is rarely of any mechanical importance.

It is well to arrange beforehand to determine the specific gravities of the waters during a cruise at as nearly as possible the same temperature; the results are then comparable, and with the minimum of reduction. It is my practice, in such circumstances, to take the mean of the temperatures at which observations have been made, and reduce all the observations to their value at this temperature. This gives the best account of their relative salinities amongst each other. To make them comparable with other observations, they must be reduced to one of the usual common temperatures, as, for instance, 15·56° C. (60° Fahr.). It will be remembered that we are dealing here with densities, that is, the weight in grammes of 1 cubic centimetre of the liquid, and if we reduce our densities to 15·56° C., we shall have the weight of a cubic centimetre of the water at that temperature. In some countries, especially in Germany, it has been the custom to reduce results to their values at 17·5° C. referred to the density of distilled water at the same temperature. The temperature 17·5° C. is not open to any objection, but we ought to adhere to densities, where our unit is the weight of a cubic centimetre of distilled water at 4° C. The common temperature to which the densities are to be reduced is, in itself, not a matter of any importance. The consideration which should guide us in its choice is to reduce to a minimum the alteration produced in the observed value by the reduction. As the usual temperature at which observations are made is the atmospheric temperature of the locality, we have to consider what is likely to be the most frequently occurring temperature. Many more observations are made in warm seas than in cold ones, and the area of the warm seas is greater than that of the cold ones. It would therefore seem reasonable to select a high temperature rather than a low one.

A sea-water of density 1·02600 at 15·56° has the following densities at other temperatures:

Temperature C.	0°	3°	10·5°	15·56°	19·6°	23·2°	26·7°	29·8°
Density	1·02818	1·02800	1·02700	1·02600	1·02500	1·02400	1·02300	1·02200
Difference of temperature		3°	7·5°	5·06°	4·04°	3·6°	3·5°	3·1°

As observations must always be made in a place protected from currents of air, it may be taken that 3° C. will be as low a temperature of observation as is likely to occur. Similarly, 29·8° C. may be taken as a higher extreme. At 3° C. the above water has a density of 1·02800, and at 29·8° C. its density is 1·02200, the difference being 0·00600; the mean of these two, 1·02500, occurs at 19·6° C., and if we assume that observations are likely to be made with equal frequency above that temperature and below it, 19·6° C. would appear to be a suitable temperature for general reduction. At the same time, the area of sea surface having a higher temperature than this is much greater than that having a lower temperature, and as the surface temperature of the sea is the principal factor determining the temperature of the air in a ship, it would seem reasonable to put the common temperature rather higher than lower. I have gone over the list of density determinations of surface water in the "Challenger," and only eighteen were made at temperatures above 29·0° C. in the seas round New Guinea. Subsequent experience in the "Buccaneer" in the Gulf of Guinea and on a passage from Panama, across the corresponding homologous feature of the Pacific, the great Central American bight, to San Francisco, showed that in these parts of the ocean during one-half of the year the temperature of the surface water and of the air is generally above 29° C.

In the following table are given the mean temperatures at which the density of surface water was determined in the different oceans:

Ocean	Number of observations	Sum	Mean temperature
South Pacific	163	3426·8	21·02°
North Pacific	117	2853·0	24·38
Pacific	280	6279·8	22·43
Molucca sea	70	1933·5	27·62
	350	8213·3	23·46
Atlantic	232	4946·7	21·32
	582	13160·0	22·61
Antarctic	77	702·7	9·12
Inland channels	21	276·8	9·99
Total	680	14139·5	20·78

The average temperature at which all the "Challenger" surface observations were made is 20·94° C. If the Antarctic cruise is excluded, the average in Atlantic and Pacific taken together is 22·87° C., the average in the Pacific alone being 23·97° C., and in the Atlantic alone 21·32° C. In exploring equatorial waters very few observations would be made at a temperature below 25° C., and, except under special and not particularly agreeable circumstances, very few above 29° C., and between the tropics they would all fall above 20° C. Looking to the much greater extent of the waters represented by the Atlantic and Pacific part of the cruise, it might seem to be more convenient to take the temperature resulting from their consideration than that found when the Antarctic is included; but it must not be forgotten that, though smaller in extent, the cold regions of the surface of the ocean are of great interest from a physical and chemical point of view, and it would seem that for a common temperature to which to reduce all densities, 20° C. would be the most convenient. The large number of densities already existing which are reduced to the constant temperature 15·56° C. are transformed by subtracting 0·00110; and these which are expressed as specific gravities at 17·5° C., referred to that of distilled water of the same temperature as unity, are transformed by subtracting 0·00189 for average ocean water. These corrections, applied to the reduction tables now in use, would at once fit them equally well for a reduction to a common temperature of 20° C., and with these tables the density of the water *in situ* is also at once found.

To recapitulate: In determining the density of the sea-water in an expedition only the absolute-weight hydrometer should be used. The samples of water should be stored in the laboratory where the observations are going to be made, and they should have sensibly the same temperature as the air while the observations are being made. If the motion of the ship is at any time too violent for it to be convenient to make the observations, then a sufficient supply of bottles should be at hand to keep the samples until the motion becomes less without interfering with the collection of other samples.

When the water is in the cylinder, its temperature is carefully taken with a trustworthy thermometer, which must be divided into tenths of a Centigrade degree. The thermometer is then removed, and the hydrometer immersed and loaded with small weights, until the water-level rises to one of the lower divisions of the scale. It is unnecessary to point out that the water in the cylinder must be at rest, that the stem of the hydrometer must be sheltered from wind, that everything must be clean, and that the ordinary precautions usually observed in every physical or chemical laboratory are to be observed. Having obtained the first reading, further small weights are added by steps of 0·1 gramme until at least nine observations have been obtained. Sometimes it is convenient to use the 0·05-gramme weight near the top or bottom of the stem. Care is taken that the stem of the hydrometer is wetted for a distance of one or two millimetres, but not more, above the division where the hydrometer is going to float. This is an essential precaution for ensuring precision. When the last observation, which must be the one nearest the upper extremity of the stem, has been made, the small weights are removed, and the hydrometer lifted out and put in safety, and the temperature again taken with the thermometer. It should not differ from the temperature found at the beginning by more than 0·3° C., and in work making any pretensions to accuracy it should not exceed 0·1° C. If a difference of temperature amounting to 0·3° C. is observed, and the temperature itself is above 20° C., then the mean temperature must not be taken and used for the nine observations, but the 0·3° C. must be distributed over them, and the temperature which the water had at the time of each observation used. The difference appears at once in the fifth place of decimals.

Whether at sea or on land, I always log the time in my laboratory work. A series of observations with the hydrometer as above described takes on an average about twelve minutes; but in that time at least nine quite independent observations have been made of the density of the water. If a sufficient number of observations have already been made with the hydrometer in distilled water of the same or nearly the same temperature, they may be used for giving the specific gravity of the water. If they have not, or if the determination is of especial importance, then a precisely similar series of observations must be made in distilled water of the same temperature, and variations of temperature amounting to 0·3° C. are then inadmissible.

When the corresponding series of observations has been made in distilled water, and they have had the small stem correction applied so as to give the displacing weight in distilled water at the exact stem divisions observed in the sample water, we have nine pairs of readings, each pair giving the weights of equal volumes of distilled water and of the sample, and therefore each pair giving by their ratio an independent determination of the specific gravity of the sample referred to that of distilled water of the same temperature as unity. The mean of the nine observations gives a result which, according to the doctrine of probabilities, should have a precision three times greater than that of a single observation. Although much may be done to avoid a large range of temperature of observation, there will always be some difference in the temperatures at which the specific gravity of the various samples is observed, and as a similar variation would take place in the unit to which they are referred, we effect a reduction to their value, taking the density of distilled water at 4° (as unity). This is obtained by multiplying the figures obtained as above for the specific gravity by the weight of one cubic centimetre of distilled water at the temperature of observation. This is one of the physical constants which have been determined with the greatest care and presumably with the greatest precision, and therefore, if reductions are to be admissible at all, this one can be made with the least fear of error. The specific gravity multiplied by this weight of unit volume of distilled water of the same temperature gives the *density* of the sample water at that temperature, that is, it gives the weight of one cubic centimetre of it in grammes. It may also be correctly described as the specific gravity of the water at the temperature, that of distilled water at 4° C. being unity.

The density at temperature of observation can now at once be reduced to its value at whatever is chosen as common temperature, and at the temperature which it had *in situ*. Naturally, these reduced values are affected by whatever uncertainty attaches to the tables used. For purposes of control, it is well, in an expedition, to preserve very carefully considerable samples of typical waters, and, as opportunity offers, to determine their specific gravity against distilled water at different temperatures. There is no reason to suppose that the precision of these determinations would stand in any way behind that of the observations on which the tables are founded, and as they would have been made on the actual waters which are under consideration, they should have the preference.

It will be obvious from these remarks how necessary it is, in an expedition, to have a supply of perfectly pure distilled water, to make parallel observations under the conditions on board. Also, as before recommended, typical samples of the waters collected should be kept for careful determination of their density with the same hydrometer at different temperatures, and especially at or near the temperature taken as the common temperature of reduction.

In the course of over twenty years' work with absolute-weight hydrometers, and the determination of the constants of about fifty different instruments, a curious fact has established itself, namely, that the rate of change of volume of the instrument with change of temperature is within certain limits variable. The limits are very narrow, and the phenomenon is to be detected only by very careful determination of the displacement in distilled water. I at first thought it might be due to distortion of the instrument with change of barometric pressure; but many series of observations with different instruments showed that the effect has no connection with barometric pressure. For this reason, in the case of normal determinations, parallel observations in the liquid under determination and in distilled water of the same temperature should always be made.

The more use is made of an absolute-weight hydrometer,

Stem division 0 10 20 30 40 50 60 70 80 90 100

BUCHANAN IN S.S. "LEIBNITZ."

Hydrometer J.Y.B. No. 14.

Nett totals	..	156·9	109·6	74·6	14·7	9·8	−27·4	−35·3	−35·1	−170·0	−97·8
Number of observations	..	58	35	30	11	48	40	25	16	41	38
Mean nett value ..	..	2·70	3·13	2·48	1·34	0.20	−0·69	−1·41	−2·19	−4·15	−2·57
Maximum value ± ..	..	9·0	10·5	7·0	4·0	5·3	4·7	5·7	7·0	10·5	8·0

THOMSON IN S.S. "SILVERTOWN."

Hydrometer J.Y.B. No. 20.

Nett totals	..	−61	−225	−290	−161	431	592	32	−54	−208	−7
Number of observations	..	74	169	152	108	136	190	130	169	137	3
Mean nett value ..	..	−0·82	−1·33	−1·91	−1·50	3·17	3·1	0·25	−0·32	−1·52	−2·3
Maximum value ± ..	..	14	11	14	13	15	13	12	10	13	3

THOMSON IN S.S. "SILVERTOWN."

Hydrometer J.Y.B. No. 11.

Nett totals	..	−5	−25	24	48	56	−6	−107	−37	—	—
Number of observations	..	4	85	87	45	77	67	95	52	—	—
Mean nett value ..	..	−1·25	−0·30	0·28	1·07	0·73	−0·09	−1·13	−·072	—	—
Maximum value ± ..	..	4	6	8	8	7	9	6	9	—	—

BUCHANAN IN STEAM-YACHT "PRINCESSE ALICE."

Hydrometer J.Y.B. No. 12.

Nett totals	..	−16·6	3·7	28·0	−7·9	0·8	−10·3	11·8	5·8	−10·8	1·8
Number of observations	..	14	17	24	5	9	27	18	18	31	7
Mean nett value ..	..	−1·19	0·22	1·17	−1·30	0·09	−0·38	1·04	0·32	−0·35	0·26
Maximum value ± ..	..	5·3	4·2	6·1	5·9	1·2	3·9	5·8	4·0	5·2	4·7

the more valuable does it become. The repeated observations in distilled water increase the knowledge of its displacement, and when many observations have been made in distilled water or in sea-water, and the single observations of each series are compared with their mean, a calibration of the stem is obtained which no other process can give. This is exemplified by the table on p. 77, where the observations made on different expeditions have been classified according to the stem readings and their difference from the means in units of the fifth decimal place.

In the case of hydrometer No. 14, it will be seen that the densities observed on the lower half of the stem are too high, and those on the upper half are too low. This points to the probability that the stem is slightly tapered, being thicker in the lower half than in the upper. It also furnishes a correction which could be applied to future observations. The other instruments also show probable very slight inequalities in the calibre of their stems.

For observing on board ship, I find that the method used in the "Challenger," of placing the cylinder on a swinging table, gives better results than any other. In the "Princesse Alice," a ship of not more than 600 tons, the motion at sea is always considerable, but in ordinary circumstances the maximum amplitude of the motion of the floating hydrometer was not over 3 millimetres. The vibration period of the hydrometer always interferes with that of the ship and swinging table, producing moments of rest, and my experience is that a moderate rate of motion is an advantage. The individual readings of a series made under favourable circumstances at sea generally agree more closely with the mean than is the case with a series made under similar circumstances on land. The limit to the amount of motion with which trustworthy observations can be made does not depend on the hydrometer, but on the observer. When the motion goes beyond a certain amount of violence, the observer's attention is entirely taken up in looking after his own stability, and in preventing his coming into collision with the swinging table.

In the "Princesse Alice," I frequently compelled myself

to observe when the motion was violent, and then kept the waters for observation under more favourable circumstances. I rarely found any sensible differences in the results; but the labour of making the observations in bad weather is very great, and has an irritating effect. In the "Challenger" there was no difficulty in deciding whether the observations should be proceeded with or not, because there was no difficulty in making them in weather that admitted of the main-deck port, which lighted and ventilated the laboratory, being kept open. When it was shut, the darkness put a stop to such observations independently of the motion.

It is, perhaps, not wholly unnecessary to point out that to obtain good results with a method such as this the observer must have a certain amount of dexterity and patience, but more particularly he must approach the matter with the desire to succeed. There is never any difficulty in making unsuccessful experiments.

One of the most important physical features of the ocean is the motion of its waters, and as it directly affects the course of a ship, it has at all times come under the notice of the mariner, and it forms an essential factor in navigation. With this end, it is the custom on board ship to keep two parallel series of determinations of the position, which are entered in the log from day to day. The one gives the position at one or more times of the day as fixed by observations of the heavenly bodies and of a correct time-keeper; the other gives the position by linear measurement of the distance run through the water, and determination by compass of its direction. The former gives the absolute position on the sphere at the moment of observation; the latter gives the position at the same moment on the supposition that the water of the ocean through which the ship has passed has no proper motion. The distance between the two positions divided by the time from the previous fixture of position is logged as current. Tidal components of the current are eliminated by the fact that the complete period of the tidal ebb and flow is very nearly twenty-four hours, and the currents produced by it in that interval must nearly balance each other.

It is by such observations that the main features of oceanic currents have been ascertained and delineated. In sailing ships and steamers of moderate speed valuable indications can be obtained, especially where the currents are strong. As the strongest oceanic currents—those in the neighbourhood of the equator—occur where the winds are feeblest and calms most frequent, much very valuable information has been received from the observations of sailing ships on long voyages from one hemisphere to the other. It was a frequent experience to be becalmed for days in the regions known as the Doldrums; but it was the almost invariable experience that, though not moving through the water, the position of the ship on the globe had altered often by as much as 50 or 60 miles in the twenty-four hours. Here the greater part of the change of position was due to current, and the amount of it as logged by an experienced navigator was a trustworthy record of the average current prevailing at the place during the interval of time between two observations. It was not the least of the many services that we owe to Maury, the founder of oceanography, that by collecting and discussing thousands of ships' logs, he produced the first reliable chart of the ocean currents. Modern steamers run so fast that the difference between their position as by observation and by dead reckoning, though important for their commanders, is of little use for our purposes.

As the "Challenger" was to all intents and purposes a sailing ship, for she was never under steam except when actually sounding and dredging, or occasionally when going into and out of harbour, good results were obtained in this way of the average current every twenty-four hours; and in the tabulated meteorological observations published in the second volume of the Narrative in the "*Challenger*" *Reports*, an entry will be found each day of the current logged. But in the first year of the cruise, a practice was occasionally adopted which, unfortunately, was departed from later, namely, to anchor one of the ship's boats either by the sounding-line or by the dredge-rope, and, from the boat thus stationary, for one of the navigating officers to make careful observations of the current, both at the surface and at certain depths below it

These observations were chiefly made in the months of August and September, 1873, on the cruise from the Cape Verde islands to Fernando Noronha. From the point of view of oceanic circulation, this is the most interesting portion of the Atlantic, and the results obtained were important and novel. The following are worthy of being quoted:

"*'Challenger' Reports*, Narrative I. p. 192.—On August 16, 1873, at Station 100, lat. 7° 1′ N., long. 15° 35′ W., depth 2425 fathoms, in the Guinea current, the dingy was anchored by the sounding-line, and the surface current was found running N. 70° E. half a mile per hour. The current-drag at 50 fathoms indicated a set of 0·45 mile per hour N. 17° E.; at 100 fathoms, N. 15° E., 0·3 mile per hour, and at 200 fathoms, N. 17° E., 0·2 mile per hour. On the 19th, at Station 101, lat. 5° 48′ N., long. 14° 20′ W., depth 2500 fathoms, the cutter was anchored by the trawl, and the surface current found running N.E. 1·3 mile per hour. On the 21st, at Station 102, lat. 3° 8′ N., long. 14° 49′ W., depth 2450 fathoms, the dingy was anchored by the lead-line, and the surface current was found running N.W. 1·25 mile per hour. On the 25th, at Station 106, lat. 1° 47′ N., long. 24° 26′ W., depth 1850 fathoms, the cutter was anchored by the trawl, and the surface current at 10.30 A.M. was found to be running west (true) 2 miles per hour; but in the afternoon its velocity had decreased to 1 mile per hour. The current-drag at 10 A.M., at 75 fathoms, showed no current; at 50 fathoms, a current of ½ a mile per hour; and at 15 fathoms, ¾ of a mile per hour, all to the west, thus showing how very superficial the equatorial current was. On the 26th, at Station 107, lat. 1° 22′ N., long. 27° 36′ W., depth 1500 fathoms, the cutter was again anchored by the trawl, and the surface current found to be running west (true) 1·5 mile per hour; and it continued to run at that rate throughout the day, instead of slackening in the afternoon, as on the 25th. Also at Station 71, lat. 38° 18′ N., long. 34° 48′ W., depth 1675 fathoms, a variation in the surface current over the day was observed, and was ascribed to tidal influence."

These observations gave absolute values for the current at the surface and at some depths below it in mid-ocean. They showed that the rate and direction of the current varies considerably with the distance from the surface, and that the current at the surface varies with the time of day, and is probably subject to a tidal influence. Although in the later part of the voyage the current observations from an anchored boat were not persevered in, no opportunity was lost in fixing the position of the ship astronomically at short intervals while she was preserving the same position in the water at the various stations. Some remarkable results were thus obtained in the equatorial regions of the Central Pacific.

"'*Challenger' Reports*, Narrative I. p. 772.—The axis of greatest velocity of the equatorial current was on the parallel of 2° N., where its speed amounted to 3 miles per hour. Such an exceptional velocity has, so far as is known, only been recorded once before, viz., by the French corvette "Eurydice," in August, 1857. The astronomical observations taken at frequent intervals showed even a greater velocity than 3 miles per hour. By these observations it appeared that the vessel was in still water between the equatorial and counter-equatorial currents on September 2, in lat. 5° 54′ N., long. 154° 2′ W. From this position to lat. 4° 32′ N., long. 147° 28′ W., the velocity of the equatorial current was ¾ mile per hour S. 53° W.; from thence to lat. 3° 55′ N., long. 148° 10′ W., its velocity was 1½ mile per hour; thence to lat. 3° 32′ N., its velocity was 1¾ mile per hour; thence to lat. 2° 34′ N., long. 149° 9′ W., its velocity was 3 miles per hour S. 76° W.; thence to lat. 2° 10′ N., long. 149° 34′ W., its velocity was 4 miles per hour; thence to lat. 1° 0′ N., long. 150° 30′ W., its velocity was 3 miles per hour S. 85° W.; thence to lat. 0° 25′ N., long. 151° W., its velocity was 2½ miles per hour; thence to lat. 0° 43′ S., long. 151° 32′ W., its velocity was 1¾ mile per hour S. 81° W., and then it gradually decreased."

Observations of this kind require good will, good eyesight, good instruments, and a good use of them all. The ocean water, which was thus proved to be moving at the rate of 4 miles per hour, appeared to the eye to be motionless. There was no trouble on its surface, yet it is a fact which may astonish the casual observer, that there are very few of the apparently most violent mountain-torrents which get over the ground at as great a speed. Of this I have assured myself by many measurements in Switzerland and other countries.

A time-honoured method of measuring the movement of oceanic waters is by throwing overboard floating-bodies, generally empty corked bottles containing a paper with the date and position of their starting-point. If suitably ballasted so as to be protected from the influence of the wind and be exposed to the action of the current alone, very valuable results can be obtained in this way. The most remarkable are those obtained by H.S.H. Albert, Prince of Monaco. His floats were especially made for the purpose, and were thrown over in series on a definite system from his schooner yacht "Hirondelle" in the course of his scientific cruise round the

Azores. The result of his work is to prove the existence of a circular drift current round the central basin of the North Atlantic. In districts where the currents are strong, much valuable information can be obtained from floats. In the "Buccaneer," when exploring the Gulf of Guinea, Mr Little, the first officer, threw a bottle overboard every day at 1 P.M., and the few that were picked up and returned show very conclusively the different motion of these waters in different months.

The tidal influence on oceanic currents which was shown by the observations in the "Challenger" has since received wide confirmation, notably by the Americans in the Gulf Stream. It is impossible, in a short paper like the present, to go into all the work that has been done in this way, and that of the Americans alone fills volumes. We have chiefly to consider what has still to be done, and how to do it.

A remarkable paper on this subject has been communicated to the Congress by Captain Anthony S. Thomson, C.B., R.N.R., with whom I was associated in the "Buccaneer" in the exploration of the Gulf of Guinea, where several attempts were made to determine the currents, both surface and under currents, in that interesting region. Captain Thomson's paper falls into two parts. The first deals with the probable causes of ocean currents, and it is quite possible that many will hesitate to accept them unreservedly. The second part, however, deals in detail with the method by which a satisfactory survey of the currents existing in any locality far from land can best be made. Here Captain Thomson's record and experience give an especial value to what he may say on this subject. The paper will be read to the Congress, and it will be convenient to keep any remarks which I may have to make on the subject for the discussion on it. I will only point out here that observations on ocean currents, in the sense here meant, cannot with profit be directed by any one but a thoroughly experienced seaman and navigator, and unless he is one, like Captain Thomson, who loves navigation as a fine art, he will rarely have the perseverance to go through with the frequent astronomical observations and the close attention to all the details of compass and log which are necessary. Besides, a complete practical knowledge of the craft of the seaman is required for the work connected with the boats and buoys and cables.

It is always advisable to begin operations by anchoring a buoy as a fixed point to which all the observations are referred. Its absolute position is determined astronomically with the greatest care and repeatedly throughout the whole series, in order to be quite sure that it has not drifted. Anchoring the ship is not advisable, because she is wanted to follow the drift-buoys with current-drags at different depths, and to make other observations which are impossible when anchored in a stream-way. Large current-drags suspended by wire from suitable buoys, whose direction and speed are followed from the ship, are the simplest means of investigating under currents. Several forms of current-meter have been constructed in late years, and those of Admiral Magnaghi of the Italian navy, and of Lieut-Commander Pillsbury of the U.S. navy, have been used with effect by their inventors. Never having had practical experience of them, I am unable to give any opinion on their working. They are delicate and expensive pieces of apparatus, but the object to be gained is worth spending money on. Ascertaining the true direction of an under current is very important and very difficult, but if the investigation of the equatorial currents is seriously taken up, there can be no doubt that the best instrument will develop itself in the work, and will persist.

Although there is demonstrably a transfer of cold water from high latitudes along the deeper layers as far as the equator, it is usually assumed that the motion is very slow—so slow, indeed, as to give none of the appearances which would be produced by a current. Yet in many parts of the open ocean we meet with rocky bottom, and it must be kept clear of sediment by some agency, the most natural being a current. When the rocky bottom of the ocean comes up to moderate depths, as in the oceanic shoals which I had the opportunity of examining in the s.s. "Dacia" in 1883, these currents, and the tidal element in them, are very evident. In archipelagoes like the Canary islands, which are separated by channels having often a minimum depth of 1200 fathoms and more, the crests of these ridges are swept bare of sediment, and are hard rock, generally calcareous and manganiferous.

At great depths it is difficult to determine the direction and rate of motion of the water, but the existence of motion can often be detected by the behaviour of the sounding wire or line when carrying out a deep sounding, and by the movements which it is necessary to give the ship as a compensation. The person making the sounding must be thoroughly acquainted with his ship as well as with the manipulating of the wire, and there must be no unfavourable accidental circumstances of weather. In the Gulf of Guinea, the sea being calm and no wind, I on several occasions met with a difference in the motion of the water when the lead passed to a greater depth than 1300 or 1400 fathoms. The motion which had to be given to the "Buccaneer," which I knew thoroughly, cannot be described so as to produce conviction, but to the person who had to order them, they admitted of no uncertainty of interpretation. In order to get quantitative results in such deep water, an accurate and easily worked current-meter must be had.

In the observation of ocean currents, as in that of all the other physical features of the sea, co-operation by several parties in different vessels, in different but neighbouring localities, is very useful. The advantages of it are well shown in the results of the Swedish expeditions into the Baltic, with which the name of Professor Pettersson is so honourably connected. Could a similar fleet be despatched to the equatorial regions of the Atlantic, and carry out combined operations in the equatorial currents, each series extending over at least seven days and repeated at different times of the year, we should very soon know a great deal more about oceanic circulation, and we should find out a number of things which would at first astonish us, as we should find that our present theories have no place for them. One of the most striking observations made on board the "Buccaneer" was that of the strong easterly current met with at a depth of not more than 25 fathoms from the surface, and extending to a considerable depth. This observation was confirmed much further to the west when the cable was being laid from Fernando Noronha to Senegal, and is referred to in Captain Thomson's paper. This current consists of comparatively dense water; the counter equatorial or Guinea current, which sets to the eastward at the surface, consists of comparatively fresh water. This current is evidently as important a factor in oceanic circulation as the westerly running surface current at the equator. It requires to be traced. Work of this kind cannot be done in a hurry, and it costs money. Were only one or two of the owners of large yachts to follow the spirited and enlightened example of the Prince of Monaco, and to combine for the thorough sifting of such a fundamental problem in the physics of the globe as the one here indicated, they would have great satisfaction themselves in the prosecution of the work, they would confer a lasting obligation on science, their names would be for ever connected with the solution of a great natural problem, and there would be value for the money expended. For the present purpose, only steam-yachts and such as are of considerable size are of use. Considering only yachts belonging to private owners, there appear in Lloyd's register 1895—

8	steam-yachts	of over	1000	tons		
50	,,	between	1000	and	500	tons
28	,,	,,	500	,,	400	,,
61	,,	,,	400	,,	300	,,
59	,,	,,	300	,,	200	,,
115	,,	,,	200	,,	100	,,

The supply of vessels, therefore, is sufficient. Two out of the fifty-eight yachts of over 500 tons, carrying captains with the necessary navigating qualifications and an interest in the work, and crews prepared for boat and other work, detailed for three months' work in the winter, would make a beginning; and if the same yachts did not repeat the experiment, they would speedily find imitators.

No. 3. [*From the Proc. of the Royal Geographical Society*, 1886, *Vol.* VIII. *pp.* 753–769.]

ON SIMILARITIES IN THE PHYSICAL GEOGRAPHY OF THE GREAT OCEANS

OSCAR PESCHEL, in his charming work on *New Problems of Comparative Geography*[1], dilates on what had been called by Agassiz geographical homologies. He compared the study of the structure of the earth with that of the structure of the animal body, the science of geography with the science of anatomy. It is from studies of comparative anatomy that much of our knowledge of the early history and development of animals is derived, and it was hoped that analogous studies of land and sea would yield similar results. The establishment of geographical homologies having a certain amount of parallelism with anatomical homologies has been an important step in this direction. Thus, the wing of an eagle, the arm of a man, and the foreleg of a horse, are homologous features, that is, they are similar organs in corresponding positions. When we contemplate the features of the globe, we find the continental land of the world forming an almost continuous belt round the north pole, and separated from it by an area of chiefly sea. From this circumpolar belt the land stretches southwards, and appears to have a tendency to form three separate continental masses. Of these, the American continent is in all respects well defined, and stretches furthest in a southerly direction. The other two show their separate individuality in their southern portions, Africa and Australia; in their northern portions they are *twinned* (to use a mineralogical expression), forming the joint continent of Europe and Asia. Since ever maps have existed, the similarity in the form of the southern extremities of these continents has struck observers, and they are classed amongst remarkable

[1] *Neue Probleme der vergleichenden Erdkunde:* von Oscar Peschel 4te Auflage. Duncker and Humblot. Leipzig. 1883.

geographical homologies. They are similar features in corresponding localities.

A remarkable group of similarities of this kind is to be found in the arrangement of enclosed seas lying to the northward of the three southern continents. To the northward of South America there are the Gulf of Mexico and the different basins of the Caribbean Sea; to the northward of Africa there are the Mediterranean with its different basins, and on the north-east the Red Sea; and to the northward of Australia there are the well-known seas of the Eastern Archipelago. These seas are bounded on all sides by islands and insular groups, and they are in continuous connection with two oceans, the Pacific and the Indian. The African seas are bounded entirely by continental land and communicate directly with two oceans; but in the limited sense that one sea, the Red Sea, communicates with the Indian Ocean by a single channel, and the Mediterranean Seas with the Atlantic, likewise by a single channel. Finally, the American seas are all in continuous communication with only one ocean, the Atlantic, the continental barrier towards the Pacific being continuous.

It is not unworthy of remark that the great depths (over 4000 fathoms) of the Atlantic and the Pacific Oceans occur immediately to the northward of these groups of seas, and in the western sinus of the northern portions of both oceans; while the greatest depression of the continental land, the region of the Dead Sea, is found similarly situated with regard to Africa. The analogy here, however, does not hold good all through, because it is a mere accident of climate that this area does not form a large and not excessively deep freshwater lake.

The enclosed seas just referred to lie on the margin of debatable ground between the continental and elevated area and the oceanic or depressed area of the earth's surface. The bed of the great ocean, including in this designation the great mass of water which surrounds the globe in its southern part and extends its important branches into the Pacific, the Indian, and the Atlantic oceans northwards, is necessarily a complementary feature in the earth's architecture as compared with the shape of the continental elevations. The one dovetails into the other. Again, in north polar regions we have a detached oceanic area, while in south polar regions we have a detached continental area. These are similarities of form, and have to do with the earth as a solid.

There is another class of similarities which are produced by climate re-acting on the surface of the globe and producing certain alterations on it. We find similar alterations of this kind in corresponding localities.

It may be well in a sentence or two to recall the characteristics of intertropical climate. As the belt between the tropics is the only region of the globe within which the sun is ever vertical, and it is always vertical over some part of it, it is the most highly heated part of the earth's surface. Acting on the air, the earth, and the water, the sun's heat renders them all specifically lighter, less dense. In the case of the air and the water, the particles of which are mobile, it produces disturbances of equilibrium, which are adjusted by the movement of the particles producing in the air, winds, and in the sea, currents. Further, as the water is capable, under the influence of heat, of changing from the liquid to the aeriform state, there is a continual exchange going on between the water and the air, producing on the one hand, rain, and on the other, the less evident phenomenon of evaporation. All places on the same parallel of latitude are situated similarly with regard to the prime cause of climate, namely, radiation *from* the sun and *into* space. But the sensible effect of the sun's rays depends in a great measure on the nature of the substance which they strike. In passing through the atmosphere they hardly warm it at all; falling on the sea they warm the water moderately, and on the land they heat the ground intensely. Hence one reason why distribution of sea and land modifies climate.

Another and secondary, but not less efficient cause, is the obstruction offered by land masses to ocean currents, and in a modified way to winds. The effect of the sun's bestowing the greatest amount of heat in the equatorial region, is to produce the north-east and south-east trade winds which blow from higher latitudes towards the equator, suffering a deflection towards the west due to the changing peripheral velocity with changing latitude. Where these winds meet, there is the equatorial belt of calms and rains. The regularity of these conditions is disturbed by the fact above mentioned, that where we have land the sun's rays raise its surface to a higher temperature than in the case of their falling upon sea. The consequence of this is that the air draws in towards these highly heated areas, and modifies the primary trade wind, producing the secondary phenomenon of monsoons. As land predominates in the northern hemisphere and water in the southern, these phenomena are more evident in the northern than in the southern hemisphere, and they are particularly evident in the Indian Ocean. The trade winds during the whole of their course are travelling from regions of lower temperature to those of higher temperature; hence, even though they were saturated with moisture at the beginning, they would, as they continued their course and their temperature rose, increase their capacity for taking up moisture. Hence their great characteristic quality—they are *drying* winds[1].

The south-east trade wind passes chiefly over water, the north-east trade blows almost equally over land and over water. Its effects on land are principally visible in its progress over Western Asia and Africa. From Siberia to Morocco we have winds mainly from the north. Their actual direction is sometimes to the eastward, sometimes to the westward of north, according to the preponderance of local influences. They are of a highly drying character, and this is put clearly in evidence by the chain of inland drainage areas and absolute deserts which occupy this broad belt of continent. Thus we have the desert of Gobi, the inland basins of the Aral and the Caspian, the salt lakes of Asia Minor, and the deserts of Arabia and North Africa.

The drying power of the wind is also shown by the fact that the Mediterranean, which lies in its course, but receives the drainage of quite half of Europe which lies in the rainy regions of the south-west winds, and that of a considerable portion of Equatorial Africa by the Nile, is essentially the receptacle of an inland drainage area, quite as much as the Caspian is.

[1] See Contents, p. xvii.

In North America, from its configuration, the extent of country traversed by this wind is not so great as in the old continent, but still we have the well-known alkali deserts on the west side of the Rocky mountains. In the southern hemisphere we have an exactly similar phenomenon. The interior of Australia, except on the windward side, occupied by the colonies of Victoria, New South Wales, and Queensland, is a waterless desert; in South Africa we have the Kalahari desert; and in South America we have the rainless regions of the Argentine Republic and the Atacama desert. Here we have, then, a feature due to climate, which repeats itself in corresponding localities in the Old World and in the New, in the northern hemisphere and in the southern.

But the trade winds produce not only dry places on the land, they produce also comparatively dry places on the sea. The areas of maximum density or salinity of the surface water are shown in the maps for the different oceans, and it will be seen that they are situated in localities corresponding with those of the deserts on land.

In concentrating the surface water in the regions over which the trade wind blows, it produces a powerful mechanical means of propagating the heat, obtained by the water at the surface from the sun, downwards into the colder depths below. All the year round, the water exposed to the influence of the trade wind is being energetically evaporated, while the rainfall in the same region is quite insignificant. The temperature of the surface varies with the season. When the temperature is rising, the water becomes specifically lighter, notwithstanding the concentration which is going on, and it has no tendency to leave the surface. When the temperature is falling, the water is still being concentrated at nearly the same rate, and the concentration and fall of temperature combine to render the water specifically heavier. It thus acquires the power of sinking through the water of same temperature but less saltness below it, and diffusing its higher temperature in the colder layers below. There is thus a diurnal and an annual oscillation which make the surface temperature felt at greater depths. The tendency is to propagate the lowest daily and the lowest annual temperature downwards, and the energy with which this takes place depends on the excess of salinity of the surface waters over that of the waters beneath, and on the range of daily and especially annual temperature. The greater the range of salinity and of surface temperature, the greater will be the oscillation produced, and the deeper will its effect be felt. Thus in the North Atlantic everything conspires to collect heat and propagate it downwards, and as a consequence we find a higher temperature at 1000 fathoms in the Sargasso Sea than in any other part of the globe. In the North Pacific the range of density is small, and the subsurface water is comparatively cold.

In passing over the surface of the ocean, the winds not only remove water in the form of vapour, they also impart motion to the water immediately under their influence. The effect of this is to produce a general motion of the denser intertropical water towards the equator and towards the west, which is intensified near the equator and forms the well-known westerly equatorial current.

In the Atlantic the westerly running waters are collected in the western sinus of the ocean, and the head of water thus produced is relieved by the overflow of the Gulf Stream. In the Pacific the westerly running waters meet no continental obstacle, and pass freely into the eastern waters of the Indian Ocean, subject to a partial reversal during the season of the south-west monsoon.

Everything conspires to produce an exaggerated heating effect in the waters of the western portion of an intertropical ocean. It is supplied with water which has already been heated further to the eastward, and this water, in virtue of its density, has the property of propagating the high temperature to greater depths. A secondary consequence of a leeward position in the ocean, due to this characteristic of such waters, is the presence of coral formations in the western regions of the Atlantic and Pacific, and, owing to the mixture of conditions, in both the eastern and western regions of the Indian Ocean.

Continental homologies, or similar features in corresponding localities, are found on the western as well as on the eastern sides of the continents. One of the most striking is the resemblance of the Gulf of Guinea on the African coast with the great Central American bight stretching from Cape St Lucas at the extremity of the Californian Peninsula, by Panama, to Cape Blanco at the mouth of the Guayaquil river, and with the unnamed bight in the Indian Ocean bounded continentally by the north-west coast of Australia, and insularly by the chain of islands stretching from the Peninsula of Malacca to Australia. Oceanically these bights are homologous. It is in them that the beginnings of the westerly-running equatorial currents are to be found, and perhaps more important still, it is in them that the easterly-running counter equatorial currents end. They are to be found in each of the three oceans, and generally on the northward side of the axis of the westerly-running current. In the Atlantic the counter equatorial current is best known by its eastern portion, the Guinea current.

The observations here recorded of the Guinea current were made on board the steamship "Buccaneer," at the invitation of the owners, the Indiarubber, Guttapercha, and Telegraph Works Company, of Silvertown, and were carried out during a survey for a telegraph cable from Sierra Leone to St Paul de Loanda. In order to render the results more intelligible, some of them have been extended in diagrammatic form. Diagram A shows the relation between the density of the water at the surface, and at 50 fathoms below it, to the distance off shore. The diagram begins on the 1st of January, when the ship had steamed over 200 miles out to sea from Conakry. Here the surface water, though differing from pure ocean water of the same latitude by its low density, is still the densest surface water met with in the period, the 1st to the 20th of January, embraced in the diagram. From this point the ship was run in shorewards, sounding frequently, and the parallelism of the density and the distance lines is very remarkable. As far as Cape Palmas it is so close that the distance from shore could almost be told from the density of the surface water. After rounding Cape Palmas, the density of the surface water shows itself less sensitive to the distance from shore, which may be due to the fact that, until near Accra, the route lay within the bight where there was little easterly current, but an occasional westerly eddy. To the westward of Cape Palmas the easterly current was very strong, especially inshore, where the water was comparatively fresh. It seemed then to strike across the bight, and was again found running very strong inshore between Cape Three Points and Cape St Paul; but the density of the water was comparatively high, 1·0253. After rounding Cape St Paul, the density of the water fell very much, and more markedly after leaving Porto Novo for S. Thomé.

The line of density at 50 fathoms is almost horizontal. The water at this depth is quite independent of surface influences. All through the Gulf of Guinea the water is found to have a maximum density at from 30 to 50 fathoms from the surface. Both at greater and at less depths the water is fresher. When the surface water is moving rapidly eastwards, at 50 fathoms it is either motionless or is moving in another direction.

For over 100 miles from Porto Novo, the surface water is that characteristic of the Guinea current, density about 1·024; at 150 miles, the density of the water rapidly falls, and for 200 miles the ship's course was through water of density below 1·022, indicating plainly the great influence of the neighbouring river Niger. The ship's course here was almost exactly parallel to the western coast-line of the Niger delta, and the very fresh water ceased when the coast-line took an easterly trend at the Nun entrance of the river. This belt of fresh water, 200 miles broad, teemed with diatoms.

The density of the surface water as far as the island of S. Thomé remained moderately uniform at about 1·023. About 100 miles south of S. Thomé the surface density rose rapidly to over 1·025 off Cape Lopez, and 100 miles further south it fell below 1·022 off the mouth of the river Ogowé. Between this and the mouth of the Congo the surface water rises in density to above 1·026. Off the mouth of the Congo the fresh water was again met with, a stretch of over 200 miles, on our

DIAGRAM A. Density at 60° F. or 15·55° C. of the Sea-Water off the Coast of Guinea

The unit of Density is, in all cases, that of Distilled Water at 4° C.

S.S. "Buccaneer." January 1886.

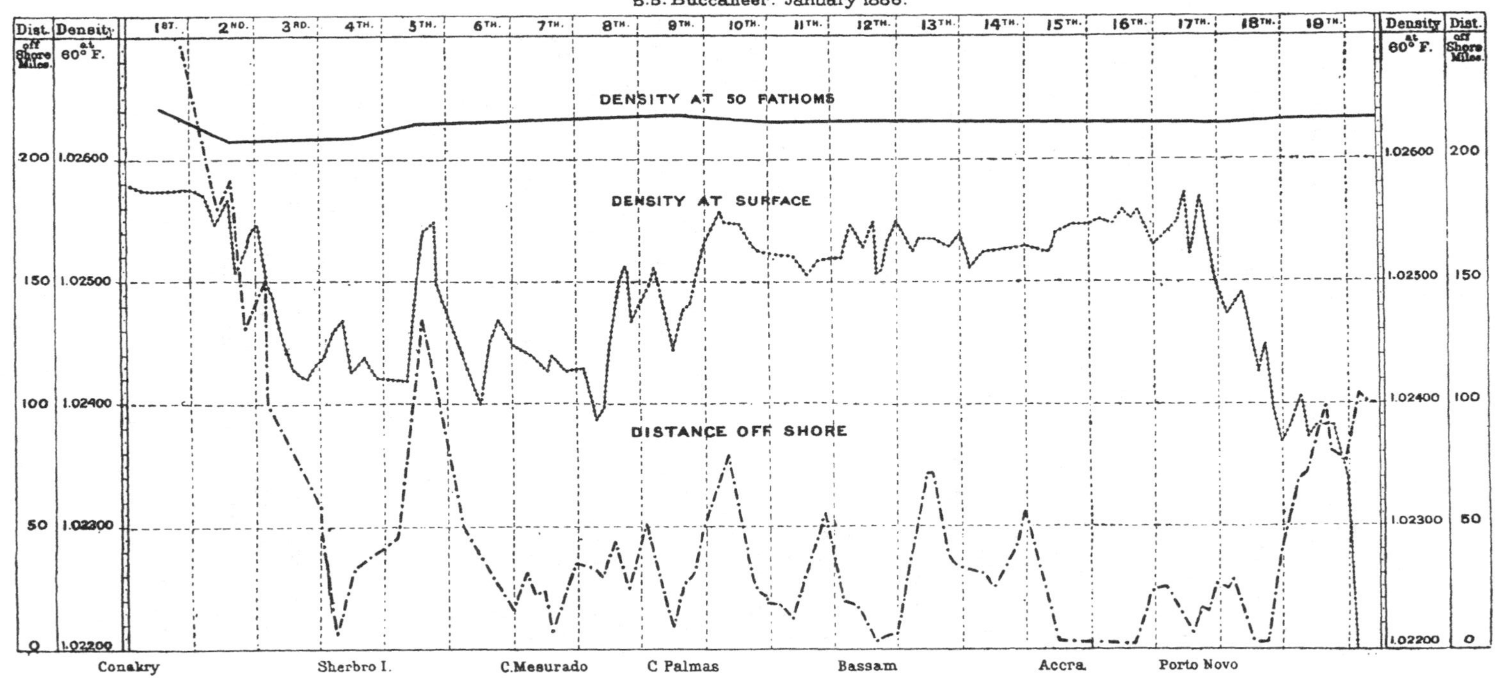

line, having water of under 1·022 density. South of the Congo and as far as Loanda the surface density rises above 1·026 and is oceanic and free from river influences.

On this line of section the part south of S. Thomé is occupied by ocean water, with river water superposed in places; the part north of the islands has a layer of intermediate Guinea current water interposed between the underlying ocean water and the overlying Niger water. Along this line the dependence of the temperature of the subsurface layers on the distribution of salinity in the water was evident. North of the island of S. Thomé we have large areas of river water on the surface with comparatively fresh Guinea current water below. South of that island the supply of river water is probably quite as large, but it lies on the ocean water of temperate tropical regions with a high surface salinity. While, therefore, we find the actual surface temperature higher to the northward of S. Thomé, the subsurface water is warmer to the southward of it. North of S. Thomé, a temperature of 65° F. is found at an average depth of about 35 fathoms; south of that island it is found as deep as 60 fathoms at a station where the density (at 60° F.) of the water from the surface downwards was over 1·026. After passing the mouth of the Congo, equatorial features disappear, and at a station near St Paul de Loanda, the cold deep water was found approaching the surface, a temperature of 65° F. being found at 15 fathoms. Here we find the beginning of the cold water of the windward shore of the South Atlantic.

Diagram B is a meridional section on long. 14° W. down to a depth of 200 fathoms, compiled from observations on board the "Buccaneer" in March 1886, and the "Challenger" in April 1876. The range is from 11° N. to 12° S. From 9° N. to 11° S. the temperature of the surface is above 80° F., and the isothermal of 80° travels at depths varying from 10 to 25 fathoms. From 10° N. to 4° S. the isotherm of 70° maintains a nearly constant level of 25 to 30 fathoms; to the southward of 4° S. it dips deeper. The isotherm of 65° follows a nearly parallel course. That of 60° comes within 30 fathoms of surface at 4° S., and dips rapidly towards the south, more gently towards the north.

DIAGRAM B. This diagram shows by means of isothermal lines the distribution of temperature in the water lying between the surface and a depth of 200 fathoms, along the meridian of 14° W. and between latitudes 11° N. and 12° S. It is compiled from observations made in the "Challenger" in April 1876 and in the "Buccaneer" in March 1886

The thick curved lines are those of equal density at constant temperature and represent the salinity of the water

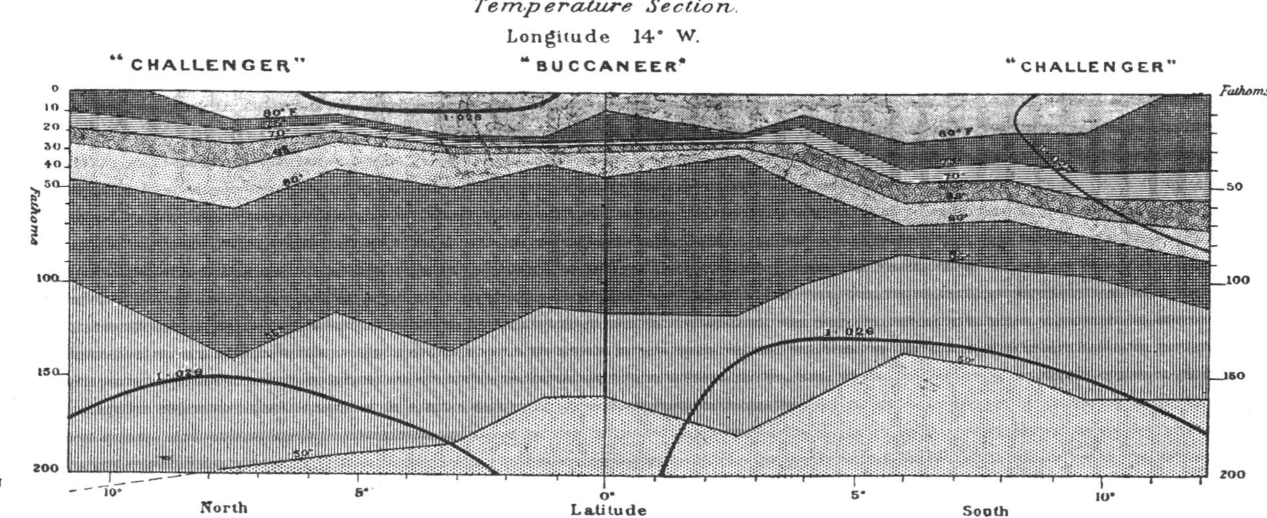

This diagram shows well the distinguishing features of an equatorial ocean in the warm season of the year. In the cooler season the surface temperature is below 80°, and the temperature gradients in the layers near the surface are gentler. At this season, with the sun nearly vertical over the equator, and a thick layer of water of temperature over 80° F. on the surface. the gradients are very steep. This was found both by the "Challenger" and the "Buccaneer," in which ship several of the "Challenger's" temperature stations were repeated. Thus station 43 of "Buccaneer" was made on 7th March, 1886, in the same position as No. 346 of "Challenger" on 6th April, 1876, namely, lat. 2° 42′ S., long. 14° 43′ W. The "Challenger" found a higher surface temperature, 82·7° against 81·7°; but the layer of water above 80° F. was thicker in March 1886 than in April 1876. The following temperatures were observed:

Depth	.. ..	Surface	10	20	25	30	50
Temp.	"Challenger"	82·7°	82·0°	78·1°	71·5°	62·2°	56·5° F.
	"Buccaneer"	81·7°	81·2°	81·0°	65·2°	61·8°	57·3° F.

It is remarkable that the "Buccaneer" found a fall of 16° F. in 5 fathoms between 20 and 25 fathoms; the steepest gradient found by the "Challenger" was about 9·3° in 5 fathoms between 25 and 30 fathoms. Similar gradients and even steeper ones are found in our fresh-water lakes about midsummer, when the surface heating is proceeding most rapidly. In July of this year (1886) a fall of 5° in *one* fathom was observed in Loch Lomond at a depth of 12 fathoms[1].

In the diagram the thick lines are those of equal density at constant temperature (60° F.). The line of 1·027 shows the penetration of dense trade-wind water, and the effect of high surface density in facilitating the propagation of heat downwards is well shown by the dipping of the isotherms. The surface outcrops of the 1·026 line to the north of the line indicate the average position of the equatorial freshet. On this line neither the "Challenger" nor the "Buccaneer" experienced any sensible easterly current, and it is well known that the Guinea current experiences marked variations with the season.

In the "Buccaneer" it was the custom every day at 1 P.M.

[1] See Paper No. 21, p 312.

to throw overboard a bottle containing a note of the position and a request to the finder that the paper might be sent home with a record of the time and place of discovery. Up to the present date four of these bottles have been reported as found. Two of these, *A* and *B*, were thrown overboard off the Kroo coast; *A* on the 6th January, in lat. 6° 40′ N., long. 12° 32′ W., a distance of 42 miles from the coast; and *B* on the 8th January, in 4° 49′ N., 9° 38′ W., a distance of 30 miles off the coast. *A* was picked up by a native who watched it coming ashore in the breakers, in lat. 5° 47′ N., long. 0° 35′ E., about 10 miles from Addah, on 5th March. *B* was picked up on Bey Beach in lat. 6° 8′ N., long. 1° 17′ E., about 40 miles further east than *A*, on the 2nd March. *A* had thus travelled 862 miles in 58 days, the average rate per day being 15 miles, and *B* had travelled 690 miles in 53 days at an average rate of 13 miles per day. They were washed ashore, one on one side and the other on the other side of Cape St Paul where the tide-rips and discoloured water show the meeting of different current systems.

The change in the character of the current at different times is shown by the fact that in travelling north from the equator towards Sierra Leone in March, the "Buccaneer" experienced no easterly current, and a bottle thrown overboard on this passage was washed ashore on the Kroo coast, whereas if the same current had prevailed in both March and January the bottle would have been carried round Cape Palmas and into the Gulf of Guinea. In connection with the absence of easterly currents off the coast may be taken the very remarkable under-current which is found setting in a south-easterly direction with a velocity of over a mile per hour at three stations almost on the equator, and to the northward of the island of Ascension. For the double purpose of examining the currents and of obtaining a large specimen of the bottom, the "Buccaneer" *was anchored in* 1800 *fathoms* of water by means of an ordinary light anchor fitted with a canvas bag to receive the mud which would otherwise fall off the flukes on its being weighed. While the ship was lying thus at anchor, the surface water was found to have a very slight westerly set. At a depth of 15 fathoms there was a difference, and at 30 fathoms the water was running so strongly to the south-east that it was impossible to make observations of temperature, as the lines, heavily loaded, drifted straight out, and could not be sunk by any weight the strain of which they could bear.

In the Pacific the counter equatorial current in the open ocean was well observed by the "Challenger" on her voyage from Hawaii to Tahiti. Her observations are represented in Diagrams C and D. The former gives the surface current in miles per day along a mean meridian of 150° W., easterly current being laid off above, and westerly current below, the meridian line, along which degrees of latitude are measured. The latter represents the distribution of *true density*, that is, the density of the water due to its salinity and temperature combined, but at constant pressure, in the water from the surface to a depth of 200 fathoms. As these lines of equal density coincide nearly with corresponding isothermal lines, the approximately corresponding isothermals are indicated in the diagrams. The thick lines intruding unconformably into the diagram represent density at constant temperature (60° F.), which is proportional to salinity. The horizontal scale of degrees of latitude is the same for both diagrams. The current diagram is carried to lat. 7° S., while the density diagram stops at 3° 48′ S. The easterly current was found between the parallels of 5° and 10° N., there being two streaks of *maximum* velocity, one between 7° and 8° N., and the other between 9° and 10°. In the former the mean daily set was 54 miles; in the latter it was probably quite as high, but it could not be accurately determined, as the ship passed from westerly to easterly current in the course of the 24 hours, and the observed current of 20 miles represented the difference of the two. The streaks or axes of strong easterly current are sharply defined by areas of abnormally low surface density. The whole of the area of easterly running water has a comparatively low density, but where there is a sudden acceleration of its velocity there is a correspondingly sudden drop in its density, so that the existence of a strong easterly current in equatorial regions may be guessed with great probability by the use of the hydro-

DIAGRAMS C and D. Both these diagrams are oriented along the same meridian, 180° W.

Diagram C shows the rate and direction of travel of the surface water, in miles per day. Diagram D shows, by means of lines of equal *true density*, the degree of statical balance which exists in the water lying between the surface and a depth of 200 fathoms.

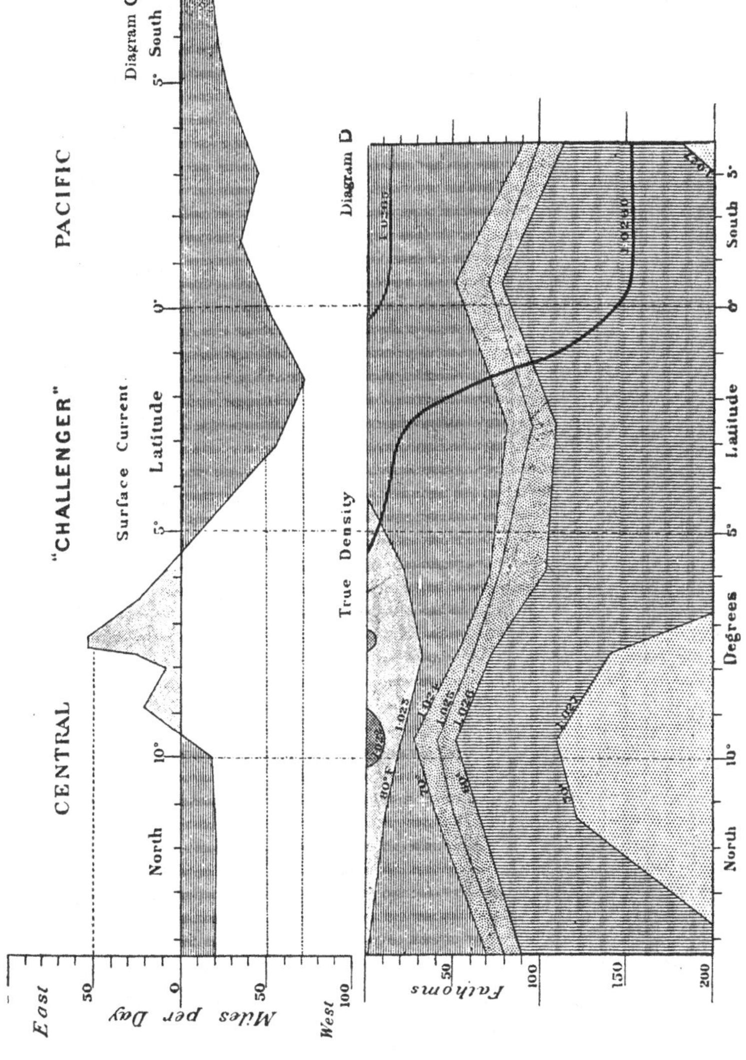

meter. Thus, in the region of general easterly current, the specific gravity (at 60° F.) of the surface ranged between 1·0255 and 1·0258, but where the current was strongest it fell to 1·0249 in lat. 9° 28′, and 1·02475 in lat. 7° 26′ N. It rose rapidly immediately the westerly running water was reached, being 1·02621 in lat. 2° 34′ N.

The diagram shows also in a very marked way the protective action of the fresh surface water in preventing the penetration of heat into the lower layers of the water. A temperature of 60° F. is found here (lat. 9° 10′ N.) at a depth of 50 fathoms from the surface, while in the westerly running current, a little further south (lat. 2° 34′ N.), the same temperature occurs at a depth of over 100 fathoms. The effect is even more striking if the mean temperature of corresponding layers is compared as in the table, p. 103. Besides stations 267 and 270, the figures for station 278, in lat. 17° 12′ S., with the island of Tahiti in sight are given. No current was logged here, but the effect of high surface salinity, gradually falling as the depth increases, in propagating heat downwards, is made particularly evident.

If the figures in the tables (p. 103) be studied it will be seen that the temperature follows the density, that is, the further the high density penetrates the deeper do we find a high temperature. At station 278 the high surface density is found to persist almost undiminished down to 100 fathoms, while at No. 270 it does not persist much beyond 25 fathoms. Accordingly, at 278 we find a mean temperature between 80 and 150 fathoms, 10·3° higher than at 270. From the surface to 80 fathoms there is very little difference in the amount of heat in the water. Besides a greater range of salinity in the water of different layers at station 278 than at station 270, there is certain to be a greater annual range of surface temperature at lat. 17° S. than at lat. 2° N., and this will also be helpful in propagating heat downwards. Further, at the equatorial station the concentrated surface water is being hurried away westwards at a rate of 70 miles in 24 hours, while at lat. 17° S. it was not found to have any very apparent motion. At station No. 267 there is a layer of very low salinity, of probably not more than 20 fathoms in thickness, yet this so completely screens the water below, that the temperature of the layer between 20 and 80 fathoms is 17° lower than at the neighbouring station No. 270. The surface water at both stations is moving away rapidly, and the deeper and colder water, which rises to supply a portion of the deficiency, cools the surface water at both stations; but at station 267, however

Mean temperature of different layers of water.

No. of Station (Challenger)	Latitude	Current in 24 hours	Mean temperature Fahr. of water between			
			Surface and 20 fathoms	20 and 80 fathoms	80 and 150 fathoms	Surface and 150 fathoms
278	17° 12′ S.	none logged	78·2°	76·3°	70·1°	73·8°
270	2° 34′ N.	W. 70′	78·5°	75·9°	59·8°	68·7°
267	9° 28′ N.	E. 19′	77·3°	58·9°	50·4°	57·4°
Difference		270–267	1·2°	17·0°	9·4°	11·3°
		278–270	−0·3°	+0·4°	10·3°	5·1°
		278–267	+0·9°	17·4°	19·7°	16·4°

Observed density (at 60° F.) of water at different depths at the same stations.

No. of Station	Density at 60° F. of water at				
	Surface	25 fathoms	50 fathoms	100 fathoms	200 fathoms
278	1·02696	1·02694	1·02697	1·02678	1·02593
270	1·02621	1·02591	1·02572	1·02567	1·02572
267	1·02490	1·02552	1·02552	1·02578	1·02567

much the surface water be cooled, it still floats on the surface, while at No. 270, on being cooled to the same temperature as the water immediately below it, it sinks through it, carrying this temperature deeper.

The study of the currents of equatorial regions would

well repay the trouble of the investigation. The counter equatorial current is particularly interesting, and its dynamics obscure. Its range is very superficial, and its physical conditions can be studied without the elaborate and costly equipment required for the research of oceanic depths.

With the change of season the counter equatorial current shifts its latitude, but within narrow limits. It seems never to be met with south of the equator and rarely south of lat. 4° N. Its extreme fluctuations are between 4° N. and 12° N., and for the greater part of the year it occupies the zone between 7° and 9° N. where it has impressed its mark on the distribution of temperature in the water below it. In Diagram B, where the lines of *true density* coincide closely with the isothermal lines, we see the isothermal of 50 (coinciding with 1·027) approach within 110 fathoms of the surface at the station in lat. 9° 28′, and it falls away to below 200 fathoms on both sides. The thick line passing unconformably through the above shows the extent to which the concentrated waters of the south-east trade wind district penetrate. To the left hand or north of this line all the water is at a density (at 60° F.) below 1·026. The water is diluted both by diffusion of the rain water above and the rising of abysmal water from below without mixture with any concentrated surface water.

In the Western Pacific on the voyage from Admiralty Island to Japan in March, the "Challenger" passed through no similar easterly current, but here it is probably completely checked by the strength of the north-east monsoon and may be much intensified at the other seasons by the south-east monsoon.

The temperature observations showed the enormous amount of highly heated comparatively dense water flowing at this season steadily westward towards the Indian Ocean. In its progress from the meridian of 150° W. to that of 145° E. the westerly stream has taken up an enormous amount of heat. The velocity of the western current is not so great, but its axis lies in the same latitude, namely between 2° and 3° N. The surface temperature is over 83° F., and at 90 fathoms the temperature of the water is still above 80°. Between these depths the water has on an average a temperature about

5° higher at the western than at the central station. The mean temperature of the layer between the surface and 140 fathoms, where the temperature curves cut each other, is 69·8° at the central station, and 77·5° at the western. The current per day logged at the western station was 30 miles, while it was 70 miles at the central one.

To the north and to the south of the equatorial bights of the western shores of Africa and America we have a remarkable similarity in the distribution of temperature in the coast waters. The transition from equatorial heat to extratropical cold is very marked; on the North American shore, at Cape St Lucas, the southern extremity of the Californian peninsula; on the North African, at Cape Verde on the South American shore, at Cape Blanco; and on the South African, at Cape Frio[1]. These areas are marked ⊙ on the map. The very low temperature of the coast water is particularly remarkable on the Morocco coast, and it extends round Cape Spartel into the Straits of Gibraltar. Here the temperature of the surface water is usually 10° F. lower than

[1] See Chart of the world between pp. 106, 107.

In this chart the direction of motion of the water in the different localities is indicated by arrows. The observer is supposed to be facing the north and looking down on the surface of the sea; the current flowing from west to east is shown by an arrow floating on the surface of the water and pointing towards the east; a current flowing from east to west is shown by an arrow floating on the surface of the water and pointing towards the west. When the motion of the water is vertical and is rising from below to the surface it is shown by the symbol ⊙ which means that the arrow which shows the direction of the current is pointing directly towards the observer and shows only its point. When the water is descending from the surface to greater depths this is indicated by the symbol ⊕ which means that the arrow which shows the direction of the currents is pointing directly away from the observer and exhibits only its butt with the feathers in section.

It will be seen that the cold water on the west coast of South America reaching to Cape Blanco in latitude 4° S., which has been attributed to a surface current from the Antarctic Ocean and called the Humboldt current, is a vertical rising of cold water from the ocean depths contiguous to the coast line and has no horizontal component capable of influencing navigation. The same holds with regard to the cold water to be found along the coast of the eastern states of North America which was erroneously attributed to a surface current from the Arctic ocean and was called the Labrador current.

on the European side of the Straits. Similarly, off Mogador the temperature of the surface water inshore is 60° F., while it is 70° F. 20 miles out at sea. The delightful climate of Tangiers and the low temperature of Mogador are directly due to the existence of this water on the shore. The same comparatively cool water is found all the way down the African coast as far as Cape Verde[1].

On the west coast of South Africa there is not much experience of sea-water temperatures, but the characteristics of the climate on shore are a low temperature, much cloud and mist, no rain, and barren waterless country. On the Atlantic side of the Cape of Good Hope the surface water inshore has a temperature about 10° F. lower than the water on the Indian Ocean side of the Cape, and all the way northwards as far as St Paul de Loanda the barrenness of the land is more than compensated by the fruitfulness of the sea, which everywhere abounds with fish, indicating cold water.

The abnormally cold water found on the west coast of South America has been well known since the beginning of the century, principally by the works of Humboldt. In the months of April and May of 1885, I made the voyage along the coast, from Valparaiso to San Francisco, and carefully observed the temperature and density of the surface water several times a day.

My voyage being made in the months of April and May, the season was autumn in the southern hemisphere, and spring in the northern. The water ought therefore to have been about its highest temperature during the cruise along the coast of South America. As on the coast of Morocco, it was observed that the lowest temperature was always found closest inshore, also that the cold inshore water was green, while some miles out it was warm, and of the deep ultramarine peculiar to tropical ocean water. From Coquimbo to Pisagua the temperature averaged 60° F., being as low as 57° just outside Coquimbo, and being actually 60° F. in Pisagua anchorage, in lat. 19° 30′ S. At Arica it was 67° F., and crossing the bight, where the course lay in open water, the temperature

[1] See Contents, p. xix.

rose to 73°, but fell again on closing the shore, being as low as 59·5° when 5 miles off Independencia Bay, in lat. 14° S. Further along the coast it averaged about 61°, and in Payta anchorage, in lat. 5° S., it was 63·5° F., though the heat on shore was intolerable. On leaving Payta the ship rounded Cape Blanco, and in a distance of 35 miles, during the night, the temperature rose from 65° to 75°. In rounding Cape Blanco the ship had entered equatorial waters. The characteristics of the South American coast waters were similar to those of the Morocco coast—the water cool, and the land desert. There is no stream or current sufficient to affect the passages of ships to the southward of Cape Blanco. Northward of Cape Blanco currents are met with running sometimes in one direction, and sometimes in another.

From the Guayaquil river the ship left the coast and steamed straight to Panama, and the surface temperature varied from 75° to 80° F., which, for equatorial water, is comparatively cool. The climate of Panama is remarkable. From January to May the temperature of the water is much lower than that of the air. Mr Ward, first officer of the s.s. "Colima," in which I made the passage from Panama to San Francisco, kept very careful meteorological logs, and I am indebted to him for much information about the coast, which he kindly communicated to me. At the end of February 1885 the mean (of four days) temperature, at midnight, of the air was 77° F. and of the water 75·5°, and at noon that of the air was 83·2° and of the water 74·8°, so that at noon the temperature of the water was 8·4° lower than that of the air. At other seasons of the year the warm counter equatorial water extends southward of Panama and communicates to it the same oppressive climate to be found further north along the coast of Central America. From Panama as far as Cape Corrientes the temperature of the surface water was above 80° F., and nearly always above 84° F., at times as high as 88° and 89° F. It gradually declined after rounding the cape to 77·4° when anchoring at Mazatlan. On getting under way in the afternoon the temperature of the surface water was 75·8°, and the cold water must have come very close to the surface, for water taken from the wash when the ship was going astern had a temperature of 72·8°. All the way across the mouth of the Gulf of California the water was very constantly at a temperature of 78° F., with a deep ultramarine colour. Both these characteristics began to change as the mountainous coast of Lower California was approached. At 2 P.M. the temperature fell to 75·8°, but recovered again to 77·0° at 4 P.M., when the ship was less than two miles off shore, and to the eastward of Cape St Lucas. The change now became very evident to the senses, and at 4.35 P.M., after the rocky promontory had been passed, the temperature was 73·4°, and at 5.45 P.M., while still close inshore, the temperature of the water had fallen to 64·4°, and its colour was quite green. On leaving the shore the temperature rose to 66·8°, and next morning at 8 A.M. it had fallen to 59°, when about two miles off Marguarete Island. The number of sharks off Cape St Lucas, and indeed far out into the mouth of the gulf, was quite extraordinary. Quantities of large fish with the glistening scales of salmon were jumping in the water close to the shore round Cape St Lucas. The temperature of the water remained constantly at about 60° F. until, after passing Santa Barbara, the ship rounded Cape Concepcion, when the temperature fell to 52° F. It was 50° at the entrance of San Francisco harbour.

These sharp transitions are found only close inshore, and they have usually been attributed to surface currents from higher latitudes. This explanation is at variance with the observations of navigators on the coasts, who do not notice any currents which would be strong enough to bring water many hundreds of miles under a burning sun without sensible rise in temperature. The occurrence of these coast areas of abnormally cold water is explained when we recognise that they are the windward shores of the oceans. The trade winds blow from them towards the equator, and in doing so mechanically remove water, which has to be supplied from the readiest source. This source is the deep water lying off the continental coast, which is supplied by a gradual drift of cold water from high latitudes. Hence, though the low temperature of the coast waters referred to is due to the cold of high latitudes, it is not supplied by a long surface current, but by a short vertical one. This view is very strongly supported not only by the temperature of the water, but by its other characteristics, especially colour. The outside ocean water is of an intense ultramarine blue; the coast water off Mogador had the clear olive-green colour met with constantly in Antarctic seas. The same is observed on the west coast of North and South America, and it would be of the highest interest to have these waters investigated from a biological point of view. No waters in the ocean so teem with life as those on the west coast of South America. A bucket of water collected over the side is turbid with living organisms, the food of countless shoals of fish, who, in their turn, afford prey for innumerable schools of porpoises. One remarkable school which accompanied the ship for some time consisted entirely of females, each accompanied by a calf following in her wake and mimicking her every movement. Along with abundance of life this coast unites facilities for investigating it. At every port there are plenty of shore boats anxious for a fare, and with a tow net and a few bottles a naturalist might make a rich collection of the shore-water fauna of the coast in one trip from Valparaiso to Panama.

The characteristics of inshore water of the windward coasts of the oceans are, an abnormally low temperature, usually a comparatively low density, and always a pronounced and peculiar green colour. These characteristics are found most marked in the water between the shore and the 100 fathom line. The water, as it drifts out to sea, rapidly loses its low temperature, and its green colour is bleached. In the year 1883 when, through the kindness of the Silvertown Company, I had the opportunity of carrying out some researches in their ship "Dacia," I dredged a number of shells from the Dacia Bank, lying between Teneriffe and Madeira. These shells were all stained with a green colour, which came out and coloured the spirit in which they were preserved. I submitted this coloured spirit to my friend Professor Hartley, of Dublin, and he determined the colouring matter to be a species of chlorophyll. Some of the coloured spirit of the "Challenger" collection was then sent by Mr Murray, with the result that Professor Hartley detected chlorophyll in it also. The diatoms so abundant in the surface waters of Antarctic regions give the water and the ice exactly the colour which is to be found in the coast waters of windward shores. It would be a matter of great interest to determine, by careful observations and investigation, if the green colour of these coast waters is directly or indirectly due to the presence of chlorophyll.

A most remarkable confirmation of the view that the cold water on the windward shores is due to a submarine source has been quite recently supplied by the observations of Captain Hoffmann of the German man-of-war "Möwe," on a voyage from Zanzibar to Aden. Zanzibar was left on 28th June, 1886, Cape Gardafui was passed on 5th July, and Aden reached on 11th July, 1886. The season was therefore that of the strongest south-east trade wind in the southern part and of the south-west monsoon in the northern part of the Indian Ocean. Captain Hoffmann kept as close to the coast as possible, chiefly with a view of checking and correcting the sailing direction for this little-frequented coast, but also in order to test the view which he held that the inshore passage at this season was not so risky as was commonly held. He found this view completely justified, both wind and sea being much less and more uniform. The daily epitome of weather during the few days spent on this coast was—sea-breeze from midday to midnight, fresh enough to enable progress to be made under steam and sail, then a light land-breeze till 8 A.M., and in the forenoon calm. The sea-breeze would be a fair wind for the "Möwe," and its lightness may be gauged by the fact that the sails alone were not thought sufficient for making passage, especially when assisted by the very strong northerly current. On one day 240 miles were made good, although the highest speed logged was only 6·5 knots. This gives an average of 3·5 knots per hour in the twenty-four hours; and, as all these currents are affected by tidal influences, its maximum speed may have been much greater. Captain Hoffmann[1] says:

[1] *Ann. Hydr.* (1886), XIV. pp. 395 and 500.

At Cape Warschek (lat. 2° 30′ N.) strong cross currents were observed, and it seemed that the along-shore current here swept out to sea. The remarkable change in temperature which here commences seems to be connected with the change of current. Up to this point the temperature both of air and water had remained sensibly the same (77° F.) as at Zanzibar. As soon as the strong current had ceased, the temperature of the water fell rapidly between the parallels of 4° N. and 8° N., and attained at Ras al Khyle the abnormally low minimum of 59° F. As a consequence of this the temperature of the air fell. With a clear sky the thermometer did not rise at midday above 68° F., so that exposure to the direct sun was in no way disagreeable. The horizon was hazy, and at night there was heavy dew. The sea had a deep olive-green, often almost black, appearance; close to the coast it was clear green. In the seas of normal temperature the colour of the water was always deep blue. In every other region of the sea one would conclude from these observations that there was here a polar stream. Here it will be impossible to avoid admitting a rising of the water from greater depths. It is proved thus, in the great depths of the ocean here, as well as in other regions, the water is cooled down to near the freezing point.

We have seen that the season when Captain Hoffmann passed along this coast was that of the south-west monsoon, and at this season the Somali coast is a pronounced windward shore, and exhibits the same characteristics as the corresponding coasts of Morocco and western South America.

A very important example of this class of homologies remains to be mentioned. It occurs on the south-east coast of South America, and it is due to the influence of the great world-wind of the southern hemisphere, which blows strongly throughout the year from west to east in the latitudes known to seamen as the "roaring forties."

This is the most powerful and the most constant mechanical driving agent on the surface of the globe. In the circle which it makes of the globe it has only one windward shore, namely, the Patagonian Coast of South America. Here I observed the occurrence of abnormally cold water in January 1885 as far north as lat. 30° S., and it is shown on the chart of homologies.

It is important that this phenomenon should have a name; and I propose to call it an *Uprise*; to mean, that, in the locality where it occurs, water is found on the surface of the ocean which persistently has a temperature lower than that corresponding to the temperature of the air and the climate of the locality. The one occurring on the western side of the South Atlantic, which has just been described, may therefore be appropriately called the *Patagonian Uprise.*

Similarly the other homologous phenomena which have been noticed in this paper may be named:

The Peruvian Uprise on the west coast of South America.

The Californian Uprise on the west coast of North America.

The New England Uprise on the east coast of North America.

The Morocco Uprise on the north-west coast of Africa: and

The Somali Uprise on the east coast of Africa.

Downthrows. The corresponding phenomenon of abnormally warm water being found at the surface and descending to greater depths may appropriately be called a Downthrow. Important examples of downthrows are found on the homologous coasts of Brazil and Australia, which are characterised by the formation of Barrier coral reefs.

Perhaps the most important as well as the most interesting downthrow is that of the Sargasso Sea, which forms a vast lake in the middle of the North Atlantic Ocean. The distribution of temperature in this sea formed the subject of a paper which was published in the *Proceedings of the Royal Society* in the year 1875 and was reprinted as Paper III in the first volume of my *Scientific Papers,* published by the Cambridge University Press in 1913. It is due to the downthrow in this region that the water, at a depth of 1000 fathoms from the surface, has a higher temperature than is to be found in any other part of the open ocean at the same depth.

No. 4. [*From Nature, June* 11, 1885, *Vol.* XXXII. *pp.* 126–130.]

RELATION BETWEEN THE TEMPERATURE OF THE SEA AND THAT OF THE AIR ABOVE IT

BEING obliged to proceed to South America at the beginning of the year 1885, I took with me a thermometer and a hydrometer in order, if circumstances were favourable, to provide myself with occupation during the somewhat long and monotonous voyage. Thanks to the kindness and courtesy of Captain Brown, of the s.s. "Leibnitz," who took a lively interest, and assisted me greatly in carrying out my observations, the voyage was neither long nor tedious.

The "Leibnitz" sailed from Southampton on January 16, 1885, and made the passage direct, without touching at intermediate ports, to Monte Video, where she arrived on February 8, after a very favourable voyage. The route lay through the most interesting meteorological districts of the Atlantic, and my principal object at starting was to make as many observations of the temperature and the density of the surface-water along the route as possible. With these I combined observations of the temperature of the air, and frequently also of the wet-bulb thermometer. Observations were begun on January 21 in lat. 34° N., and continued up to the morning of arrival in the River Plate.

I have put together the simultaneous observations of the temperature of the air and the water with those of the wet-bulb thermometer, as they possess some interest of their own; the observations of density are kept for a future opportunity, as the reductions in connection with them are not quite finished.

The *thermometer* used for all the observations was divided into individual degrees of the Centigrade scale, and was of the ordinary form of German manufacture, with a paper scale.

The degrees were 1·6 mm. apart, so that there was no difficulty in estimating tenths of a degree. Its zero was verified on board by immersing it in pounded ice, and found correct. The ice was well pounded in a clean towel, and a tall tumbler was filled with it; the thermometer was then thrust into it and allowed to remain till sufficient ice had melted to fill up the interstices, producing a perfect magma of ice and water down to the bottom. The mercury remained constant on the zero line. The temperature of the air was 25° C.

Having plenty of pounded ice at my disposal I poured off the water which had formed by melting, and replaced it by sea-water, containing 35·65 grammes salt per kilogramme, and then immersed the thermometer; it fell rapidly below zero, and remained constant at – 1°. I then strained away the sea-water from the ice and replaced it by a mixture of equal volumes sea-water and distilled water: the thermometer fell to – 0·45°, and remained constant for some time at that temperature. When the ice was mixed with distilled water alone, the thermometer again stood at 0° C. These experiments were made to verify some observations of Pettersson, quoted in his investigations into the nature of ice formed from waters of different degrees of salinity, in connection with the voyage of the "Vega." He there says, referring to the melting temperature of different kinds of ice, that pure fresh-water ice, when immersed in sea-water, melts at a temperature considerably below 0° C. Writing from memory, I think he puts the melting point at from – 1° to – 2° C. Having both the ice and the sea-water ready at hand, I repeated this remarkable experiment. The result showed that Pettersson's observation is quite correct, and that the lowering of the melting point is roughly proportional to the salt held in solution. When equal volumes of the sea-water and distilled water of the same temperature were mixed, there was no change of temperature. I do not remember if Dr Pettersson furnished an explanation of this remarkable phenomenon, and I am unable to supply one myself, but it must necessarily affect the validity of conclusions as to the composition of sea-water ice drawn from its melting point. When the "Challenger" was in Antarctic waters I made a number of observations on the melting point of ice collected from broken pieces of the pack, and found that it began to melt a little below – 1° C. I concluded that either it was one solid substance or a mixture of several solids. But if pure ice melts at a different temperature according to the medium in which it is placed, then this reasoning is faulty, for inclosed brine would have much the same effect as inclosed salt or crystalline hydrate.

Temperature of the water. The water was collected in a small bucket, well clear of the side of the ship, and on the opposite side from that through which the condensing water of the engine is discharged. Its temperature was determined as soon as the sample was brought on board.

As the ship left the Channel in the middle of winter, and proceeded nearly due south, the temperature of the water rose rapidly at first. Observations were begun on January 21, in lat. 34° N., and between this latitude and lat. 10° N. the rate of rise was very steady, averaging 0·36° C. per degree of latitude. From lat. 5° N. to 15° S. the temperature is very uniform and high, averaging 26·86° C. After passing lat. 15° S. the temperature falls, and begins to show greater variations, as the shallow water on the Abrolhos Bank is approached. The average temperature of the water over this bank was 25·56° C. After passing Cape Frio, and between the parallels of 25° and 30° of south latitude, the variations of temperature were considerable and often abrupt; the maximum observed in this part was 26·7° C., and the minimum 24·3° C. As the higher temperature generally accompanies a greater salinity, it is probable that these variations are due, not to any terrestrial source, such as large rivers, but to an oceanic cause, the less saline and colder water of the deeper ocean strata being thrown up against the coast, and mixing imperfectly with the hot and dense surface water. In lat. 30° S. the influence of the River Plate makes itself distinctly felt by a general rapid fall of temperature. As the ship got into soundings, with the change in colour and other properties of the water, the temperature fell rapidly to between 23° and 24° C., and to 22° C. in six fathoms off Flores Island close to Monte Video. The minimum temperature observed in this part was 20° C. at 2 A.M. between Lobos Island and Maldonado Point.

Excluding the latter part of the voyage between the River Plate and lat. 15° S., where the conditions are a good deal affected by purely local causes, the surface water shows well-marked diurnal maxima and minima of temperature. From lat. 9° N. to lat. 2° N. the ship passed through the equatorial belt of calms and rains, which separates the regions of the north-east and the south-east trade winds from each other. It is characterised by a calm sea, a cloudy sky, and heavy rains. Here the temperature was subject to very little diurnal variation (0·3° C.). On approaching St Paul's Rocks, a few miles north of the equator, the clouds cleared away completely, and there was a calm sea, a clear sky, and a very powerful sun. The result was a comparatively great rise of temperature in the afternoon; and yet the greatest difference between any neighbouring maximum and minimum in this region was only 1·1° C.

The maximum temperature of the sea surface observed during the voyage was 27·4° C. (81·3° F.) at 2 P.M. on January 31, in lat. 7° 35′ S., the Brazilian coast being about 100 miles distant. The temperature of the water will be further considered in connection with its density; at present its connection with the temperature of the air will be more particularly considered.

Temperature of the air. Along with the temperature of the water, that of the air during daylight was determined. It is probably very rare, in any part of the ocean, to find the mean temperature of the air agreeing exactly with that of the surface water, and in many places the differences are considerable. In order to be able to compare the temperature of the air with that of the water, it is necessary that both should be determined with equal accuracy. The temperature of the water is easily and accurately determined by agitating the thermometer in a bucket of it freshly collected. With the air it is somewhat different. Having only one thermometer with me, I was obliged to use it for all purposes, and I could not hang it up in a thermometer box, even if I had had one, and had deemed it advisable to do so. On board ship, however, I am convinced that it is quite impossible to fix a thermometer box in such a position as always to secure such an exposure as to justify the assumption that the indications of the thermometer may be taken as the true temperature of the air. Even on shore and under the most advantageous circumstances, the temperature of the thermometer in the atmosphere of the best constructed box is too much dependent on the temperature and capacity for heat of the material of the box for it to be assumed always to be identical with that of the air outside, at the moment of reading. I was obliged, therefore, to adopt the method of whirling the thermometer, at the end of a short string, in the air, in whatever part of the ship happened at the moment to afford the most favourable conditions, and reading it when it had assumed a constant temperature. The temperature of the air is thus determined in exactly the same way as that of the water, namely, by agitating a thermometer in it, and the comparison of the two is therefore likely to lead to trustworthy conclusions.

Temperature of wet-bulb thermometer. The series of observations with this instrument is not so complete as that with the dry thermometer, but they possess some interest. The method of observation was the following: the temperature of the air having been determined by whirling the thermometer in it, a bucket of sea-water was fetched and its temperature taken; the thermometer was then exposed, with its bulb still wet with sea-water, to the breeze in a suitable part of the ship, and its temperature observed when it became constant. The exposure of the instrument requires some care. The bulb must be quite free from grease, which can be readily secured by washing it with soap and water. It is then dipped into the water and allowed to drip for a second. It is then held somewhat inclined to the direction of the wind and to the horizon, and rotated gently on its axis so that the bulb be kept covered with a continuous film of water which is locally thickened by gravity, which tends to form a drop on the lower side of the bulb. The reading of the thermometer is observed while it is being rotated. Had I intended from the beginning to make a series of wet and dry bulb observations, I should probably have used fresh water from the first. I began to expose the thermometer, merely in order to have an indication whether the atmosphere was saturated or not, and I expected, in the damp equatorial regions, to find the atmosphere so heavily saturated as to be incapable of producing any sensible lowering of the thermometer with damped bulb. For this purpose it seemed to be quite sufficient to expose the thermometer wet with sea-water. Having begun with sea-water the observations were continued with it. A few comparative observations were made in order to determine the effect of replacing the sea-water by fresh water. On February 2, after a shower, the temperature of the air was 25° C. When wet with sea-water the temperature of the thermometer was 23·5° C., and with rain-water 23·1° C. Similarly, at noon on the same day, the following temperatures were observed: dry bulb, 26·1° C.; wet bulb (sea), 24·5° C., wet bulb (rain), 24·2° C. The air, at this time, appeared to the sensation to be damper than at any other time, and yet, when suitably exposed, there was a difference of nearly 2° C. between the wet and dry thermometers.

There is an advantage in having the bulb of the thermometer wet with a continuous film of water, instead of being surrounded with damp muslin, namely, that it more nearly resembles the surface of the sea, which is exposed to the influence of the atmosphere. Observations with the wet thermometer were not made as regularly as those with the dry instrument, and no observations were made with either of them after dark, owing to the difficulty of securing proper exposure and reading the instrument with a lantern, without heating it.

The temperature of the air and of the water were taken generally every two hours from 6 A.M. to 6 P.M., but the intervals between the observations were not always the same. These observations showed that only on two days, January 31 and February 1, between lat. 6° S. and 12° S., did the mean *day* temperature of the air exceed that of the surface water. On these days the temperatures were taken every two hours from 6 A.M. to 6 P.M., and the means of the groups of seven observations gave: on January 31, air, 27·13° C.; sea, 26·90° C.; difference, 0·23° C.; and on February 1, air, 27·26° C.; sea, 26·96° C.; difference, 0·30° C. These differences would have been reduced in amount if the observations had been carried on through the night, though, from the very high temperature of the air just before sunrise on February 1 and 2, they would not have been reduced to zero.

In the table (p. 121) all the simultaneous observations of temperature of air and water made during the voyage, except those of the last day, when approaching the mouth of the River Plate, are collected in small tables for each day. The time of day is given in hours, from 0 to 24; the temperatures are in Centigrade degrees; *t* denotes the temperature of the sea surface, $t - T$ the difference between it and the temperature of the air, and $T - T'$ the difference between the readings of the thermometer in air with its bulb dry and when it is wet with sea-water. At the head of each table is given the meteorological district of the ocean through which the ship was passing, as "north-east trade winds," "equatorial calms," and the like; also the day of the month (1885) and the latitude and longitude at noon of the day. The means at the foot of each table are simply the arithmetical means of the numbers in each column; and their meaning and value are at once apparent on inspecting the column.

With the two exceptions above-named, the temperature of the sea was always found higher than that of the air, over the day, and only very seldom was it exceeded by that of the air at the hottest time of the day. Had the observations been carried on through the night, the contrast between the two temperatures would have been much greater. On January 31 and February 1 the conditions were somewhat exceptional. On the former of these days the ship passed into the northerly monsoon, which prevails all down the Brazilian coast during the southern summer. Like the similar monsoons in the northern hemisphere, it is caused by the proximity of a large mass of land, which gets intensely heated by the vertical rays of the sun. On January 30 the wind had been light south-easterly; during the night it fell calm, and at sunrise a light easterly wind sprang up, which gradually drew around towards the north and blew all day, with just sufficient force to travel exactly at the same rate as the ship (11½ knots); consequently, during the whole of the day the atmosphere on the deck was motionless, with a very powerful sun beating on it and heating up everything, so that it was impossible to find any place where the air could be got, coming fresh on board, without having been exposed to the influence of the highly heated deck and fittings. It is therefore certain that the air-temperatures are somewhat above the truth.

It is probable that, when the true temperature of the air can be ascertained, it will be found to be usually below that of the sea surface. The cause of this is, I think, to be found in the relative dryness of the atmosphere over the ocean. If the observations with the wet-bulb thermometer be considered, it will be seen that the least difference of reading between the dry-bulb and the wet-bulb thermometers was 1° C. on January 28, when the ship was in the middle of the equatorial belt of calms and rains. In this region perfectly saturated air might be expected, and with instruments exposed in the usual form of box I have no doubt that here, and in the very oppressive weather of the northerly monsoon, the two instruments would have given identical readings. The readings of the air-temperature on January 28 were perfectly trustworthy, as the sky was thickly overcast with dense rain-clouds all day there was thus no risk of overheating; the readings with the wetted bulb were equally satisfactory, so that the results of the observations on that day may be taken to represent fairly the normal state of things in the "Doldrums." The temperature of the sea varied from 26·3° to 26·6° C., the mean of five observations during the day being 26·42° C. The mean temperature of the air during the day was 0·92° lower than that of the sea, or 25·5° C., and the temperature of the wet-bulb thermometer 1·3° lower still, or 24·2° C. It will be seen that, on the two exceptional days, January 31 and February 1, the difference between the wet-bulb and the dry-bulb thermometers is greater than would be expected from the oppressive damp feeling of the air; it is therefore all the more likely that the dry-bulb readings are too high as indicated above. However, it is important to observe that in all the regions passed through, whether in the westerly winds of the North Atlantic or the equatorial calms, or the monsoon of the South Atlantic, the temperature of the wet-bulb thermometer is always very markedly below that of the dry-bulb thermometer. In fact, such is the mobility of the atmosphere that it rarely has the opportunity of saturating itself; and if the effect which must be produced when this air meets the surface of the water be considered, it will, I think, afford some explanation of why at sea the temperature of the air, even by day alone, is usually markedly below that of the sea surface.

TABLE *giving the temperature of the sea surface (t), the difference between it and the temperature of the air ($t - T$), and the difference between the readings of the dry-bulb and the wet-bulb thermometers in air ($T - T'$), at different hours of the day.*

N.W. Winds — Jan. 23. Lat. 26° 24′ N. Long. 21° 21′ W.

Hours	t	$t - T$	$T - T'$
	°C	°C.	°C.
8½	18·9	2·1	
10	19·3	1·1	
12	19·2	0·7	
14	19·3	0·8	
16⅓	19·7	1·7	
17½	19·6	1·9	
Mean	19·33	1·38	

N.E. Trade Wind — Jan. 24, Lat. 22° 5′ N. Long. 23° 6′ W.

Hours	t	$t - T$	$T - T'$
	°C.	°C.	°C.
8¼	20·5	1·5	
10½	20·7	1·0	
12	20·9	0·8	
14	21·1	1·1	
16	21·2	1·2	
17½	21·2	1·3	
	20·93	1·15	

N.E. Trade Wind — Jan. 25, Lat. 18° 6′ N. Long. 24° 39′ W.

Hours	t	$t - T$	$T - T'$
	°C.	°C.	°C.
8¼	22·0	1·4	1·6
10	22·0	0·8	
12	22·2	0·7	
14	22·2	1·0	
16	22·7	1·7	
17¼	22·3	1·2	
	22·23	1·13	

N.E. Trade Wind — Jan. 26, Lat. 13° 46′ N. Long. 26° 6′ W.

Hours	t	$t - T$	$T - T'$
	°C.	°C.	°C.
8	23·2	1·0	3·0
10	23·8	0·8	—
12	24·0	0·7	1·8
14	24·0	0·9	2·1
16	24·2	1·1	
17¼	24·2	1·0	
	23·90	0·92	

N.E. Trade Wind — Jan. 27, Lat. 9° 25′ N. Long. 27° 21′ W.

Hours	t	$t - T$	$T - T'$
	°C.	°C.	°C.
6	25·4	1·5	—
8	25·6	0·9	2·7
10	25·7	0·9	—
13	26·0	1·0	3·0
16	26·0	1·0	
17½	26·0	1·0	
	25·79	1·05	

Equatorial Calms — Jan. 28, Lat. 5° 16′ N. Long. 28° 32′ W.

Hours	t	$t - T$	$T - T'$
6	26·4	6·7	1·7
9	26·3	1·0	—
12	26·3	1·1	1·4
15	26·6	1·0	1·1
18	26·5	0·8	1·0
—	—	—	—
—	—	—	—
Mean	26·42	0·92	1·30

Equatorial Calms — Jan. 29, Lat. 1° 15′ N. Long. 29° 16′ W.

Hours	t	$t - T$	$T - T'$
6	26·5	0·6	1·4
9	26·7	0·8	—
12	27·0	0·3	1·8
15	27·0	0·0	2·5
18	26·8	0·2	2·3
—	—	—	—
—	—	—	—
	26·80	0·38	2·00

S.E. Trade Wind — Jan. 30, Lat. 2° 53′ S. Long. 31° 5′ W.

Hours	t	$t - T$	$T - T'$
6	26·3	0·3	1·6
9	26·4	−0·2	1·9
12	26·8	0·0	2·4
15	27·0	+0·4	2·2
18	26·8	0·7	1·8
—	—	—	—
—	—	—	—
	26·67	0·24	1·98

Northerly Monsoon — Jan. 31, Lat. 7° 5′ S. Long. 33° 2′ W.

Hours	t	$t - T$	$T - T'$
6	26·5	0·5	1·7
8	26·5	−0·2	2·3
10	26·8	−0·6	2·6
12	27·1	−0·7	2·8
14	27·4	−0·3	2·6
16	27·1	−0·3	—
18	26·9	—	2·3
	26·90	−0·23	2·38

Northerly Monsoon — Feb. 1, Lat. 11° 18′ S. Long. 34° 53′ W.

Hours	t	$t - T$	$T - T'$
6	26·8	0·1	2·4
8	26·9	−0·1	2·2
10	27·0	−0·3	2·7
12	27·0	−0·8	—
14	27·0	−0·7	2·8
16	27·0	−0·3	2·3
18	27·0	—	2·0
	26·96	−0·30	2·40

Northerly Monsoon — Feb. 2, Lat. 15° 30′ S. Long. 36° 53′ W.

Hours	t	$t - T$	$T - T'$
6	26·9	−0·1	2·3
8	26·9	−0·1	2·0
10½	26·7	+1·7	1·5
12	26·9	0·8	1·6
—	—	—	—
16¼	26·3	0·3	—
18	25·5	−0·2	—
Mean	26·54	0·40	1·85

Northerly Monsoon — Feb. 3, Lat. 19° 48′ S. Long. 38° 42′ W.

Hours	t	$t - T$	$T - T'$
6	24·3	0·1	1·8
8	25·7	0·8	2·1
9	25·4	0·6	—
12	25·2	1·2	—
14	25·9	1·3	1·8
16	26·0	0·1	1·6
18	25·9	—	—
	25·48	0·60	1·82

S.E. Trade Wind — Feb. 4, Lat. 24° 0′ S. Long. 40° 33′ W.

Hours	t	$t - T$	$T - T'$
5½	25·5	1·5	—
8	26·0	1·1	2·9
10	25·8	0·2	—
12	25·7	0·5	2·9
14	25·8	0·8	2·8
16	26·0	1·0	—
—	—	—	—
	25·80	0·85	2·87

S.E. Trade Wind — Feb. 5, Lat. 27° 8′ S. Long. 44° 1′ W.

Hours	t	$t - T$	$T - T'$
5½	24·4	0·6	—
8	24·5	0·2	2·8
10	24·5	−0·3	3·7
12	25·2	−0·1	3·3
14	26·4	+1·4	4·7
16¼	26·7	2·0	3·5
18	26·3	2·0	4·3
	25·43	0·97	3·72

S.E. Trade Wind — Feb. 6, Lat. 30° 13′ S. Long. 47° 52′ W.

Hours	t	$t - T$	$T - T'$
6	24·7	0·6	2·8
8	25·0	0·7	2·7
10	24·8	1·1	2·7
—	—	—	—
14	25·2	0·9	3·0
16	24·7	0·7	2·9
18	24·8	1·1	3·7
	24·87	0·85	2·97

If we consider the film of water immediately at the surface of the sea, having the atmosphere on the one side of it and the bulk of the water on the other, it is strictly comparable with the film of water surrounding the bulb of the thermometer, when exposed to the atmosphere in the way described above; and the air playing upon it must produce exactly the same effect in the one case as in the other. The evaporation lowers the temperature of the aqueous film, which proceeds to extract heat from the neighbouring bodies—namely, in the one case the air and the bulb of the thermometer, and in the other case the air and the layer of water immediately below the surface film. If we imagine for a moment the surface film separated from the bulk of the water below it by a diaphragm impervious to heat and exposed to the atmosphere so as to suffer evaporation and lowering of temperature, then on the removal of the diaphragm it would immediately sink away from the surface and its place would be taken by warmer, and therefore less dense, water from below. In the case of sea-water this effect would be slightly intensified by the concentration produced by evaporation. But while the water below supplies some of the heat rendered latent by the evaporation of the water, the air above it supplies its share, and is cooled. In both cases the heat thus lost is made good by the direct radiation from the sun. Through a moderately dry atmosphere the rays pass with comparatively little heating effect, but are largely absorbed on entering the water. Consequently the loss of heat which the water suffers by evaporation at the surface of separation is made good more abundantly than that sustained by the air; and the difference in power of absorption of radiant heat exhibited by these two substances is thus sufficient to keep up a permanent difference of temperature between the water and the air immediately above it.

Starting with air and water at the same temperature, we may imagine the process as taking place in three acts. First, the water at the surface evaporates, and the air on the one side, and the water on the other, are cooled; second, in order to make up for the heat thus rendered latent and lost, the sun shines upon both alike, but the water absorbs a larger proportion of the heat of its rays than the air does; and finally, a portion of this excess is removed from the water by the simple contact of the air at its surface. *The nett effect of these causes is to produce a permanent excess of temperature of the surface water of the sea over that of the air above it, provided that the air is not completely saturated with moisture.*

From what I have seen and experienced in the regions visited by the south-west monsoon in the east, I cannot doubt that there are often cases where the most carefully exposed wet-bulb and dry-bulb thermometers would show identical readings, and the atmosphere is completely saturated with vapour of water. Thus it is probable that the temperature of the air would not be inferior to that of the water. Further, when, on the eastern coasts of Asia, the south-west monsoon blows out of the China Sea and penetrates far into the North Pacific, off the coasts of Japan it attains a latitude of naturally lower temperature than that from which it proceeded, so that much of the water with which it was laden, and which is held diffused through it as a true gas, is condensed and remains suspended in it, producing a visible haze, which obscures the horizon and moistens all solid objects exposed to it. Here the conditions are reversed, and instead of the air losing heat to evaporate the water, it receives the heat liberated by the condensation of the steam removed from waters of lower latitudes. Such conditions are, however, certainly local, and there can be little doubt that, as a rule, the temperature of the surface water of the sea is higher than that of the air. The temperature of the air depends on that of the water which tends to warm it, and on the degree of its own dryness, by virtue of which the water has a tendency to evaporate into it and, by extracting heat from it for this purpose, to cool it.

It is obvious that local circumstances such as currents may produce differences between the temperature of the air and the water, but such cases are not here under consideration[1].

[1] See Contents, p. xxiii.

No. 5. [*From Nature, July* 21, 1910, *Vol.* LXXXIV. *pp.* 87–89.]

THE COLOUR OF THE SEA[1]

APROPOS of the report (*Nature*, March 10) of Lord Rayleigh's lecture dealing with the parts played by reflection and transmission of light in the production of the integral impression of colour on the eye of an observer looking at the sea from the deck of a ship, I should like to be permitted to make some observations on the proper colour of the water of the ocean, as it is a subject which has occupied my attention, off and on, during the last forty years.

During the voyage of the "Challenger" I began to log the colour of the water in February, 1874, when she was working in the neighbourhood of the Antarctic circle. My attention was there directed to it by the frequent and abrupt passage of the ship from water of the clear indigo colour of the ocean of temperate latitudes to the deep olive-green water which is a distinctive feature of these icy regions.

The green colour is due to the abundance of diatoms. These are so plentiful and so preponderant that, besides putting their stamp on the surface, they furnish a distinct type of oceanic deposit, the *diatom-ooze*. The green colour of the water is due, not only to the living diatoms, but also, and perhaps to a greater extent, to the excretions of the animals for the subsistence of which the diatoms furnish the ultimate food supply. The crowds of penguins and other birds to be met with in these seas stain all the ice green where they have rested. The water, inhabited by diatoms and affected by diatomaceous débris, has a deep olive-green colour which is characteristic, and this I accepted as one colour-type of the water of the ocean. It is seen best in the water the transparency of which is not interfered with by too great a crowd of the

[1] See Contents, p. xxiii.

diatoms themselves. Water belonging to this type of colour is not confined to polar latitudes; it is met with in a certain class of homologous districts of the warmer ocean, in tropical and even in equatorial latitudes.

When we quit the edge of the polar ice and steer equatorwards, the surface water assumes a pronounced indigo colour, and this persists until we pass the fortieth parallel. If we start from the equator and sail polewards, the colour of the surface water persists as a pure and brilliant ultramarine until the thirtieth parallel is passed. The passage from the ultramarine to the indigo, and *vice versa*, is usually very rapid, and the area of mixture is restricted. No one who has once sailed in the ultramarine waters of the intratropical ocean and has observed, as well as seen, its colour, can ever mistake any other colour for it. If he has doubt as to whether the water through which he is passing is ultramarine or not, he may be sure that it is not. The ultramarine and the indigo are the two great colour-types to which the mass of the surface water of the deep sea belongs, and, with the olive-green, they make the three fundamental colour-types which are required, and are sufficient for the adequate logging of the colour of the surface water of the ocean.

The water of the Mediterranean belongs to the ultramarine type, but it always appears to me to have a harder tone than the soft and brilliant ultramarine of the intratropical ocean.

With regard to the method of judging the colour of the water, much unnecessary difficulty is made. The first precaution to be observed is to take up a position where the greatest amount of light can reach the eye after passing through the water, and the smallest amount after being reflected from its surface. There is generally little difficulty in accomplishing this on one side or the other of the ship and by looking as nearly as possible vertically into the water.

The "Challenger," like other men-of-war of her date, was fully rigged, and built for sailing as well as for steaming. When under sail the propeller causes a certain amount of retardation, and to remedy this she was fitted with a "screw-well" into which the propeller could be hoisted out of the water. This proved to be a perfect observation tube for determining the true colour of the water. Its diameter was about 6 feet; it passed from the upper deck through the captain's cabin on the main deck and the ward-room on the lower deck into the water. Looked into from the deck, the sea-water appeared to be enclosed in it as the water is in a well, but with this difference, that the water, by day, was brilliantly illuminated from below. There being no clearance between the surface of the water in the well and the structure of the ship, no light could enter except through the water. No direct sky-light could reach it down the well, because the poop awning, which was practically always spread during the day, completely excluded it. The screw-well was, in effect, an artificial and perfected *Grotto di Capri*, which was carried round the world. It was perfected, inasmuch as there is a passage for boats to penetrate into the grotto from the outside, while the screw-well is entirely shut off. During the whole of the voyage the colour of the water was under observation in this very perfect apparatus.

The statement that the blue colour of the sea is nothing but the reflection of the blue of the sky was at first frequently made, even on days when the sky was completely overcast; a visit to the screw-well, especially on overcast days, never failed to convince the doubter that the water contained in its own mass sufficient colour to account for all that was perceived. When the ship was in green water the view was never advanced that its colour was due to reflection from the sky.

As ships with screw-wells long ago disappeared from the sea, it may not be superfluous to point out that what could be observed in the screw-well was altogether different from what can be seen in the wake of the screw of a modern steamer. While the screw-well was a perfect instrument for gauging the colour of the water, the determination of its transparency was more conveniently made from a boat. Thus in mid-Pacific, with the aid of a "water-glass" to eliminate the disturbing action of ripples, a metal plate measuring only 4 by 4 inches, painted white and not masked by the suspending line, was distinctly seen at a depth of 25 fathoms (45 metres).

Beyond this depth it became indistinct, and became invisible at about 27 fathoms, but this was due mainly to its smallness and to its want of steadiness, being attached to the boat, which rose and fell with the swell. At 25 fathoms the plate had a pale ultramarine colour, and its edges were sharply defined. These separated the column of water, into which I looked through the water-glass, into a central column of rectangular section having a depth of 25 fathoms, and into a column, surrounding and contiguous with it, which had a depth many times greater. These columns, being juxtaposed, were placed in the way most favourable for the comparison of their colours. The colour of the central column, 25 fathoms in length, was a pure but pale ultramarine; that of the external and uninterrupted column through which the whole unabsorbed and undissipated part of the sunlight which had penetrated into the water returned to the surface was of the same tone, but of many times greater intensity. Assuming the intensity of the colour to be proportional to the length of the column of water traversed by the light, it is to be concluded that the length of the uninterrupted column which transmitted the more intense colour was many times greater than 25 fathoms. It must be noted that the glass plate forming the bottom of the small tub, which is called a "water-glass," was during the observation completely protected from direct sky-light by my head and the brim of the panama hat which, at that time, I always wore when exposed to the sun.

It has already been said that water of as pure a green as that of the Antarctic occurs in other and warmer districts of the ocean. My attention was first directed to this during the cruise of the "Dacia," which, although it occupied no more than three weeks, marks an epoch in deep-sea research. A short account of it is given in a paper by me—"On Oceanic Shoals discovered in s.s. "Dacia" in October, 1883"—and published in the *Proceedings of the Royal Society of Edinburgh*, 1885, XIII. p. 748. Perhaps the most remarkable of these shoals was the one which was named the "Coral Patch," in lat. 34° 57′ N., long. 11° 57′ W., the exploration of which, along with that of the tidal currents in the open ocean (*Proc.*

Roy. Soc., 1888, XLIII. p. 356), supplied the evidence which definitively established the fact that coral islands are not a product of subsidence.

When the survey of this shoal had been completed, in so far as the time at the disposal of a steamer engaged on a commercial mission permitted, a line of soundings was run from the "Patch" to the African coast at Mogador. Independently of the high land which is visible from the sea at a distance of many miles, the approach of the coast is indicated by a fall in the temperature of the water of the sea surface, and a remarkable change in its colour. Outside, the temperature of the surface water was 21° C., and its colour was ultramarine. After sighting the land its temperature fell, at first slowly, then rapidly, and, when at a distance of two miles from Mogador, it was only 16° C. The colour at the same time had become a pure olive-green, which maintained its transparency until close to the shore, where it became masked by the solid matter kept continually in suspension by the mechanical energy of the breaking waves.

The pure green colour of the water and its temperature, so much lower than that which could persist at the surface of the sea in the latitude of Mogador, made me for a moment think that it might be in reality Antarctic water which had found its way, at or near the bottom, into the northern hemisphere, having been diverted first to the west while in the South Atlantic, then to the east after crossing the line. But this idea could persist only for a moment, because the temperature and the density of the bottom water were found to be those characteristic of the bottom water of the eastern basin of the North Atlantic, as shown by the "Challenger" observations, and these are much higher than those of any other ocean.

The low temperature of the water showed that it could not come on the surface from the north or south or west of it, and the only source from which it could come was from below the surface. Deep water comes close to the coast, and the water at 2000 fathoms was found to have a temperature of 2·5° C., so that the supply of cold from this source was adequate, and it was available with a very small expenditure of energy.

Having arrived at the surface and following the south-westerly drift of the surface water, exposure to the sun raised the temperature of the water and discharged its colour *pari passu.* It was evident that there was here a case of the rising of deep water at the weather coast of an ocean, away from which the prevailing wind was continually driving the surface water.

From Mogador the "Dacia" proceeded to the "Seine Bank," in lat. 33° 47′ N., long. 14° 1′ W., and explored it thoroughly. Among the specimens brought up on the grapnel were masses of dead coral and shells, all having the same green colour. Some of these fragments were preserved in spirit, which quickly assumed the green colour, leaving the shells and coral practically decolourised. I sent the bottle, with the specimens and spirit, to my friend Prof. W. N. Hartley, in Dublin, who was good enough to subject them to spectroscopic examination. He wrote to me on February 15, 1884: "I have made a spectroscopic examination of the colouring matter you sent me and have no doubt that it is altered chlorophyll. I have got identical wave-length measurements for the absorption band with your liquid and a specimen of very pure chlorophyll dissolved in ether"; and he adds, "there is very little real substance in even a dark green solution."

As the year 1884 belongs now to the remote past, I recalled the matter to Prof. Hartley, and, confirming his previous information, he added: "I believe my impression at the time was that the chlorophyll was the colouring matter of a living micro-organism, and that these settled upon the shells, but when not deposited they were floating in the sea water." I am obliged to Prof. Hartley for kindly permitting me to use these private communications. Further information will be found in his paper on chlorophyll from the deep sea (*Proc. Roy. Soc. Edin.*, 1885, XIII. 130).

Prof. Hartley's report furnished a remarkable confirmation of my first impression in so far as it showed that the green water of the Mogador coast owed its colour to the same substance as did the diatom-crowded water of the Antarctic, namely, chlorophyll.

In April and May of 1885 I made a coasting voyage from

Valparaiso to San Francisco. Excepting the equatorial part, stretching from Cape Blanco to Panama and round the coast of Central America to near Mazatlan, the west coast of the American continent between the fortieth parallels is the weather shore of the Pacific Ocean. All along it cold and green water is met with, in the same way as we have seen to be the case on the Atlantic coast of Morocco. On the South American coast the green water was found to extend, with few interruptions, from Valparaiso, lat. 33° S., to Cape Blanco, lat. 4° 27′ S. As on the Morocco coast, the green colour and the low temperature of the water are found only close to the shore. At a distance of ten miles outside the colour is blue, and the temperature normal for the latitude. There can be little doubt that, as the localities where the green water occurs are geographically homologous, so the substance which produces the colour is generically the same, namely, chlorophyll.

The following particulars are taken from my unpublished journal. The only ports or anchorages where the water was blue were Huasco, lat. 28° 27′ S., temperature of the surface water 14·7° C., and Carizal, lat. 28° 5′ S., temperature 15·1° C. The occurrence in this latitude of blue water with so low a temperature is very remarkable.

At Antafogasta, lat. 23° 39′ S., the water was greenish-blue, and its temperature was 18·0° C. Between this port and Iquique the ship's course took her to a distance of nearly twenty miles from the coast, and there the colour of the water was ultramarine and its temperature 21·2° C. At Iquique the water was quite green, and its temperature 17° C. Between this port and Arica the water was quite green, even at a distance of five miles from the coast, where the temperature was 19·5° C., but on anchoring at Pisagua, lat. 19° 36′ S., the temperature of the water was only 15·2° C. At Arica, lat. 18° 28′ S., the water was equally green, but its temperature was 19·5° C. Arica lies in the angle where the trend of the coast changes from north to about north-west. From Arica the ship made a longer run to Chala, lat. 15° 49′ S., keeping at a distance of fifteen to twenty miles from the coast. Here ultramarine water was met with, its temperature rising to 23·2° C., but even at fifteen miles from this coast some green water was met with having a temperature of 18·8° C. I attributed to this the foggy state of the atmosphere which prevailed. This obscured the sun, and retarded both the heating and the bleaching of the water. In lat. 14° 8′ S., when six miles off shore, the water was quite green, and its temperature 15·1° C. Outside of Callao, lat. 12° 0′ S., the water was green, and its temperature 16·3° C.; in the harbour its temperature was 17·5° C., and its colour a dirty green, turbid and milky with sulphur, smelling strongly of sulphuretted hydrogen, and full of dead fish. Continuing northwards, off Ferrol Islands, lat. 9° 11′ S., the temperature of the water was 16·0° C., and its colour olive-green. At Payta, lat. 5° 5′ S., the temperature of the water was 17·1° C., and its colour a chalky green.

The green and cold shore water ceased abruptly at Cape Blanco, lat. 4° 27′ S., and during the passage round this cape from Payta to the entrance of the Guayaquil River, lat. 3° 9′ S., the temperature of the water rose from 17·1° to 25·2° C. From this locality a pretty straight line was followed across the equatorial current near its source to Panama, lat. 9° 0′ N. During the passage the temperature of the water varied between 25° and 27° C., and it maintained a blue colour throughout. At Panama, however, with a temperature of 27° C., the water was quite green.

A similar occurrence of cold and green water near the shore was observed on the North American coast from Cape San Lucas, at the extremity of the Californian peninsula, to San Francisco. In the equatorial waters which wash the coast from Cape Blanco, lat. 4° 27′ S., to Panama, and thence to Cape Corrientes, lat. 20° 25′ N., long. 105° 43′ W., green water is prevalent along the shore, but its temperature is very high, 28° or 29° C.

I will here refer to only one other locality, and that a well-known one, where the weather shore of an ocean is associated with green water of abnormally low temperature, namely, the east coast of North America from Florida to Nova Scotia. The cold and green water which is found on this coast, and lying between it and the western edge of the Gulf Stream, is usually attributed to the Labrador current, which is charged with the duty of bringing cold water from Baffin's Bay as a surface current round Newfoundland and down the coast to Cape Hatteras and even beyond it. The principle was the same as that which moved Humboldt to attribute the cold water, which we have described in connection with the Pacific coast of tropical South America, to a surface current from the Antarctic Ocean. In my paper "On Similarities," etc., I have shown that Humboldt's explanation postulated an impossibility. The deeper layers of the water on the coast itself are capable of supplying, as and when required, much more cold than is wanted, and that with the least expenditure of energy. The same is the case with the "cold wall." Besides the south-westerly winds of the North Atlantic, and perhaps independently of them, the Gulf Stream itself, pouring its waters in a stream of great momentum past the American coast and out into the open ocean, performs the function of a colossal jet-pump, carrying water away from the surface and leaving its place to be taken by the other water which can get there most easily. This is the cold water of the deeper layers *in situ*. It is this hydraulic cold-water service which tempers the climate of the eastern States. The labours of the U.S. Coast Survey during the last seventy years have shown that fluctuations, both regular and irregular, occur in the flow of the Gulf Stream. These necessarily re-act on the supply of cold water drawn from the deep and spread over the continental shelf. Such variations are probably the source of the accidents which occasionally occur and cause the extinction of marine life over large tracts of shoal water on that coast[1].

[1] See Contents, p. xxv.

No. 6. [*From Proc. Roy. Soc. Edin.* 1890, *Vol.* XVIII. *pp* 17–39.]

ON THE OCCURRENCE OF SULPHUR IN MARINE MUDS AND NODULES, AND ITS BEARING ON THEIR MODE OF FORMATION

In the first section of the cruise of the "Challenger," that from Teneriffe to Sombrero, the existence was established of deep-sea muds, perfectly free from carbonate of lime, consisting mainly of silicates mixed with ochreous material, principally hydrated oxides of iron and manganese, and of local concentrations of these materials in the form of nodules and of coatings or incrustations on dead calcareous matter. The qualitative composition of these concentrations was carefully determined, and it was particularly noted that whether in the form of nodules or of incrustations they were aggregations of the general materials of the bottom, and not concretions or coatings of pure hydrous oxides.

On the section between Bermuda and the Azores some very suggestive specimens were got from the bottom on 27th June 1873, when the ship dredged in 1675 fathoms in lat. 38° 18′ N., long. 34° 48′ W. A number of light-coloured concretions were brought up which were much perforated by worm-holes, the walls of which were all stained blackish brown. The substance of the concretions consisted of carbonate of lime and silicates, and the black lining of the holes was peroxide of manganese. The various specimens obtained on this occasion showed the deposition of oxide of manganese in various stages, from those which showed only specks or stains to those containing a considerable percentage[1]. The

[1] They are described in my report, *Proc. Roy. Soc.*, 1876, vol. XXIV. p. 606. It is reprinted as No. 5 of my *Collected Scientific Papers*, published by the Cambridge University Press, 1913.

most remarkable fact, however, was the close association of the oxide of manganese with the work of annelids. It was noticed here for the first time, and produced a strong conviction that the occurrence of peroxide of manganese at the bottom of the sea depended in some way or other on the organic life existing there.

After this comparatively little manganese was met with, until, on approaching the south coast of Australia, a large haul was obtained from a depth of 2600 fathoms in lat. 42° 42′ S., long. 134° 10′ E. and throughout the whole of the Pacific, when the trawl was put over in water sufficiently deep and sufficiently far from land, it rarely failed to collect abundance of manganese nodules, of all shapes and sizes, and surrounding all kinds of nuclei. Concretions also were obtained from time to time, recalling those of the North Atlantic above referred to. Thus, on the plateau of the Kermadec Islands, large lumps of a tufaceous sandstone were brought up, which were much perforated by serpular borings, and these were lined with peroxide of manganese. At the first station after leaving Japan, and on the landward side of the deep gully which runs parallel with the islands, a large haul was obtained, chiefly of pumice, tuff, and volcanic mud concretions. These were much perforated by worms, and the holes were lined with black oxide of manganese. One concretion was broken open in the plane of one of the worm-holes, and the worm was found dead in it[1]. On another portion a dead worm was found adhering, and on removing it a black stain was found below it consisting of peroxide of manganese. The connection of the peroxide of manganese with the life of these animals was very marked in this case, and continued to occupy my attention from time to time, though without arriving at any satisfactory solution, during the cruise. It must not be forgotten that an invariable feature of the nodules was that they gave off abundance of alkaline and empyreumatic-smelling water on being heated, which served further to connect them with the organic world.

After the return of the "Challenger" I did a good deal of

[1] The body of this worm was tested and found free from manganese.

dredging in the summers of several years (1877-1882) in the seas on the west coast of Scotland, and on the 21st September 1878 I brought up from the deepest parts of Loch Fyne (104 fathoms) a quantity of sandy mud, with large quantities of dead pecten shells, and along with them true manganese nodules, with all the outward characteristics of those from the greatest depths of the open ocean; and this similarity was maintained on chemical examination. The dredging anchor must on this occasion have been dropped in the very richest part of the deposit; for the mud, which had undergone no concentrating process, was found, on being submitted to mechanical analysis, to consist of rather over 30 per cent. of nodules[1]. This was a very remarkable discovery; for although peroxide of manganese was not wanting in the shallower dredgings of the "Challenger," it existed only as coatings and similar deposits and not as nodules, which were believed to be dependent for their formation on the conditions obtaining in very deep water. After this, particular attention was paid to the occurrence of manganese in all dredgings, and it was found to be abundant all round our coasts as a coating on shells, and more especially as the binding and colouring matter of worm tubes; but no nodules were anywhere found except in the deep part of Loch Fyne. Some years afterwards Mr Murray found them in great abundance on the Skelmorlie Bank in the Firth of Clyde in 10 fathoms.

In the same summer (1878) I made a number of observations in the channel off the north-east part of the Island of Arran, where the water reaches a depth of 90 fathoms. A galvanised iron bucket was used as dredge, with a weight attached behind, and one before it; so that its action was rather to skim the surface than to dig into the lower layers of the bottom. It brought up a quantity of a very fine red mud, in which manganese grains could be detected, not apparently differing from those found in oceanic red clays. In the process of levigation, when the mud was stirred up with water and the light flocculent portion poured off, the heavier portion which had settled to the bottom of the vessel had the appearance

[1] *Nature*, 1878, vol. XVIII. p. 628.

of having been cast into elongated pellets. When these were stirred up again with water they were partially broken up into flocculent matter, which was poured off, leaving again pellets as before; and this could be continued until the whole of the mud had been washed away as flocculi, produced by the breaking up of these pellets. In the case of the particular mud under description, hardly anything in the shape of sand or coarser material remained behind. The ground-fauna, chiefly ophiurids, seemed to be abundant; and the pellets above described were the casts excreted by these creatures, which subsist on what nutriment they can pick up by triturating and passing the sand or mud through their bodies. In some of these animals the triturating apparatus takes formidable proportions, as in sea-urchins; and it is probable that the sand found at low water owes its state of comminution largely to these animals and to worms, such as the ordinary lob-worm used for bait. When examining deep-sea clays in the "Challenger" I had observed the pellet formation, without, however, being able to refer it to any probable cause. Now, however, it became probable that the same causes are at work in deep as in shallow seas, and that the matter forming the bottom of the sea is being continually passed and re-passed through the bodies of the numerous tribes of animals which demonstrably subsist on the mud and its contents.

In the following season, 1879, I made an extended cruise through the greater part of the waters of the west coast of Scotland, visiting most of the deeper spots, and paying particular attention to the occurrence of coprolitic mouldings of the mud. Thus, on 16th June 1879, dredging in the deep part of the Sound of Raasay in 155 fathoms[1], "a little mud came up. It was a fine grey clay, which effervesced with acids and smelled of H_2S. On washing a quantity of it there remained the coprolitic masses and very little fine sand. There appeared to be a good deal of carbonate in a very fine state of division. There were very few shell particles visible, and the effervescence of what looked like flocculent clay was not inconsiderable." At the time I explained this flocculent carbonate as having

[1] From deck-book of steam yacht "Mallard," 1879.

been produced out of the silicates of the mud by the ground animals forming sulphide of calcium, which was transformed into carbonate by the carbonic acid of the water. On the following day another haul was got in the same locality and with similar results; it is noted that—"Sticking to the outside of the bag were many legs of ophiurids, which will account for the coprolites." When attention had once been paid to it, the coprolitic moulding of the mud, when of a suitable consistency, was found to be practically universal round our shores[1].

Shore muds, that is, the terrigenous deposits which are found all along the shores of continents, and even at great depths—generally present the characteristic appearance of a reddish surface layer, overlying a bluish substratum. This characteristic is observed in deposits even far out at sea, and, where it is not masked by large amounts of calcareous matter, is evidently due to the oxidation of the bluish ferrous salts, on their coming in contact with the sea-water, which always contains dissolved oxygen.

A very remarkable example of a blue clay—for it was too tenacious to be called a mud—was obtained in the Sound of Jura, and it was particularly noticeable for the amount of sulphides which it contained, and instructive by their

[1] Later, in the year 1886, when accompanying the expedition to survey the Gulf of Guinea in the steamship "Buccaneer," I found the same thing practically universal all along the African coast, and developed in a most remarkable manner on the coast flat within a considerable radius of the mouth of the river Congo. Here it was necessary to introduce a new designation for muds, and in this district the most frequent entries in the deck-book as to the nature of the bottom are "cop. m.," meaning coprolitic mud. These so-called coprolites were almost jet black and of the size of mice droppings, and they were covered with the same substance in flocculent form, or were free from it, according to the scour of the tide in the locality. It was best developed in comparatively shallow water, and more especially in a depth of 50 fathoms, when the large ash bucket, to the use of which as a dredge I found it convenient to revert, came up full of these coprolites, without any flocculent matter whatever. All along the coast the mud of the locality was moulded in a similar way, though it was not so striking. When the course of the cruise took us across the open ocean to Ascension, and thence northwards, we were able to trace the transition of the more earthy shore coprolites into the more mineralised and glauconitic pelagic ones.

complete disappearance on drying. It is worthy of more particular mention.

On 6th July 1879 the anchor dredge was put over in the Sound of Jura, where a depth of 120 fathoms was marked on the chart. It did not hold, and the yacht drifted, dragging it over the ground in a northerly direction before the wind and tide. Suddenly it hooked the ground, and brought the vessel up with a great strain on the cable. In heaving up it was with difficulty that the anchor was broken out of the ground; and when it was brought to the surface the bag was full of a fine, unctuous, very tenacious blue clay, with some of the reddish-brown surface mud covering it. There were a few pieces of broken shell and rock, also smooth and rounded pebbles, which seemed to occur principally in the part separating the surface mud from the blue clay, but there was very little of this kind of matter. The whole bagful, weighing more than 1 cwt., consisted almost entirely of homogeneous blue clay of a tenacity similar to the clay dug for brickmaking, and quite different from ordinary "blue muds." The clay was rather foul-smelling, and gave off abundance of sulphuretted hydrogen when treated with hydrochloric acid. It was so tenacious that it was impossible to break it up in water for the purpose of levigation, which is always very easily accomplished with ordinary muds. A considerable portion of it was dried and taken for analysis. It was found that, as soon as dry, not a trace of sulphide was to be found; but the mass of the clay was permeated with fine particles of oxide of iron, each of which represented a previous particle of sulphide. The contrast between the fresh moist clay, which was thoroughly impregnated with sulphides, and the dried clay, without a trace of them, was very striking[1].

The fact then had been demonstrated that the mud is being continually passed and re-passed through the bodies of

[1] A condensed account of my views of the part played by the sulphates of the sea water in the production of the ochreous deposits on the bottom of the ocean, and of the carbonate of lime of the shells of the Mollusca, is published in the *Reports of the British Association* (York), 1881, p. 584.

animals inhabiting the bottom of the sea. In doing so the mineral matter of which it consists comes in contact with the organic secretions of the animals, mixed with sea-water, and is ground up along with them in the milling organs of the animals.

The Reducing Action of organic matter on sulphates has long been known, and its importance as an agent in geological metamorphosis was thoroughly recognised by Bischof[1].

The effect of Trituration in promoting the chemical decomposition of silicates by water was demonstrated by Daubrée[2], more particularly in the case of Felspar. I found the observations to hold good also for Augite. Clear crystals of this mineral from the Tristan da Cunha group, when pulverised with water in an agate mortar, rendered the water alkaline to turmeric paper.

It is evident therefore that at the bottom of the sea a number of conditions occur together, which are favourable to the production of chemical change. The ground animals, in the search of food, pass the mud through their bodies, grinding it up, and bringing it thoroughly into contact at the moment of comminution with the sea-water and the digestive secretions of the animal. The action of these secretions on the sulphates in the sea-water is to produce sulphides, and the action of the sulphides on the ochreous matter of the bottom is to produce sulphides of iron and manganese. Even if the bottom were covered with felspathic or augite sand, the sulphides, acting on these silicates in the moment of partial decomposition, would convert the ochreous oxides by degrees into sulphides. That the volcanic material, lava, dust, scoriae, pumice, which forms the bulk of the unaltered material of the bottom of the ocean, is so dealt with by the animals, is evident from the specimen from the Pacific on the table, which is not a singular specimen, but rather a typical one.

Having extracted what nutriment they can from the mud, the animals reject it, containing a certain proportion of sulphides of iron and manganese. These sulphides, it is well known,

[1] Bischof, *Lehrbuch der Chemischen und Physikalischen Geologie* (1863), I. 31, 358.

[2] Daubrée, *Geologie Experimentale*, I. 268.

are exceedingly unstable in presence of water and oxygen, and if they come to lie on the surface of the mud, where they are exposed to the action of the sea-water, which always contains dissolved oxygen, they must be quickly transformed into oxides. In the oxidation of ferrous sulphide by this process there is always separation of free sulphur, which, however, is to a great extent further oxidised; but it is probable that some would persist. If then the process just described represents at all what takes place in nature, we should expect to find in the ochreous deposits (the hydrous oxides of iron and manganese) some relics of their connection with the organic world. These are not wanting. All the deep-sea muds and manganese concretions, of every diversity of form, gave without exception, when freshly collected and heated in a tube over the lamp, a large quantity of ammoniacal water. It was important to see if sulphur could be detected. And here it is well to bear in mind that in the case of a "blue mud," which may contain unaltered sulphide, the sulphur found in the dried sample will come at least in part from that sulphide, and will be due to the oxidation by the atmosphere in the process of drying. In the case of an oceanic "red clay" or manganese nodule, where no blue matter is present, any sulphur which is found may be properly ascribed to oxidation on the bottom of the sea.

Acting on these considerations, in the winter of 1880–81, a number of muds and nodules were examined with a view to the detection, and if possible the estimation, of free sulphur.

Estimation of Sulphur in Muds.

A certain quantity of the clay, dried at about 80° C., was put into a bottle with a known weight of chloroform. The stopper was tied down, and it was then put into a water-bath for about an hour at about the temperature of boiling chloroform (61° C.). It was then allowed to cool, filtered into a weighed fractionating flask, and washed twice with a little more chloroform. The chloroform was then distilled off, and the residue heated slightly and weighed.

The residue was treated with hot nitric and hydrochloric

acids, diluted, and filtered if necessary, barium chloride solution added, allowed to stand, filtered, and the precipitate of barium sulphate weighed.

In the first few samples the barium sulphate was not weighed, but the quantity of sulphur judged by the amount of barium sulphate precipitated.

At first bisulphide of carbon was used, but it was departed from, because, although perfectly pure, and leaving no trace of sulphur on evaporation, it was thought that it would be well to use a solvent not containing any sulphur. A portion of the blue clay from the Sound of Jura, which when fresh contains much sulphide, was in the dried state tested with both solvents, with the following results:

Treatment with Bisulphide of Carbon. A quantity of the clay was pounded and dried at about 80°. 50·00 grammes were put into a bottle with 236·0 grammes of bisulphide of carbon, and allowed to stand all night. A weighed portion of the carbon bisulphide was then taken out and put into a weighed flask, and the carbon bisulphide distilled off, and the residue weighed. 0·28 per cent. of sulphur was found in this way.

The sulphur dissolved completely in a small quantity of bisulphide. Next day another portion of carbon bisulphide was taken out and put into a weighed flask, distilled, and the residue weighed. This gave 0·33 per cent. of sulphur.

The bisulphide was tested to see whether it contained any free sulphur; it turned out to be very nearly pure.

Treatment with Chloroform. Another 50·00 grammes of clay were treated with 183·6 grammes of chloroform. The mixture was heated for an hour on the water-bath at about the temperature of boiling chloroform (61° C.). A portion of the chloroform was then taken out, evaporated, and the residue weighed. This gave 0·39 per cent. of sulphur.

There was very little oil present in the residue, which was nearly pure sulphur.

In the first ten samples the $BaSO_4$ was not weighed, but the residue was always oxidised and the presence of sulphur proved by the formation of sulphuric acid. When sulphur was found constantly and in appreciable quantity, I then decided to weigh it, the operation being, from an analytical point of view, an advantageous one, as the sulphate of barium weighed weighs seven times more than the sulphur to be estimated. By far the largest amount of sulphur is contained in the clay from the Sound of Jura, which, in its fresh state, contained large quantities of sulphides, which were completely oxidised on drying. The 0·197 grammes of residue may be taken to be pure sulphur, which makes about 0·4 per cent. By far the greater part, if not the whole, of this sulphur was formed by oxidation during drying. Had it been possible to collect and examine separately the reddish-brown surface layer, we should, no doubt, have found very much less sulphur, but it would have been mainly due to oxidation by the oxygen of the bottom water.

The "oil," which is extracted from all the muds along with the sulphur, and which varies a good deal in quantity, is due to the animal *débris* intimately mixed with the mud and with the materials of the nodules, which are made up, for the most part, of the materials of the bottom[1].

Nos. 2 and 3. The manganese nodules of the 12th July 1875, from the North Pacific, in lat. 37° 52′ N., long. 160° 17′ W., came from a depth of 2740 fathoms, where they appear to have been exceptionally abundant. Those of the 16th September 1875 came from a locality where they were equally abundant. The water was a little shallower, being 2350 fathoms, in lat. 13° 28′ S., long. 149° 30′ W. In both the samples of these nodules examined. the weight of the residue is considerable, but as there was a little oil in both cases it is not possible to give the percentage of sulphur.

No. 4. The mud from the Sound of Raasay, off the west coast of Ross-shire, was dredged from 150 fathoms, and consisted of very fine soft grey mud, which on washing left a large residue of coprolitic pellets.

No. 5 is a similar mud from Loch Duich, also in Ross-shire; it harboured many annelids.

No. 6 is from the station in Loch Fyne, where, for the first time, manganese nodules were obtained in comparatively shallow water. It is a sandy clay with many dead shells.

[1] For table of analytical results, see pp. 144, 145.

No. 7 is the red clay from 90 fathoms in the Firth of Clyde, off the north-east part of the Island of Arran, which has already been referred to. It is a very fine red ochreous mud, much resembling the oceanic clays. On washing, it is found to be almost completely moulded into coprolitic pellets, and supports an abundant ground fauna. Like oceanic clays, on careful washing, grains of peroxide of manganese can be isolated, and it contains over 1 per cent. of phosphoric acid.

No. 8 is red clay from lat. 18° 56′ N., long. 59° 35′ W., depth 2975 fathoms, in the western basin of the North Atlantic.

No. 9 is a coating of peroxide of manganese from an oceanic concretion, but the locality has been omitted to be noted.

No. 10 is mud from 115 fathoms in the channel between the Island of Searba and the Garvelloch Islands, about twenty miles S.W. of Oban. Peroxide of manganese is very abundant here as a coating on dead shells.

No. 11 is from the upper basin of Loch Fyne, in 60 fathoms. The mud here contains a remarkably large amount of sulphur. The upper basin of a sea loch is, as regards many of its conditions, and notably as regards the nature of the mud at its bottom, in a state intermediate between that of the open sea and that of a fresh-water lake. The mineral constituents are usually in a lower state of oxidation than outside; and this is accompanied by, and partly due to, the relatively large amount of vegetable *débris* from the land. All these circumstances may retard the disappearance of the sulphur.

Nos. 12 and 13 are globigerina oozes from the Pacific and the Atlantic respectively, their particular locality not noted.

No. 14 is from the same locality as No. 10.

No. 15 is from a position north-east of the Island of Rum, in 147 fathoms, soft grey mud.

No. 16 is a blue mud, from 2050 fathoms in the Celebes Sea.

No. 17 is a glauconitic mud from the east coast of Australia, in 410 fathoms, lat. 34° 13′ S., long. 151° 38′ E.

No. 18 is a diatomaceous mud from the Antarctic Ocean, in 1950 fathoms, lat. 53° 55′ S., long. 148° 35′ E.

TABLE *giving the results of the treatment of various samples of sea-bottom with chloroform for the extraction of sulphur.*

No.	Description of sample	Weight of sample taken (grammes)	Weight of chloroform added (grammes)	Weight of residue (grammes)	Per cent. of residue	Weight of $BaSO_4$ (grammes)	Per cent. of sulphur	Remarks
		a	b	c	$d=100\frac{c}{a}$	e	$f=13{\cdot}73\frac{e}{a}$	
1	Clay (*C*), Sound of Jura	50·0	183·6	0·1970	0·39	—	—	Very little oil; sulphur looked very pure
2	Manganese nodule, 16th Sept. 1875	78·8	228·1	0·0232	0·029	—	—	There was a little oil
3	,, ,, 12th July 1875	77·8	232·3	0·0470	0·064	—	—	,, ,,
4	Sound of Raasay, 1880	65·7	111·9	0·0230	0·034	—	—	A good deal of oil
5	Loch Duich	81·2	120·5	0·0298	0·036	—	—	,, ,,
6	Loch Fyne, 104 fathoms, 1st Oct. 1878	117·2	138·5	0·0150	0·055	—	—	,, ,,
7	Clay (*A*), Glen Sannox, 3rd Oct. 1878	84·2	153·3	0·0318	0·038	—	—	,, ,,
8	Red clay mud, 2975 fathoms	78·0	156·0	0·0222	0·028	—	—	A little oil
9	Manganese coating concretion	97·5	178·2	0·0022	0·002	0·00135	0·0002	Very little oil
10	Off Garvelloch Islands, 23rd July 1881	79·6	119·0	0·0182	0·023	0·01345	0·0023	A little oil
11	Loch Fyne, Upper Basin	78·1	140·8	0·0550	0·074	0·03945	0·01	Large amount of oil. A crystal of sulphur separated out before chloroform had distilled off

12	Globigerina ooze (Pacific)	79·5	182·3	0·0072	0·009	0·00095	0·00016	A little oil
13	,, ,, (Atlantic)	78·5	165·5	0·0250	0·032	0·00195	0·00034	Very little if any oil. Did not require to be filtered after oxidation
14	Garvelloch Islands (siftings)	90·0	152·2	0·0094	0·010	0·00915	0·0014	Very little oil
15	Off Rum, 15th Aug. 1881	125·7	183·5	0·0154	0·012	0·01195	0·0013	,, ,,
16	Blue mud, 8th Feb. 1875	79·0	135·2	0·0086	0·011	0·01895	0·0033	Very little if any oil. Did not require to be filtered after oxidation
17	Bottom, 410 fathoms	91·1	114·9	0·0042	0·005	0·00195	0·00029	Very little oil
18	Diatomaceous mud	15·6	160·0	0·0212	0·136	0·00275	0·0024	Very little if any oil. Did not require to be filtered
19	13th March 1874, 2600 fathoms	44·5	128·3	0·0090	0·020	0·00295	0·00067	Very little oil
20	Rad. ooze, 25th Aug. 1875, 2900 fathoms	44·4	170·8	0·0270	0·061	0·0102	0·0031	A little oil
21	Globigerina ooze (Atlantic)	73·8	185·2	0·002	0·0026	0·0015	0·0002	Very little oil. Did not require filtering
22	,, ,, (Pacific)	70·5	185·0	0·0025	0·0035	0·0024	0·0004	Very little oil. Did not require filtering
23	Manganese nodule, 16th Sept. 1875	58·8	128·7	0·0012	0·0020	0·0001	0·000017	Nodule dissolved with HCl in presence of $FeCl_2$ and residue treated
24	Loch Fyne, Otter House	79·8	178·1	0·0142	0·017	0·0053	0·0009	A little oil
25	Loch Ness, off Urquhart Castle	75·0	150·8	0·2805	0·374	0·0292	0·00413	Much oil and solid fat
26	Isle Oronsay, 6 fathoms, 19th July 1879..	85·4	147·5	0·0038	0·004	0·0028	0·00044	Very little oil. Did not need filtering
27	Garroch Head, 87 fathoms, 13th June 1879	89·2	153·5	0·0036	0·0043	0·0020	0·00032	Very little oil. Did not need filtering

No. 19 is a red clay dredged on the 13th March 1874, in 2600 fathoms, lat. 42° 42′ S., long. 134° 10′ E. Along with the mud a large quantity of manganese nodules was brought up.

No. 20 is a radiolarian ooze from the North Pacific, lat. 12° 40′ N., long. 152° 1′ W., depth 2900 fathoms.

Nos. 21 and 22 are again samples of globigerina ooze from the Atlantic and the Pacific respectively. These samples differ from Nos. 12 and 13 inasmuch as the Pacific sample now contains more sulphur than the Atlantic one.

No. 23 is the insoluble residue left after treating a nodule from the same locality as No. 2 with hydrochloric acid and ferrous chloride. The difference is very remarkable. In No. 2 the sulphur was not determined—that is, the barium sulphate produced by its oxidation was not weighed; but it was one of these samples which showed that the amount present was so appreciable that it was worth while determining it as accurately as possible, so that it is certain that it must have contained at least an average amount. In the case of the natural nodule (No. 2) the weight of chloroform residue per 100 grammes substance was 29 milligrammes; in the case of the extracted nodule No. 23 it is 2 milligrammes, and the weight of sulphate of barium is put down as 1 decimilligramme. In fact, the sulphur in the nodule had disappeared under the treatment.

No. 24 is from Loch Fyne, in 87 fathoms, opposite Otter House, and a little further up the loch than the station No. 6, but both of them in the outer loch, as opposed to No. 11, which is in the upper and semi-enclosed basin. The contrast between No. 24 and No. 11 is remarkable. In the upper basin the amount of the chloroform residue per 100 grms. substance was 74 milligrms., and 10 milligrms. of it was sulphur. In the outer loch there were only 17 milligrms. of residue and 1 milligrm. sulphur.

No. 25 is a very remarkable white clay from the bottom of Loch Ness, and therefore a fresh-water formation. It occurs in a small area opposite Urquhart Castle, and in various depths, often covered by a thin layer of peaty substance; but in some places, in depths of about 30 fathoms, the sounding-tube brings up the white clay alone. It was observed also in Loch Oich. It is chemically quite distinct from the marine clays, being much more acid. The amount of matter extracted by chloroform is enormous, being 374 milligrms. per 100 grms., most of which is oil or wax, but containing 4 milligrms. of oxidisable sulphur. It is not impossible that in this case the sulphur may exist as an organic compound; and the amount of oily matter in the clay is interesting in the indication which it gives of the possible mode of formation of our oil bearing shales.

No. 26 is from the anchorage of Isle Oronsay in the Sound of Sleat.

No. 27 is from a depth of 87 fathoms off Garroch Head, in the Firth of Clyde. Both in this case and in that of No. 26 the amounts of residue and of sulphur are insignificant.

Sulphur was thus detected in all these samples and determined in the greater number of them. Putting aside shallow water coast muds, the largest amounts of sulphur are found in the Celebes Sea (No. 16), in the diatomaceous ooze of the Antarctic (No. 18), and in the radiolarian ooze of the Pacific (No. 20). So far, therefore, as it goes, we have the evidence of the sulphur in favour of former organic agency. It is worthy of remark that the property of giving off alkaline water on heating has in the course of years disappeared, and in its place the nodules on being heated give off acid vapours, which, it is true, contain some ammonia, but along with an excess of nitric acid, which is without doubt due to the gradual oxidation of the nitrogenous matter. It is possible that the finely divided sulphur may diminish and finally disappear in the same way. But in 1881, there was still enough to be easily determined. Let us consider the chemical reactions more closely.

When a mud containing ferrous sulphide is treated with dilute hydrochloric acid, the sulphide dissolves with evolution of sulphuretted hydrogen, so long as there is no substance present which has a decomposing action on the sulphuretted hydrogen. If there be ferric salt either mixed with the mud or in the solution, then it is reduced to ferrous salt, with the destruction of the equivalent amount of H_2S and separation of sulphur. If the ferric salt be in excess, no sulphuretted hydrogen makes its appearance at all. The reaction is very simple:

$$FeS + 2HCl = FeCl_2 + H_2S$$

and $$H_2S + Fe_2Cl_6 = 2FeCl_2 + 2HCl + S$$

∴ by addition $$FeS + Fe_2Cl_6 = 3FeCl_2 + S$$

because the 2HCl appears on both sides of the equation, and is in fact unnecessary. A *trace* of free acid is no doubt necessary, and it is turned over and over again in the reaction of indefinite quantities of FeS on Fe_2Cl_6, after the manner of a catalytic action.

The same reaction takes place if we use ferric sulphate in place of ferric chloride.

It is evident, therefore, that if we have a sample of mud containing sulphide, and we mix it thoroughly with a solution of Fe_2Cl_6 or $Fe_2(SO_4)_3$, we shall have in the ferric salt reduced a measure of the decomposable sulphide present. The ferrous salt can be readily determined by permanganate of potash or otherwise. It will be seen from the above equation that one molecule FeS decomposes one molecule Fe_2Cl_6 with the formation of three molecules $FeCl_2$, so that the FeS in the mud is one-third of the ferrous salt found.

In order to make some preliminary experiments, a mixture of 100 grms. alum and 30 grms. ferrous sulphate were dissolved in about ¾ litre of water and precipitated with ammonia and sulphide of ammonium. The precipitate was thoroughly washed by repeated decantations, the flask being always filled up to the neck, and corked and allowed to settle. When it was completely washed the surplus water was poured off, and the precipitate, suspended in about ½ litre of water, was preserved in a well-stoppered reagent bottle. The precipitate consists of alumina and sulphide of iron, and may therefore be taken as an imitation of a simple form of mud. I made some experiments to see with what amount of agreement in the results one could titrate a number of different samples of the same mud.

Three flasks were placed side by side, and into each 50 c.c. suspended FeS mud were measured. The mixture of $FeS + Al_2O_3$ was thoroughly shaken up, then run into a narrow graduated cylinder, holding 50 c.c., which was emptied into the flask and then washed once into it with distilled water. To each of the flasks was then added 10 c.c. of the reddish-brown but still acid, ferric sulphate solution, and the contents shaken. In a few seconds the black colour of the sediment had disappeared entirely, being replaced by a yellowish-red precipitate, which disappeared for the most part on the addition of dilute sulphuric acid. Water was then added to bring up the volume to 250 c.c., and the titration was effected with permanganate of potash solution $\left(1 \text{ litre containing } \frac{KMnO_4}{50} \text{ grms.}\right)$. The three portions of 50 c.c. required each 11·6, 11·6, and 11·7 c.c. permanganate respectively. We see then that a suspended precipitate can be measured off about as accurately as a dissolved salt.

It is evident, then, that if we have a mud containing FeS and other ferrous compounds decomposable by HCl, we can determine first the FeS by adding Fe_2Cl_6 and titrating a portion with permanganate; then the other ferrous compounds, by adding HCl and titrating another portion with permanganate, due account being kept of the weights and volumes used. In order to try the method in practice, three soundings were made;—on 30th September 1881 in the Sound of Raasay, off Croulin Island, 120 fathoms; and on the 1st October 1881 in Loch Duich, in 49 and 51 fathoms. The first of these represents more or less the conditions in the open sea of coast waters; the last two represent the conditions in a semi-enclosed loch basin. The Sound of Raasay mud was a light grey mud, with no offensive qualities. Both samples from Loch Duich were very foul smelling. All three samples were tightly stoppered up in their wet condition, and examined on 20th and 21st October 1881 in my laboratory in Edinburgh. I unfortunately had no suitable ferric solution afloat with me so as to treat them immediately. In the three weeks that both muds from Loch Duich were kept in bottles, the surface layer got completely oxidised, and on opening the bottles the smell was gone; but, on breaking through the surface layer, the unaltered black mud was exposed with all its original qualities, including its peculiar odour.

The following was the method used in the case of the Loch Duich mud from 49 fathoms. Two portions of the damp unaltered mud were weighed out; one portion, 6·724 grms., was dried at 100° C., and the other, 7·881 grms., was treated with deep red Fe_2Cl_6 in a 100 c.c. flask, which was then filled up to the mark with water. 50 c.c. of this solution, containing 3·94 grms. damp mud, were acidified with sulphuric acid and titrated with permanganate $\left(\frac{KMnO_4}{50} \text{ grms. per litre}\right)$, using 1·9 c.c. To the remaining 50 c.c. with sediment (the volume of which may here be neglected) were added 4 c.c. of strong hydrochloric acid (12·5 HCl grms. per litre), filled up to the mark, and allowed to settle. 50 c.c. of this solution, containing 1·97 grms. damp mud, were further acidified with sulphuric acid and titrated with the same permanganate, of which 1·7 c.c. were used. A litre of the above permanganate oxidises 5·6 grms. iron from the ferrous to the ferric state. In the first operation, 50 c.c. solution used 1·9 c.c. permanganate, therefore the whole amount of mud, 7·881 grms., when treated with ferric chloride, would require 3·8 c.c. = 0·0213 grm. iron.

After treatment with hydrochloric acid a quantity of solution equivalent to 1·97 grms. wet mud required 1·7 c.c. permanganate, so that 7·881 grms. mud would require 6·8 c.c. when treated with both HCl and Fe_2Cl_6, which represents 0·0381 grm. iron. Therefore, total iron found by

Permanganate in HCl + Fe_2Cl_6 solution	..	0·0381 grm.
Deduct Iron found in Fe_2Cl_6 solution	..	0·0213 ,,
Leaves Iron present as Ferrous Salt extracted by Hydrochloric Acid		0·0168 ,,

Of the 0·213 grm. iron found in the first solution we have seen that only one-third is to be reckoned as belonging to the mud, and to be taken as forming FeS, so that in 7·881 grms. wet mud we have 0·0071 grm. iron present as sulphide, equal to 0·0112 grm. FeS, and 0·0168 grm. iron present as ferrous oxide extracted by hydrochloric acid, equal to 0·0216 grm. FeO.

The 6·724 grms. wet mud weighed when dried at 100° C., during which it was oxidised as well as dried, 2·011 grms., equal to a loss of 70·1 per cent. Therefore the dry mud is 29·9 per cent. of the damp mud taken. The 7·881 grms. damp mud therefore represent 2·3564 grms. dry mud, and therefore we find that the mud taken as dry contains 0·47 per cent. FeS and 0·92 per cent. FeO in some other easily decomposable combination.

The other samples were treated in the same way, and in the Loch Duich mud, from 51 fathoms, 0·94 per cent. FeS + 0·65 per cent. FeO were found. It is remarkable that the amount of FeS should be so small in such offensive muds.

In the outside mud from 120 fathoms in the Sound of Raasay only 0·05 per cent. FeS and 0·1 per cent. FeO were found.

In connection with this mud, which contained some shell *débris*, the method was found to be less applicable than to muds free from calcareous matter. The reason is obvious; because, on adding a neutral ferric solution to a mud containing carbonate of lime, precipitation of the ferric oxide by the lime immediately commences. This would not really interfere with the reaction, because the FeS would reduce the precipitated Fe_2O_3 all the same, and the ferrous salt can still be determined by permanganate; but in truly calcareous bottoms this action is troublesome, and the method will require special study in this direction. In the semi-enclosed basins of the sea lochs, which, as has already been observed, form a transition between the open sea and fresh-water lakes, the bottom resembles more nearly that of the fresh-water lakes, in the absence of mollusca, and in the abundance of organic matter of vegetable origin, than that of the open sea with its abundant and varied ground fauna. It differs from those of fresh-water lakes in being bathed by sea-water largely impregnated with sulphates. Consequently it is in the inner basins of sea lochs that the conditions for a constant production of sulphides are present, while the same conditions are hostile to the presence of calcareous organisms. Hence it is in these basins that the greatest quantities of sulphides are found, and it is in their muds that the above method is most applicable.

The sulphuretted muds, however, are so alterable by atmospheric influences that it is essential that they should be treated immediately on collection. For this purpose weighed wide-mouthed bottles with good stoppers should be provided. When a specimen of mud is brought up from the bottom, a sample of it is immediately taken with a spatula and put into one of these bottles containing a known quantity of ferric chloride solution, at least sufficient to completely cover the sample of mud. Another sample, as nearly similar to the first as possible, is taken and stoppered in another bottle for drying. In this way a large amount of valuable information might be gained; but it will be evident from the nature of the case that the actual figures obtained in any one particular case are affected by a considerable possible error.

In the month of June 1881 I carried out a number of laboratory experiments bearing on this subject, using the sulphides of different metals of the iron group. These bodies were all prepared in the same way, namely, by precipitating the sulphates with sulphide of ammonium, and washing by decantation in stoppered bottles, always filled up quite full. A quantity of hydrated ferric oxide was also prepared by precipitating ferric chloride with ammonia and washing. All of these precipitates, when thoroughly washed, were preserved suspended in distilled water in well-stoppered reagent bottles.

Ferrous Sulphide and Ferric Oxide. When quite neutral these substances do not re-act on one another, at least at once. But if the water has the slightest acid reaction, reduction of the sesquioxide and production of sulphur take place rapidly. A mixture of Fe_2O_3 and FeS in water and quite neutral was corked up and allowed to stand for five days, when the sediment was found to be separated into two sharply-defined layers—the upper red, consisting of the oxide, and the lower black, of the sulphide. When brought together, therefore, in presence of nothing but distilled water, there is no appreciable resultant action.

Manganous Sulphide can be preserved perfectly under distilled water in well-stoppered bottles filled to the neck. A considerable quantity was prepared in the summer of 1881, and, when thoroughly washed, it was put away in three separate bottles. The contents of only one bottle were used for experimental purposes, and the upper part of it got coloured immediately black with oxide of manganese, from the oxidation of the flakes of sulphide which adhered to the surface of the upper part of the bottle, left dry when some of the water and precipitate had been poured out. This took place at the time, and was to be expected. The two other bottles, which were filled up with the manganous sulphide at the time of preparation, have never been opened since, though they have all the time been exposed to the light, and are exactly in the condition in which they were when bottled nine and a half years ago. There is no trace of oxidation.

Manganous Sulphide and Hydrous Ferric Oxide. Both substances are used, suspended in distilled water. If the ferric oxide be cautiously added to the sulphide of manganese, and both suspended in water, the red patches are seen to disappear, and the general colour of the suspended matter becomes rather lighter in colour than the MnS, and there is no formation of FeS. If further additions of Fe_2O_3 be made, red flakes deposit themselves. They do not appear to be unaltered Fe_2O_3, but are exactly like the "red cherty particles" of manganese bottoms. On still further additions of Fe_2O_3, the colour changes quickly, though not instantaneously, to black, with, however, a large admixture of white particles, the two being easily seen to be perfectly distinct. There is also a quantity of precipitated sulphur which remains floating in the liquid long after the heavy matter has subsided.

Prosecuting this line of experiment, I made three mixtures in suitable flasks.

No. 1 contained MnS and Fe_2O_3, the MnS being in excess. There was formation of red cherty particles, but nothing black.

No. 2. The same substances, but containing the Fe_2O_3 in excess; the mixture quickly turned black.

No. 3. The same as No. 2, only it was made up with warm water, and it turned black almost at once.

These experiments were repeated and with the same results. The above flasks, Nos. 1, 2, and 3, were corked up and allowed to stand over night. No. 1 contained numerous black particles, as well as red cherty ones, and an excess of MnS as well as sulphur. Nos. 2 and 3 were much as they had been the night before, except that the white particles had almost entirely disappeared, as also all red particles. The reactions are considerably accelerated by heat.

On examining the contents of each of these flasks, no peroxide of manganese was found, but large quantities of sulphide of iron. The likeness in the red flakes to the cherty particles of the bottom muds in the manganese districts of the South Pacific, and of the kernels of some manganese nodules, was very striking. It is not improbable that the first action of the MnS on the Fe_2O_3 may be accompanied by the formation of mixed oxides of iron and manganese; but there is much to be done in this direction in the strictly quantitative investigation of the interaction of the insoluble, but not inert, compounds of this as well as of other groups of metals.

Ferric Sulphate and Manganous Sulphide. Experiments were now made, using the iron as a ferric salt in solution, and for this purpose ferric sulphate was used. It was made as nearly neutral as possible by addition of ammonia. The MnS was, as before, suspended in distilled water.

On adding ferric sulphate to excess of MnS, the formation of FeS is immediate.

On adding a large excess of $Fe_2(SO_4)_3$ the FeS is decomposed, there is formation of basic salt, and on dissolving it with H_2SO_4 the solution contains large quantities of ferrous sulphate.

On experimenting with solution of ferrous sulphate it was found that excess of MnS precipitates the iron completely as FeS, acting exactly like an alkaline sulphide.

The rationale, therefore, of the above reaction is very simple. Thus

$$Fe_2(SO_4)_3 + MnS = 2FeSO_4 + MnSO_4 + S \ldots (1)$$

and

$$2FeSO_4 + 2MnS = 2FeS + 2MnSO_4 \ldots\ldots (2).$$

Therefore adding (1) and (2) we have

$$Fe_2(SO_4)_3 + 3MnS = 2FeS + 3MnSO_4 + S \ldots (3)$$

and

$$2Fe_2(SO_4)_3 + 2FeS = 6FeSO_4 + 2S \ldots\ldots\ldots (4),$$

and

$$(3) + (4) = 3Fe_2(SO_4)_3 + 3MnS = 6FeSO_4 + 3MnSO_4 + 3S \ldots (5)$$

or

$$\frac{(5)}{(3)} = Fe_2(SO_4)_3 + MnS = 2FeSO_4 + MnSO_4 + S \ldots (6),$$

which is identical with (1) and by adding more MnS we get the conditions of equation (2), and so on, repeating the cycle.

Hence, if we add MnS to excess of $Fe_2(SO_4)_3$, we should

get reduction of the ferric salt without formation of FeS. On adding excess of MnS, we get formation of FeS, and then on adding excess of $Fe_2(SO)_4$ we get back to the same state of things as at first.

The reaction of equation (1) can be obtained by very cautiously adding small quantities of suspended MnS to a very large excess of $Fe_2(SO_4)_3$. Still there is always local formation of FeS which disappears on mixing, so that the reaction is really that of the whole cycle. The action, therefore, of MnS on soluble iron salts is in the first instance to reduce whatever is in the ferric state to the ferrous, and then at once to precipitate the ferrous salt as sulphide, a manganous salt taking the place of the ferrous salt in the solution.

When added in great excess to solutions of nickel sulphate, manganous sulphide precipitates it as NiS. When added to solution of sulphate of zinc, it either does not precipitate it at all or only very slightly at ordinary temperatures. Sulphide of zinc was not found to precipitate manganese sulphate solution.

As the result, then, of the observations and experiments which have been recited I was led to believe that the principal agent in the comminution of the mineral matter found at the bottom of both deep and shallow seas and oceans is the ground fauna of the sea, which depends for its subsistence on the organic matter which it can extract from the mud.

In order to fit them for collecting their nutriment in this way the animals have been fitted with different forms of masticating or milling apparatus, so as to thoroughly deal with the matter which they pass through their bodies. It has been shown that most silicates are decomposed to a certain extent when ground or pulverised under water; so that the mere mastication of the sand or mud in presence of pure water would have a decomposing action on the silicates which it contains. This action is much assisted, in the case of marine animals, by the fact that the water which they pass through their bodies along with the sand is charged with sulphates. These are easily reduced to sulphides by the action of the organic matter of the secretions of the animals. The resulting sulphide at once suffers double decomposition with any oxide of iron or manganese which is present as such in the mud, or may be being set at liberty from silicates under the decomposing influence of trituration under water. The sulphides of manganese and iron so formed are in course of nature extruded by the animals, and if exposed to the sea water on the surface of the mud are quickly oxidised, the manganese taking priority. The mud below the surface layer, in localities where ground life is abundant, remains blue, being protected by the oxidation of what is above it.

At the bottom of the ocean the mineral matter is thus exposed to a reducing process due to the life of the animals which inhabit it, and to an oxidising process due to the oxygen dissolved in the water. Other things being equal, the redness or blueness of a mud or clay depends on the relative activity of these processes. They also exercise a controlling or modifying influence on one another. For, although marine animals are much less sensitive to variation in the amount of oxygen in their atmosphere than terrestrial animals, it is certain that there must be a limit to the deficiency of oxygen which each animal can support; and when this limit is approached, its reducing activity is diminished, or, it may be, extinguished. The water in the course of circulation is being continually renewed, and, meeting with a diminished amount of freshly reduced matter, it is able to push the oxidation of the mud to a greater depth. It is easily conceivable that in many of the deep parts of the ocean the amount of ground life may be so limited that the water has no difficulty in oxidising at once its ejecta; and these conditions would be favourable to the formation of a red clay or chocolate mud according to the preponderance of iron or manganese.

While dealing with this subject it is proper to refer to Darwin's book on *Vegetable Mould and Earthworms*, which was published in 1881. His masterly investigations, in the kindred department of the part played by earthworms in the formation of the terrestrial soil, strengthened me much in my belief in the soundness of the views above developed as to the formation of marine muds. Indeed, to a certain extent he extends his views himself to the case of marine muds. At page 256, after noticing that it is due to the milling action of the gizzards of worms that the supply of exceedingly finely divided mineral matter, which is removed from the surface of every field by every shower of rain, is constantly renewed, he adds in a note: "This conclusion reminds me of the vast amount of extremely fine chalky mud which is formed within the lagoons of many atolls, where the sea is tranquil and waves cannot triturate the blocks of coral. The mud must, so I believe, be attributed to the innumerable annelids and other animals which burrow into the dead coral, and to the fishes, Holothurians, etc., which browse on the living corals." Darwin further gives an approximate numerical result or estimate of the work of earthworms which is interesting. At page 258 he says: "Nor should we forget, in considering the power which worms exert in triturating particles of rock, that there is good evidence that on each acre of land which is sufficiently damp and not too sandy, gravelly, or rocky for worms to inhabit, a weight of more than 10 tons of earth annually passes through their bodies and is brought to the surface."

But this does not exhaust the mysterious efficacy of mechanical comminution when pushed to a high power.

The activity of the gizzard in the ground-fauna of both sea and land in supporting life by promoting chemical action between the materials passed through it, which otherwise are inert towards each other, is an example in Nature which was unconsciously imitated in Art by one of the greatest discoveries in the history of Medicine, namely that of the Dynamization of drugs by trituration, due to Samuel Hahnemann.

Taking the ***Aurum Solubile*** of the alchemists as his finger-post, he showed that metals and other insoluble substances, by trituration to a certain degree of fineness, become soluble in water, alcohol and other neutral liquids.

The practical result of this discovery is that alcoholic solutions of most of the metals in the one-millionth and higher dilutions have been officinal in homœopathic pharmacies for the last hundred years.

No. 7. [*From Proc. Roy. Soc. Edin.*, 1877, *Vol.* IX. *pp.* 287–289.]

NOTE ON THE MANGANESE NODULES FOUND ON THE BED OF THE OCEAN

THE manganese nodules occur in greater or less quantity all over the ocean-bed, and most abundantly in the Pacific. They occur in all sizes, from minute grains to masses of a pound weight, and even greater, and form nodular concretions of concentric shells, round a nucleus, which is very frequently a piece of pumice or a shark's tooth. Their outside has a peculiar and very characteristic mammillated surface, which enables them to be identified at a glance. When freshly brought up they are very soft, being easily scraped to powder with a knife. They gradually get harder on exposure to the air.

The powder, heated in a closed tube, gives out water which re-acts alkaline, and has an empyreumatic odour. Heated with strong hydrochloric acid, it liberates abundance of chlorine, and the residue which remains is white, consisting of silica, clay, and sand, the sand being the same as is found in the bottom mud from the same locality. Their composition varies greatly, different nodules containing different quantities of mechanically admixed mud, and the number of different elements found in them is very large. Copper, iron, cobalt, nickel, manganese, alumina, lime, magnesia, silica, and phosphoric acid have been detected in a large number; but I have not as yet been able to make a complete analysis of any of them. I have, however, made a few determinations of the most important component substances. For this purpose the outside and densest layers of the nodules were selected, and portions of them were pulverised and dried for ten or twelve hours at 140° C. The amount of chlorine liberated on treatment with hydrochloric acid was determined by Bunsen's method, and the iron was determined by titration with stannous chloride. The samples analysed were from four different localities.

Nos. 2, 4, and 5 were from the same place, No. 2 being the matter collected round a shark's tooth as nucleus; Nos. 4 and 5 being the outside rinds of ordinary nodules.

The results are given in the following table, the numbers being in many cases the means of several observations:

Locality		No.	*A*	*B*	*C*	*D*	*E*	*F*	*G*	
Lat.	Long.		Insoluble Residue	O	MnO_2	MnO	Fe_2O_3	Al_2O_3	H_2O	Na_2O
13° 52′ S.	149° 17′ W.	2	17·55	6·13	33·30	27·18	—	—	—	—
"	"	4	15·30	5·92	32·23	—	23·86	—	—	—
"	"	5	15·30	6·49	35·28	—	24·85	—	10·2	—
37° 52′ N.	160° 17′ W.	6	36·24	6·49	24·41	—	20·16	3·83	7·70	5·98
42° 42′ S.	134° 10′ E.	7	17·98	7·54	41·11	33·53	18·04	2·55	7·31	—
22° 21′ S.	150° 17′ W.	8	21·74	5·19	28·20	—	24·52	7·67	8·54	8·5

A is the residue which remains undissolved after treating the mineral with strong hydrochloric acid, evaporating to dryness and redissolving. In No. 5 it contains 85·16 per cent. silica, and in No. 6, 82·27 per cent.

B is the "available oxygen" determined by Bunsen's method.

C is the MnO_2 equivalent to the available oxygen.

D is the MnO found by weighing as Mn_3O_4.

E is the Fe_2O_3 found by titration with $SnCl_2$.

F is the alumina found by subtracting the Fe_2O_3 found in *E* from the weight of the precipitate with acetate of soda.

G is the water expelled on ignition; it is obtained by deducting two-thirds of the oxygen found in *B* from the loss of weight by ignition.

It will be seen from the results given in the above table that the nodules from different localities vary greatly in composition, though in the same locality they have similar composition, irrespectively of the nature of the nodules. The insoluble residue contains, besides silica and clay, sand of the same mineral nature as is found in the bottom at the same locality. The manganese is present wholly as MnO_2, and the iron as Fe_2O_3. In No. 6 there is 0·3 per cent. of cobalt; this metal, along with copper and a little nickel, is present in all of them. Zinc was not found in any of the above specimens.

No. 8. [*From Nature, Oct.* 10, 1878, *Vol.* XVIII. *p.* 628.]

MANGANESE NODULES IN LOCH FYNE[1]

On September 21, 1878, I anchored the steam yacht "Mallard" near the mouth of Loch Fyne, in 104 fathoms, for the purpose of making physical and chemical observations on the water of this, the deepest part of the Firth of Clyde. When the anchor was got up a large mass of clay and shells was found sticking to one of the flukes. It was gently dried, and on examining it I observed a number of nodular concretions, which, on being freed from the surrounding clay, presented a finely mammillated black surface, were easily cut with a knife, giving a brownish-black powder, which liberated chlorine from strong hydrochloric acid, and possessed all the properties of peroxide of manganese; in short, they were identical with the manganese nodules which we found in the "Challenger" to form so important a constituent of the sea-bottom in the greatest depths.

One half of the dried mud was carefully broken up and searched through, the nodules being collected by themselves and also the shells. It was thus separated into three portions, which were weighed, with the following results:

Manganese nodules	142·7 grammes	30	per cent.
Shells	35·0	7·5	,,
Sandy clay	289·0	62·5	,,
Total ..	466·7 ,,	100·0	,,

The manganese nodules, therefore, made up thirty per cent. of the weight of the mud. Compared with those frequently met with on board the "Challenger," the nodules were small. In the sample examined there were eighty-three nodules weighing 142·7 grammes, hence the average weight was 1·7

[1] See Contents, p. xxviii.

grammes. Their volume was found to be 58 c.c., so that the average volume was 0·7 c.c., and the specific gravity 2·46. Their form was roughly spherical, the largest, which was somewhat elongated, measured 13 × 9 × 6 millimetres, the average diameter of them all being 11·4 millimetres.

Of the eighty-three nodules so obtained I have split twenty-two. When subjected to this treatment they are found to differ in constitution from the majority of those obtained on board the "Challenger." Although they had not been exposed to any heat they were hard and sandy to the knife, and when treated with strong hydrochloric acid, they left a large amount of mineral (chiefly quartz) sand. This difference, however, is explained by the different kind of bottom from which they were obtained. In dissolving up nodules which had come from "red clay" in 2500 or 3000 fathoms, I always found the same mineral sand left as on treating the clay in the same way. But the amount of sand was always quite insignificant as compared with the clay; hence the nodules were easily cut with the knife. They, however, got harder on keeping. In Loch Fyne the bulk of the mud consists of quartz sand, giving the nodules the appearance of sandstone, whose binding material is made up to a great extent of peroxide of manganese, and hence the gritty feeling to the edge of the knife.

Where a hard nucleus has been found it has always been a piece of rock from the neighbouring shore, but in most instances (in sixteen out of twenty-two examined) the ordinary arrangement has been reversed, the nodule consisting of a soft rich nucleus of peroxide of manganese, surrounded by a black sandy rind, the whole enveloped in the characteristically mammillated black skin.

I hope very shortly to be able to report more fully on them; in the meantime, I have only been able to verify their nature by finding abundance of a higher oxide of manganese, easily recognisable quantities of cobalt, and the presence of water, which, on being expelled by heat, has an alkaline reaction and an empyreumatic odour, properties in which they agree with those which I had occasion to test on board the "Challenger."

Their position in the mud, with dead shells above, below, and on all sides of them, will, when carefully studied, no doubt throw much light on their age and method of formation. I have observed two nodules firmly attached to the interior of shells, one having evidently been directed in its growth by the shape of the shell.

In endeavouring to procure a further supply I dropped anchor in about the same depth, but about a hundred yards further down the loch, and I obtained about the same amount of mud, but it contained very much more shell and no nodules. Also in Kilbrennan Sound, between Arran and Cantyre, in a depth of eighty-five fathoms, there was much shell and pebble, but no nodules. So far, therefore, this occurrence appears to be very local.

No. 9. [*From Trans. Roy. Soc. Edin., Vol.* XXXVI. *Part* 2, 1891, *p.* 459.]

ON THE COMPOSITION OF OCEANIC AND LITTORAL MANGANESE NODULES

THE following analyses were made some years ago, principally with the object of ascertaining the state of oxidation of the manganese in the nodules. The nodules examined came from three different localities, two of them oceanic and the third littoral. Samples marked I, II, and III are from nodules brought up in the trawl on board the "Challenger," on 13th March 1874, in lat. 42° 42′ S., long. 134° 10′ E. The depth of the water was 2600 fathoms, and the temperature of the bottom water 0·2° C. The density of the bottom water was 1·02570 at 15·56° C. Being from a high southern latitude, and therefore near the source of surface aeration, the water is highly charged with atmospheric gases, especially oxygen. It contained, per litre, 18·4 c.c. of mixed nitrogen and oxygen, of which 31·81 per cent. was oxygen, and 27·33 c.c., or 53·7 milligrammes, loosely-bound carbonic acid. The position of the station is about 400 miles south-west of the nearest part of the Australian coast, and about 500 miles west of Tasmania. It was the deepest water observed in the Antarctic voyage between the Cape of Good Hope and Melbourne. The haul was a very abundant one, and a few notes which I made at the time may be interesting: "The water was found unexpectedly deep, the bottom being red clay, with some Foraminifera. The bag of the trawl came up quite full of this mud, with many animals and a large number of manganese nodules. These were of all shapes, and with the characteristic mammillated surface, which in some was accentuated to such a degree as to give them a botryoidal appearance, like specimens of Psilomelane. Many of them were perfectly spherical, others formed groups of spheres. One of these spherical nodules was found, on being broken, to contain a hard kernel of a mineral, giving a powder of the colour of bichromate of potash, with a conchoidal fracture and resinous lustre. Round this the spherical shells of manganese were gathered, and could be easily broken off with the fingers. Another nodule was noticed with the same yellow resinous-looking substance in the centre, but it mixed with the manganese forming part of the substance, and could not be detached from the surrounding shells. It has a light wine-yellow colour by transmitted light, and polarises light. Many flat pieces were observed, with horizontal stratification and botryoidal surface. Whether flat or spherical, the manganese was put on in layers, separated by very fine sheets of the mud of the locality. There was one nodule which had formed round the fragment of another, and therefore older nodule, the distinction between the two being well marked by the inclination of the mud sheets of the kernel to those of the shell. There were many where the clay or mud formed a large percentage of the mass, either as interbedded layers or as pockets, and in some of these pockets Foraminifera were to be seen. Amongst the collection were two ear-bones of whales and a very fresh shark's tooth, covered with the incrustation. The occasional occurrence of icebergs at the surface was made probable by the presence of two pieces of granite, the one with a very thin covering, and the other with over one-eighth of an inch thickness of manganese. On the fracture of one of the pieces, it was evident that the manganese had filtered into the interior of the stone, colouring the quartz a beautiful amethyst purple."

This haul was remarkable in many ways; and not the least in being the only important haul which we got in the vicinity of continental land, and with no volcanic islands near. The samples taken for analysis were—

I. The outer rind or shell of a spherical nodule, which was detached without any trace of the kernel. It was 10 millimetres thick.

II. A similar piece, detached from another spherical nodule, but with traces of the kernel attached. It was also 10 millimetres thick.

III. A piece of a horizontally stratified concretion, with botryoidal upper surface. It was 15 millimetres thick.

Samples marked IV and V are from one nodule. IV is part of the outer rind, 13 millimetres thick, consisting of concentric layers, most of them mottled with reddish yellow spots, and separated by fine seams of purple-brown oxide of manganese, without yellow specks. V is the kernel of the same nodule. It is harder than the rind, from which it easily splits away. It is nearly spherical, with a radius of 16 millimetres. At the centre is a small colourless piece of mineral matter.

This nodule was one of an enormous haul made by the "Challenger" on the 12th July 1875 in the North Pacific, in lat. 37° 52′ N., long. 160° 17′ W. The depth of the water was 2740 fathoms. The temperature of the bottom water was 1·0° C., and its density at 15·56° C. was 1·02573. The gaseous contents of the water were—17·7 c.c. mixed nitrogen and oxygen, of which 16·95 per cent. was oxygen, and 21·48 c.c., or 42 milligrammes, carbonic acid per litre. It will be seen that the water contains very much less oxygen than was contained in the bottom water off the Australian coast. In fact, in this water the dissolved oxygen has been reduced to almost exactly half the amount which it contained when it left the surface. All over this district, where manganese greatly abounds, the dissolved oxygen has been reduced from 34 per cent. of the mixed gases, as at the surface, to from 16 to 22 per cent. at the bottom.

The position of this station lies midway between the Aleutian Islands and the Sandwich Islands, being 1000 miles distant from the nearest island of either group; it is 1600 miles distant from the nearest point of the North American continent in the same latitude. Hence, although there were many stations further from land than this, it can claim to be quite beyond the reach of any continental influence. The station from which samples I, II, and III come, although comparatively close to the Australian coast, may be said also to be practically beyond the sphere of its influence, as, owing to its climate, it is almost destitute of drainage.

The *Littoral Nodules*, samples *M*, *N*, *P*, *Q*, *K*, *R*, etc., are

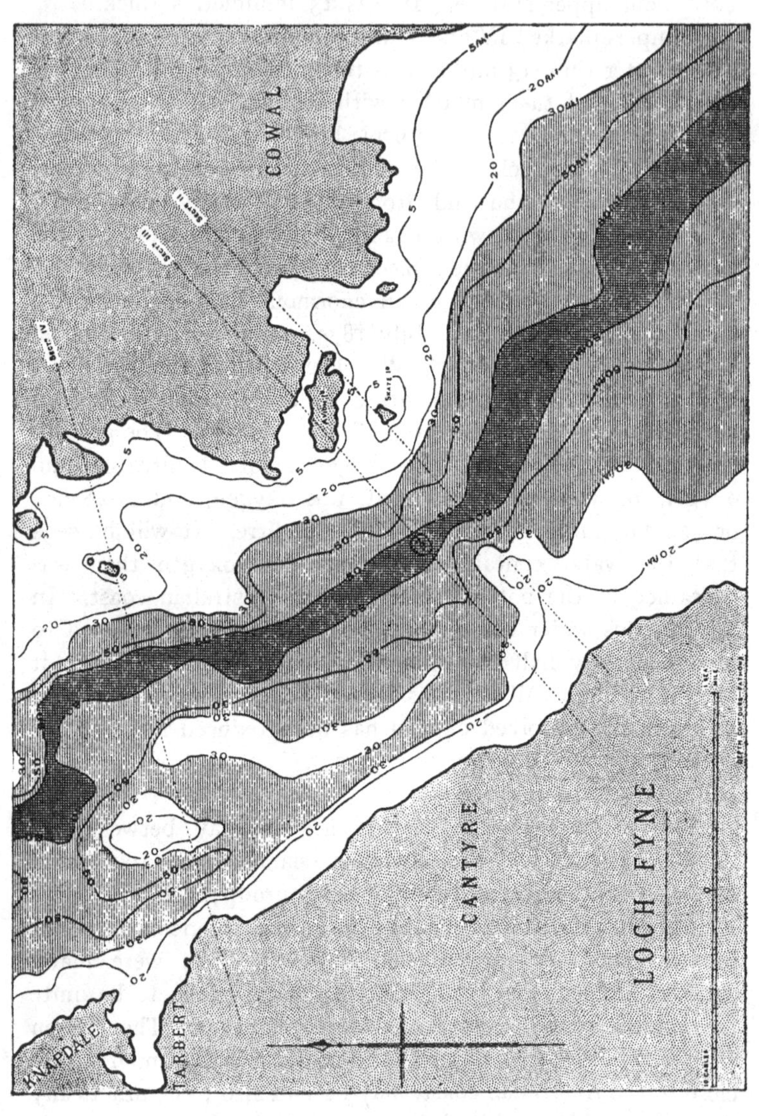
COWAL
CANTYRE
LOCH FYNE
KNAPDALE
TARBERT

from Loch Fyne, one of the most important arms of the Firth of Clyde. They are from a lump of mud which came up on the fluke of the deep-sea kedge-anchor of the steam yacht "Mallard" after a series of temperature observations in the deepest part of the loch on the 21st September 1878.

The Firth of Clyde is the name given to the most remarkable group of fiord-like channels and sea lochs in the British Islands. They form a compact basin or depression, and it has been named after the principal stream which empties itself into it—the river Clyde. If we draw a straight line through the Craig of Ailsa and Sanda Island, at the extremity of the Peninsula of Cantyre, we shall have delineated the Firth seawards, and a line of soundings along this line discloses an almost perfectly uniform depth of water of from 23 to 25 fathoms. If we run a line of soundings at right angles to this line, we shall find the water deepening as we retire from the line whether we go northwards

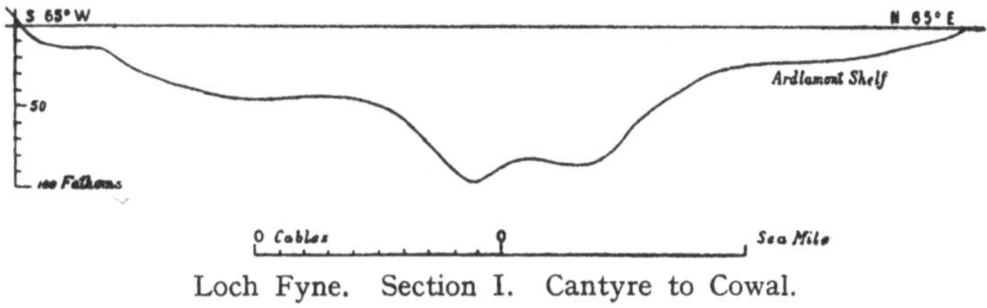

Loch Fyne. Section I. Cantyre to Cowal.

or southwards. If we go northwards we find the depth increase gradually and steadily as we pass the Island of Arran, whether by the main channel on the east, or by Kilbrennan Sound on the west; so that Arran stands, as it were, on an inclined plane sloping northwards from the mouth of the Firth, attaining a depth of 90 fathoms off the N.E. point of the island, and continuing as a deep trough into Loch Fyne, with a maximum depth of 104 fathoms[1] close to Skate Island and about three miles from the entrance of the loch. This deep trough only occupies a portion of the width of the loch, as is shown by the chart and sections. At its mouth the loch measures four nautical miles across, but the width of the portion of it over 50 fathoms in depth is only one mile. Both the loch, as a whole,

[1] See Map.

and the deep channel contract until they reach a minimum of sectional area, where this deep spot is. Here the whole width across from Skate Island to the south shore is only 1·6 mile, and of this one-half, or 0·8 mile, is occupied by a shallow tongue, with about 20 fathoms, projecting from the south shore. The deep channel, with over 50 fathoms, is here contracted to little over 0·3 mile. It will be seen from these data that dredging in the deepest water is difficult, because it implies drifting; and as the deep channel, besides being narrow, makes here a sort of elbow or sinuosity, as if there had been foldings in a vertical plane, it would seem to be impossible, as I generally found it to be in practice, to drift for any distance in any direction without rapidly getting into shallower water.

The chart is taken from Admiralty Charts Nos. 2133 and 2321, on the scale of 2 inches to the sea mile. The areas from the shore on both sides of the loch to the contour line of 30 fathoms is left unshaded; that lying between the contours of 30 and 80 fathoms is lightly shaded and areas enclosed by the contour of 80 fathoms are darkly shaded. The great constriction at Skate Island, with the deepest spot lying immediately on one side, is well shown both in the map and

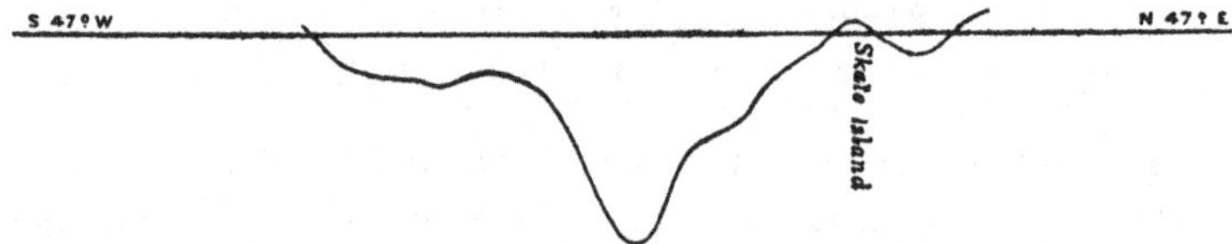

Loch Fyne. Section II. Cantyre to Cowal through 20-fathom bank and Skate Island.

in Sections II and III. In Section II the area of the whole section is 52,190 square fathoms; that of water lying at a greater depth than 30 fathoms is 17,900[1] square fathoms.

[1] In marine charts the unit of distance is always the nautical mile, which is equal to one minute of arc of a great circle of the globe, and it is subdivided into 10 cables of 100 fathoms each; so that the nautical mile is 1000 fathoms. Hence, in applying the decimal system to geographical measurements, the fathom is the natural unit so long as we retain the subdivision of the circumference of the circle into degrees and minutes as at present. The metre and kilometre are very inconvenient and clumsy in this respect

In Section III these quantities are 61,500 and 21,500 square fathoms. For comparison with these we have Section I at the mouth of Loch Fyne and Section VI at the widest part, towards Loch Gilp. In the former of these the total area is 154,820 square fathoms, and the area of water over 30 fathoms is 61,920 square fathoms; in the latter these figures are 129,200 and 67,500. Sections IV and V show transition stages from the narrowest to the widest portions of the landward portions

Loch Fyne. Section IIA. The same as Section II on natural scale.

of the loch. Section IIA is constructed on a uniform scale of lengths and depths, and gives the natural proportions at the narrowest points.

Loch Fyne. Section III. Cantyre to Cowal through deepest spot parallel to Section II.

As the locality where the nodules are found is very restricted, it is important to fix its position as nearly as possible. Now, there is not much difficulty in fixing a vessel's position in a direction up and down the loch. Skate Island and adjacent

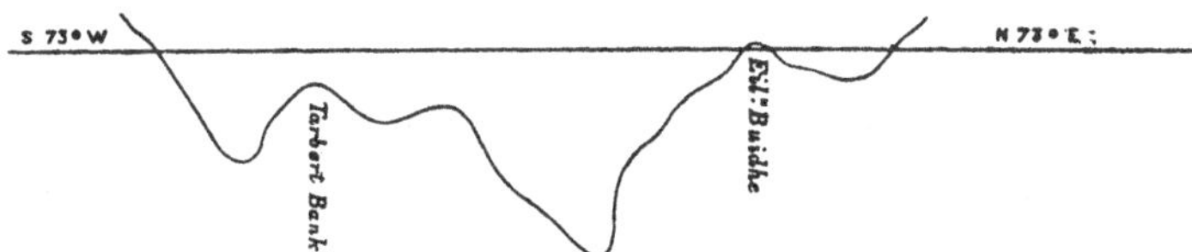

Loch Fyne. Section IV. Cantyre to Cowal through Tarbert Bank and Eiln. Buidhe.

features make it possible to get good leading marks in a direction perpendicular to the coast, but to find her place on this line is very difficult, as there is no feature that gives any leading, and there is nothing near and in the right direction to take bearings from; still, with careful work, it was always possible to find the spot. Thus, on the day in question, 21st September 1878, the first sounding gave exactly 104 fathoms, and everything was done to drop the anchor in the same spot; yet, when it was let go, it took the ground in 60 fathoms, the vessel having drifted, owing to the south-westerly breeze, towards the north shore. The cable was immediately brought to the winch and the anchor hove up, though it

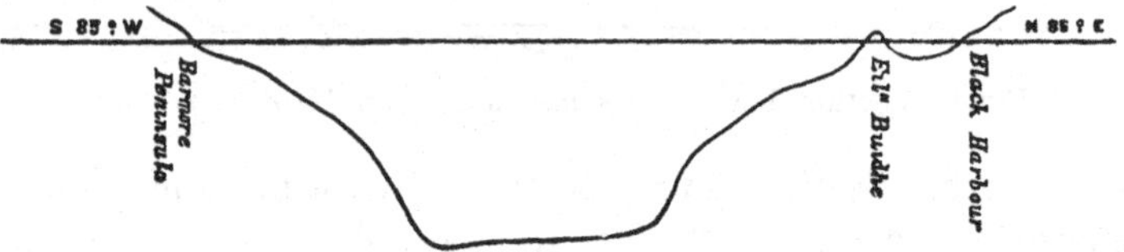

Loch Fyne. Section V. Knapdale to Cowal through Barmore and Black Harbour.

required a good deal of humouring to get it out of the ground. The anchor was not brought quite up to the bows, but, when about 5 fathoms of cable were still out, the yacht was steamed out to the proper position and the anchor immediately dropped. and the depth, as given by the wire-rope cable, was as nearly as possible 104 fathoms. More cable was then given, until 150 fathoms were out. It must be remembered that I was using a small kedge-anchor of ordinary type and weighing half a

Loch Fyne. Section VI. Knapdale to Cowal through widest part.

hundredweight, and my object was simply to anchor the vessel in the deepest water so as to be able to take temperatures and collect samples of water at leisure and without having to manœuvre the vessel. Although the anchor had not dragged, the yacht had tailed before the south-west wind so far toward the north shore that when I put over the sounding-line with a number of deep-sea thermometers, bottom was struck at 60 fathoms. On getting up the anchor, and after heaving in the extra cable which had been paid out, I sounded when the cable was up and down, and found exactly 104 fathoms, so that the anchor had held and remained in the deep place while the ship swung into water that was 44 fathoms shallower. When the anchor was brought up, a large mass of clay and dead shells, principally *Pecten*, was sticking to one of the flukes; and this was the specimen which contained the manganese nodules, which at that time were not supposed to exist anywhere out of the deepest oceans. As the anchor held firmly in the first instance in 60 fathoms, and was not examined before being lowered into the 104-fathom spot, it might be held to have picked up the sample in 60 fathoms, and kept it all the time it was holding the ship in 104 fathoms; but this is not likely. On 24th September 1878 the anchor was again dropped in apparently exactly the same spot, and it brought a quantity of clay and shells exactly like those of the 21st, but containing no nodules. On 1st October a few nodules were got in the deep trough; and dredging in 50 fathoms on the north side and in 40 fathoms on the south side none were found. On the south side the ground was very rough and rocky in 40 fathoms.

Although everything went to show that the abundance of nodules really occurred in the deep trough, it was disappointing that I could never get complete confirmation. On 15th July 1881, however, I was again successful, for the anchor-dredge brought up a large bagful of mud, containing abundance of nodules. On this occasion the position of the ship was carefully watched, showing that she dragged over about half a mile of ground, beginning in 100 fathoms and gradually shoaling to 85 fathoms, when the anchor was brought up. There can, therefore, be no longer any doubt that it is really in the deep trough that the nodules occur, and only in a very limited area of it.

The mud of 21st September 1878 was divided in two equal portions, one of which was preserved and the other was analysed mechanically. It was separated into three constituents, the nodules, the shells, and the residue, a sandy clay. The number of nodules thus obtained was 83, and they weighed 142·7 grammes, whence the mean weight of a nodule was 1·7 grammes.

Packed as closely as possible in a graduated cylinder of 37 millimetres internal diameter, they occupied a length of 130 millimetres, reaching to 146 c.c. on the graduation. I then poured in 100 c.c. water, which stood at 158 c.c. Hence the volume of the nodules was 58 c.c., and the average volume of one was 0·7 c.c., and their density 2·46. The shells were all dead, and principally *Pecten*; they weighed 35 grammes. The following was the mechanical composition of the mud:

Nodules	142·7 grammes	30	per cent.
Shells	35 ,,	7½	,,
Sandy clay	289 ,,	62½	,,
Total	466·7 ,,	100	,,

Of twenty-two nodules which were split open, sixteen contained soft nuclei of about the size of a pea, and apparently very rich in manganese. One of these nuclei was so slightly attached to the rind as to fall out when the nodule was split. The rind is always very hard and gritty, and when the oxides of iron and manganese are removed by hydrochloric acid, it falls into a mass of sand, similar to that which makes up a large proportion of the rind. This agrees with the idea that the nodules are agglomerations of the mud found *in situ*, and cemented by the ochreous oxides. One nodule (No. 24) was interesting, as showing the complete soft kernel, loose, in a cavity of mud, the rind not having as yet formed, although it was evidently forming, the mud being stained from the inner wall of the shell outwards. No. 25 was a similar nodule, only the kernel had no free space round it. The mud round it was stained yellow. This seems to be very general with the growing nodule; the oxide of iron spreads itself in front of the oxide of manganese. A little of the yellow shell gives no manganese reaction with hydrochloric acid. A nodule of the size of a pea was found, resembling the others perfectly, except that it contained Fe_2O_3, and no MnO_2. Very careful examination of the mud from which the nodules and shells had been removed showed that perfect nodules exist down to the size of a pin-head, and all through the mud there were specks showing where probably nodules had begun to form. Although so near the shore, the mud contains hardly any pebbles larger than a

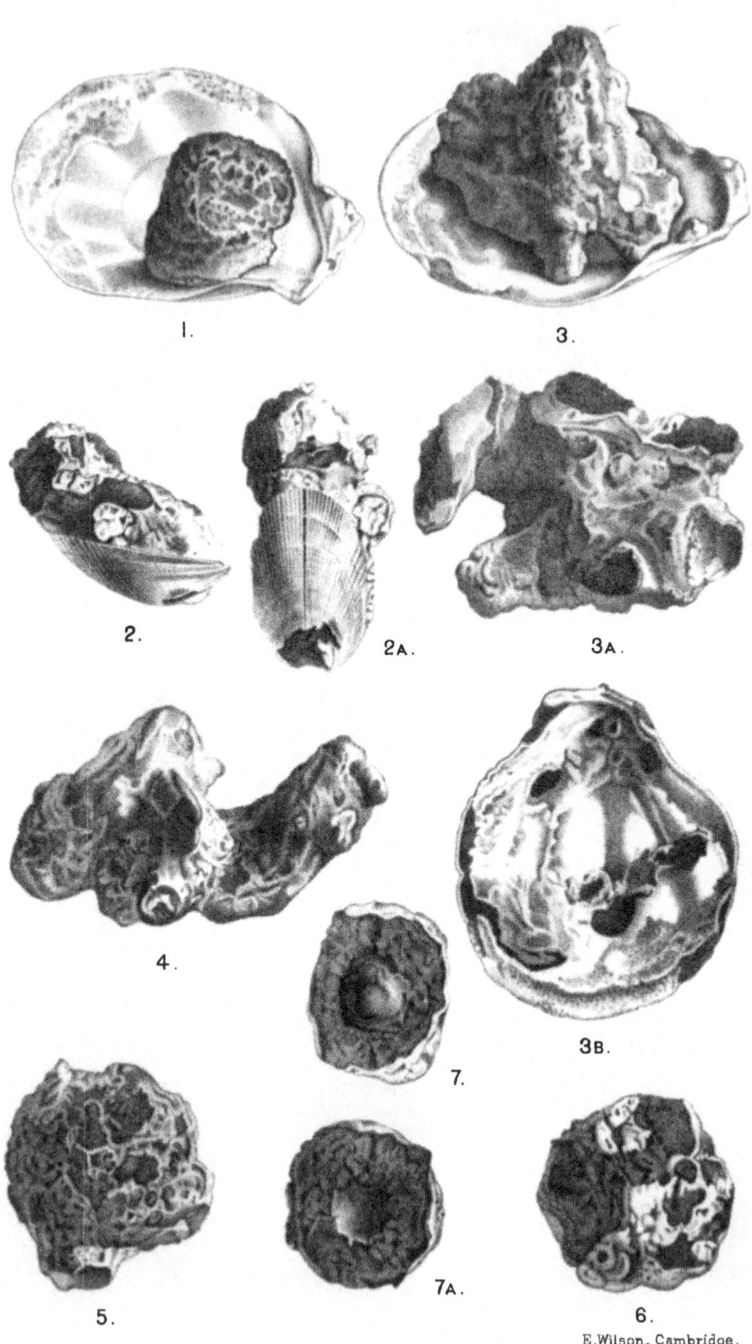

MANGANESE NODULES FROM THE CLYDE.

grain of sand. This shows conclusively that the nodules have not been washed down into their present position; and that they have been formed *in situ* is further shown by the fact that the mud round them is generally stained yellow with ferric oxide. A number of nodules were picked out for illustration and analysis.

In the Plate, fig. 1 represents a nodule which was found attached to the rim of a dead pecten shell. It is represented in one and a half times the natural size.

Figs. 2 and 2A represent, in twice natural size, a nodule entirely filling up a shell.

Figs. 3, 3A, and 3B represent, in twice natural size, (3) a nodule found attached to a dead pecten shell; (3A) the under side of the nodule, showing bases of attachment; and (3B) a view of the shell, showing surfaces of attachment.

Fig. 4, in one and a half times natural size, shows a remarkable nugget-shaped nodule.

Fig. 5, in twice natural size, shows a pear-shaped nodule, with stem of attachment.

Fig. 6, in twice natural size, shows a spherical nodule of the commonest form.

Figs. 7 and 7A show a nodule split through the middle, and with a semi-detached nucleus. The half of the kernel protrudes in fig. 7, and the corresponding cavity is visible in fig. 7A.

All these nodules were got on 21st September 1878.

The nodules taken for analysis are as follows: *M* and *N* (called Nos. 10 and 14 in original notes) were chosen as average specimens. *P* was a rather softer and *Q* a rather more stony nodule. *K* and *R* are the kernels and rinds respectively of five nodules. These nodules were split, and the kernels and rinds separated as carefully and as completely as possible. Their approximated weights were:

Kernels	1·127	grammes
Rinds	10·710	,,
Total ..	11·837	,,

S is an average sample taken from a number of nodules pounded up and mixed.

The analysis of these samples was undertaken at first only to show the state of oxidation of the manganese. In the case of the Loch Fyne nodules, it was carried further, with a view of showing the nature of the other principal constituents. The analyses under the letters *X*, *Y*, and *Z* constitute together a complete analysis, though it was not at first intended to be such, otherwise the whole of the requisite material would have been prepared and extracted at once. These analyses were made in 1879 by Dr George M'Gowan, F.R.S.E., now of Bangor, and were carried out with the greatest care and attention.

Analytical Method followed for the Determination of the State of Oxidation of the Manganese.

The analysis of these nodules was conducted as follows (any particular point relating to any one nodule will be given in detail along with the analysis of that nodule):

A. The nodules were reduced to an impalpable powder in an agate mortar, and were not dried previous to analysis.

B. In two separate portions (of about 2 decigrams each) the *available oxygen* was determined by Bunsen's iodometric method.

C. Two portions, of about 0·5 grams each, were dissolved in about 5 c.c. of pure strong hydrochloric acid, in a small covered beaker, heat being applied cautiously at first to prevent loss by effervescence of chlorine. The mixture was then evaporated to dryness over a water-bath, covered with a glass shade; the residue, when free from all excess of acid, moistened with a few drops of strong hydrochloric acid, and the *insoluble residue* filtered off, ignited, and weighed.

(*a*) *Insoluble Residue.*—This insoluble residue was kept in a corked tube, and afterwards fused with $(NaK)CO_3$ (in the proportion of about four parts of carbonate to one of residue) for the estimation of silica. The filtrate from the silica was examined *qualitatively* for the bases present.

(*b*) The *filtrate* from the insoluble residue was treated with a moderate excess of *pure* carbonate of baryta. In practice about 2 grams were required. This was added in a state of cream to the liquid in a small flask. The flask was corked and allowed to stand some hours, being shaken occasionally. The liquid was then filtered off, and the precipitate, after thorough washing, tested for manganese with carbonate of soda on platinum foil.

D. To the filtrate from the carbonate of baryta precipitate excess of sodium acetate was added, so that there should be not less than $3C_2H_4O_2$ to HCl in the solution (in practice 15 to 20 c.c. of a 1 in 10 solution of $Na\bar{A}$ were added). Before adding the acetate it is well to drop in a little acetic acid, otherwise an immediate precipitation of barium salt occurs. Bromine in excess was then added to the sufficiently dilute fluid, and the flask corked and placed in a moderately warm place for some time. Gentle warming at first greatly aids the precipitation of the MnO_2. When the manganese had begun to deposit, and the liquid in the flask was at a temperature of about 50° C., it was removed to a colder place (at the ordinary temperature of the room), and allowed to stand over-night. If necessary, a little more bromine was first added; this is regulated by practice. Next day the flasks were heated in order to drive off all superfluous bromine, a few drops of alcohol being added towards the end to get rid of the last traces. When the solution became perfectly colourless it was filtered, all the MnO_2 that came easily out of the flask being thrown on the filter and washed with boiling water. The *filtrate* was tested first for more MnO_2 with $Na\bar{A}$ and bromine, and then for nickel and copper (*vide* below).

(*a*) The *filter*, on which was the bulk of the MnO_2 precipitate, was dried. When dry, the MnO_2 was returned to the original flask (on the sides of which some MnO_2 remained sticking), the filter was ignited, and its ash also dropped into the flask. This manganese precipitate was then dissolved in a few drops of strong HCl, the chlorine boiled off, and the solution slightly diluted. It was found that the MnO_2 always brought down some baryta with it, and this was separated by precipitation with a few drops of sulphuric acid. This solution was then filtered from the $BaSO_4$; evaporated to dryness over a water-bath to get rid of all excess of hydrochloric acid; the residue taken up with water; the solution filtered again if necessary; and the manganese precipitated by *pure* carbonate of soda solution, only a very slight excess being used. The $MnCO_3$ was washed first by decantation, then dried, and ignited to Mn_3O_4.

The filtrate from the $MnCO_3$ was always kept over-night, when sometimes a second minute precipitate came down. The Mn_3O_4 was kept, and the nickel and cobalt in it determined afterwards.

(*b*) The *filtrate* from the MnO_2 precipitate with bromine was, in the case of the "Challenger" nodules, neutralised with NH_4OH, and then excess of NH_4HS added, the flask filled up and corked, and allowed to stand for one or more days. Generally a minute black precipitate, accompanied by much barium sulphate, came down. This was filtered off, ignited, dissolved in HCl, the solution evaporated to dryness in a small basin, the residue taken up with water, and preferably a drop of H_2SO_4, and filtered. In the filtrate the nickel oxide was precipitated by a drop or two of pure caustic potash. After ignition this precipitate was always tested for purity. It is better, however (and this was done in the case of the Loch Fyne nodule), to neutralise the filtrate from the MnO_2 with NH_4OH, add NH_4HS, and then acetic acid in excess, and pass H_2S gas through the gently-warmed solution (as is done in the separation of MnO from NiO and CoO). Thus, after standing over-night, minute traces of nickel came down, which remained dissolved in the NH_4HS solution.

E. The Mn_3O_4 *precipitates* were dissolved in a few drops of strong HCl in a small basin, the solution evaporated to dryness over the water-bath, taken up with water, and the solution filtered. To the filtrate, after neutralising any trace of acid present with NH_4OH, sulphide of ammonium and acetic acid were added, and sulphuretted hydrogen gas passed through the gently-warmed solution to precipitate the nickel and cobalt as sulphides. After standing over-night (the flask being full and corked) this precipitate was filtered off, and the filtrate treated in the same manner with ammonia, ammonium sulphide, acetic acid, and sulphuretted hydrogen. In this way a minute additional precipitate was obtained. The two precipitates were dried, ignited, and dissolved together in hydrochloric acid, and exactly the same process gone through again, so as to get rid of all traces of manganese. The second two precipitates of cobalt and nickel sulphides were again dried, ignited, redissolved in hydrochloric acid, the solution evaporated to dryness, the residue taken up with water, and this solution filtered. The nickel and cobalt in the filtrate were then precipitated by a few drops of pure caustic potash. This precipitated the nickel and cobalt together. The precipitate was ignited and weighed as $NiO + Co_3O_4$. It was then again dissolved in hydrochloric acid, all excess of acid evaporated off, the residue taken up with water, the solution filtered into a small beaker, and gently evaporated down. During this evaporation very often a minute quantity of silica and ferric hydrate came down, and the solution had to be refiltered. After the liquid became very concentrated—reduced to a few drops, in fact—the cobalt was precipitated as double nitrite of cobalt and potassium. A solution of 1 in 4 nitrite of potassium was used.

(*a*) The *cobalt precipitate*, after being filtered off and washed with acetate of soda, was dissolved in water or dilute hydrochloric acid, the filter ignited, and the ash added to the solution. The solution was evaporated to dryness, water added, and the solution filtered. The cobalt was then precipitated by caustic potash solution. It was ignited to Co_3O_4.

(*b*) To the *filtrate* from the potassium nitrite precipitate excess of hydrochloric acid was added, the solution evaporated, and the nickel precipitated by caustic potash. It was found necessary to redissolve this nickel precipitate and reprecipitate it, on account of the relatively very large quantity of potassium chloride which had to be washed out of it. This gave the nickel as NiO. The nickel and cobalt precipitates were afterwards tested in the dry way.

Tabular List of Nodules submitted to Analysis.

Oceanic Nodules—

Nos. I, II, and III are from a locality 400 miles south of the Australian coast, in lat. 42° 42′ S., long. 134° 10′ E. Depth, 2600 fathoms. I is from the outer rind of a large spherical nodule, and it was detached without any trace of the kernel. II is a similar piece taken from the outside shell of a smaller spherical nodule, with traces of an augitic kernel. III is a piece of a horizontally-stratified nodule with botryoidal upper surface.

Nos. IV and V are from the same nodule, which was collected on the 12th July 1875 in the North Pacific in lat. 37° 52′ N., long. 160° 17′ W. Depth, 2740 fathoms. IV is a portion of the outer shell, 13 millimetres thick, consisting of concentric layers, most of them mottled with reddish-yellow spots. V is the kernel of the same nodule, and is harder than the rind, from which it easily splits away. It was nearly spherical and with 16 millimetres radius. At the centre was a small colourless piece of mineral.

Littoral Nodules—

These are all from Loch Fyne in 104 fathoms (see page 167). *M*, *N*, *P*, and *Q* are separate selected nodules; *K* and *R* are the kernels and rinds of five nodules.

The analytical details have been collected in a series of tables, as follows:

TABLE I. *Determination of Available Oxygen by the Iodometric Method.*

Number of sample	Weight of sample taken (grms.)	Volume of hypo-sulphite used (c.c.)	Equivalent weight of Iodine (grms.)	Equivalent weight of Oxygen (grms.)	Oxygen per cent.	Equivalent per cent. of MnO_2
	a	b	$c=0{\cdot}012316b$	$d=0{\cdot}063c$	$e=100\frac{d}{a}$	$f=5{\cdot}4375e$
I (1)	0·1711	12·07	0·14856	0·009358	5·47	29·74
(2)	0·2698	19·06	0·23465	0·01478	5·48	29·80
II (1)	0·1815	13·52	0·16647	0·010486	5·78	31·43
(2)	0·1734	13·03	0·16040	0·010104	5·82	31·65
III (1)	0·1802	12·20	0·1502	0·009463	5·25	28·55
(2)	0·1620	11·05	0·13601	0·0085677	5·29	28·76
IV (1)	0·1924	10·10	0·12439	0·0078356	4·07	22·13
(2)	0·1650	8·657	0·106567	0·0067129	4·07	22·13
V (1)	0·1514	8·306	0·102246	0·00644	4·25	23·13
(2)	0·1731	9·65	0·11875	0·0073546	4·25	23·13
M (1)	0·1768	9·03	0·11113	0·007001	3·96	21·53
(2)	0·2210	11·30	0·1391	0·008762	3·96	21·53
N (1)	0·1859	5·936	0·07307	0·004603	2·48	13·48
(2)	0·2547	8·17	0·10057	0·006335	2·47	13·54
P (1)	0·1754	8·18	0·10073	0·006345	3·62	19·67
(2)	0·2117	9·77	0·1203	0·007578	3·58	19·47
Q (1)	0·1945	8·43	0·10375	0·006535	3·36	18·27
(2)	0·1815	7·83	0·09636	0·00607	3·34	18·18
K (1)	0·1116	8·94	—	—	6·25	34·00
(2)	0·1004	8·153	—	—	6·33	34·43
(3)	0·0553	4·401	—	—	6·21	33·77
R (1)	0·2286	9·96	0·12296	0·007746	3·39	18·43
(2)	0·3070	13·43	0·16446	0·01036	3·37	18·32

TABLE II. *Determination of Insoluble Residue.*

Number of sample		Weight of sample taken (grammes)	Weight of insoluble residue (grammes)	Insoluble residue per cent.	Weight of insoluble residue taken for SiO_2 estimation	Weight of SiO_2 (grammes)	SiO_2 per cent.	
							In insoluble residue	Equivalent in nodules
		a	b	$c=100\frac{b}{a}$	d	e	$f=100\frac{e}{d}$	$g=\frac{b}{a}f$
I	(1)	0·5290	0·1088	20·56	0·0958	0·0819	85·45	17·46
	(2)	0·4733	0·0961	20·30				
II	(1)	0·4802	0·0815	16·96	0·1475	0·1278	86·75	14·75
	(2)	0·6001	0·1018	16·95				
	(3)	0·5254	0·0894	17·01				
III	(1)	0·5633	0·0937	16·63	0·1291	0·1127	87·26	14·68
	(2)	0·5839	0·0983	16·83				
	(3)	0·5566	0·0937	16·83				
IV	(1)	0·5035	0·1386	27·52	0·1727	0·1442	83·47	23·09
	(2)	0·5248	0·1453	27·68				
	(3)	0·4943	0·1358	27·63				
V	(1)	0·5015	0·1001	19·95	0·1139	0·0961	84·34	16·86
	(2)	0·4825	0·0967	20·03				
M	(1)	0·4977	0·1469	29·51	0·1631	1·1388	85·10	25·24
	(2)	0·5516	0·1646	29·83				
N	(1)	0·5021	0·1395	27·78	0·1377	0·1161	84·28	23·55
	(2)	0·5321	0·1496	28·10				
P	(1)	0·4639	0·1491	32·13	0·1458	0·1217	83·44	26·83
	(2)	0·4767	0·1535	32·17				
Q	(1)	0·5045	0·1463	28·99	0·1436	0·1208	82·70	23·94
	(2)	0·5110	0·1478	28·91				
R	(1)	0·5814	0·1717	29·56	—	—	—	—
	(2)	0·5144	0·1503	29·21				

TABLE III. *Determination of Manganese, Nickel, and Cobalt.*

Number of sample	Weight of sample taken (grams.)	Weight of Mn_3O_4 + NiO + Co_3O_4 (grms.)	Per cent. of Mn_3O_4 + NiO + Co_3O_4 in nodules	Weight of mixed oxides taken for Ni and Co estimation	Weight of NiO + Co_3O_4 (grams.)	Per cent. of NiO + Co_3O_4 in nodules	Per cent. of MnO in nodules	Total weight of NiO + Co_3O_4 (grams.)	Weight of Co_3O_4 (grams.)	Per cent. of Co_2O_3 in nodules	Weight of NiO (grams.)	Per cent. of NiO in nodules
	a	b	$c = 100\frac{b}{a}$	d	e	$f = 100\frac{eb}{ad}$	$g = 0{\cdot}93(c-f)$	$h = e_1 + e_2$	i	$k = 100\frac{1{\cdot}033i}{a_1 + a_2}$	l	$m = 100\frac{b}{a_1 + a_2}$
I (1)	0·5290	0·1526	28·84	0·1524	0·0063	1·20	25·71	0·0128	0·0019	0·19	0·0116	1·11
(2)	0·4733	0·1377	29·08	0·1371	0·0065	1·38	25·76					
II	0·4802	0·1463	30·46	0·1452	0·0077	1·61	26·94	0·0077	0·0008	0·19	0·0062	1·32
III	0·5633	0·1509	26·81	0·1496	0·0066	1·21	23·91	0·0066	0·0011	0·21	0·0049	0·90
IV	0·5035	0·1056	20·97	0·1048	0·0051	0·99	18·69	0·0051	0·0001	0·19	0·0034	0·68
V (1)	0·5015	0·1076	21·49	0·2084	0·0088	0·92	19·30	0·0088	0·0031	0·34	0·0040	0·42
(2)	0·4825	0·1043	21·61				19·41					
M (1)	0·4977	0·1679	33·74	0·3525	0·00024	0·02	31·36	0·00024	0·00024	0·02		
(2)	0·5516	0·1859	33·72				31·34					
N (1)	0·5021	0·1507	30·02	0·3100	0·00024	0·02	27·90	0·00024	0·00024	0·02		
(2)	0·5321	0·1598	30·03				27·91					
P (1)	0·4639	0·1427	30·76	0·2891	0·00044	0·05	28·56	0·00044	0·00044	0·05		
(2)	0·4767	0·1464	30·71				28·52					
Q (1)	0·5045	0·1588	31·48	0·3167	0·00054	0·03	29·45	0·00034	0·00034	0·03		
(2)	0·5110	0·1611	31·53				29·29					
K	0·2673	0·1068	39·94				37·15					
R (1)	0·5814	0·1858	31·95				29·72					
(2)	0·5144	0·1631	31·70				29·49					

TABLE IV. *Oceanic Nodules.* (*Summary of Results.*)

Number of sample	Insoluble residue per cent.	Per cent.* SiO_2 in	Oxygen per cent.	MnO per cent.	Co_2O_3 per cent.	Ni_2O_3 per cent.	Formula of Peroxide
I	20·56	(17·46)	5·47	23·71	—	—	($MnO_{1·945}$)
	20·30	85·45	5·48	23·76	0·19	1·23	$MnO_{1·901}$
II	16·95	—	—	—	—	—	—
	17·01	(14·75)	5·78	—	—	—	($MnO_{1·954}$)
	16·96	86·75	5·82	26·94	0·19	1·46	$MnO_{1·908}$
III	16·83	—	5·25	—	—	—	—
	16·83	(14·68)	5·29	—	—	—	($MnO_{1·979}$)
	16·63	87·26	5·28	23·91	0·21	1·00	$MnO_{1·937}$
IV	27·68	—	—	—	—	—	—
	27·63	(23·09)	4·07	—	—	—	($MnO_{1·951}$)
	27·52	83·47	4·07	18·69	0·19	0·75	$MnO_{1·926}$
V	19·95	(16·86)	4·25	19·30	—	—	($MnO_{1·974}$)
	20·03	84·34	4·25	19·41	0·34	0·47	$MnO_{1·941}$

In calculating the "Formula of Peroxide" the MnO equivalents of the Co_2O_3 and Ni_2O are taken in the lower ones, and only the MnO found is taken in calculating those in brackets.

TABLE V. *Littoral Nodules.* (*Summary of Results.*)

Number of sample	Insoluble residue per cent.	Per cent.* SiO_2 in	Oxygen per cent.	MnO per cent.	Co_2O_3 per cent.	NiO and CuO	Formula of Peroxide
M	29·51	(25·24)	3·96	31·36	—	Decided traces of copper were present in all the samples. The traces of nickel were so small as to be doubtful	—
	29·83	85·10	3·96	31·34	0·02		$MnO_{1·562}$
N	27·78	(25·55)	2·48	27·90	—		—
	28·10	84·28	2·49	27·91	0·02		$MnO_{1·394}$
P	32·13	(26·83)	3·62	28·56	—		—
	32·17	83·44	3·58	28·52	0·05		$MnO_{1·558}$
Q	28·99	(23·94)	3·36	29·25	—		—
	28·91	82·70	3·34	29·30	0·03		$MnO_{1·507}$
K	—	—	6·25	—	—		—
	—	—	6·33	—	—		—
	—	—	6·21	37·15	—		$MnO_{1·75}$
R	29·56	—	3·39	29·72	—		—
	29·21	—	3·37	29·49	—		$MnO_{1·505}$

* The upper numbers, in brackets, refer to the original substance, the lower ones to the insoluble residue.

In connection with these Tables it may not be amiss to quote the analyses which I made of some nodules in 1876, and published in the *Proceedings* of this Society[1]. The samples analysed were from four different localities, two of which are identical with those from which the nodules I–V came.

Nos. 2, 4, and 5 were from the same place, No. 2 being the matter collected round a shark's tooth as nucleus; Nos. 4 and 5 being the outside rinds of ordinary nodules.

The results are given in the following Table, the numbers being in many cases the means of several observations:

Locality		No.	*A*	*B*	*C*	*D*	*E*	*F*	*G*	
Lat.	Long.		In-soluble Residue	O	MnO_2	MnO	Fe_2O_3	Al_2O_3	H_2O	Na_2O
13° 52′ S.	149° 17′ W.	2	17·55	6·13	33·30	27·18	—	—	—	—
,,	,,	4	15·30	5·92	32·23	—	23·86	—	—	—
,,	,,	5	15·30	6·49	35·28	—	24·85	—	10·2	—
37° 52′ N.	160° 17′ W.	6	36·24	6·49	24·41	—	20·16	3·83	7·70	5·98
42° 42′ S.	134° 10′ E.	7	17·98	7·54	41·11	33·53	18·04	2·55	7·31	—
22° 21′ S.	150° 17′ W.	8	21·74	5·19	28·20	—	24·52	7·67	8·54	8 5

A is the residue which remains undissolved after treating the mineral with strong hydrochloric acid, evaporating to dryness and redissolving. In No. 5 it contains 85·16 per cent. silica, and in No. 6, 82·27 per cent.

B is the available oxygen determined by Bunsen's method.

C is the MnO_2 equivalent to the available oxygen.

D is the MnO found by weighing as Mn_3O_4.

E is the Fe_2O_3 found by titration with $SnCl_2$.

F is the alumina found by subtracting the Fe_2O_3 found in *E* from the weight of the precipitate with acetate of soda.

G is the water expelled on ignition; it is obtained by deducting two-thirds of the oxygen found in *B* from the loss of weight by ignition.

The samples were dried for ten or twelve hours at 140° C., and therefore the percentages are higher than those in Table IV, the samples in it having been analysed air-dried.

[1] *Proc. Roy. Soc. Edin.* 1876, vol. IX. p. 287.

From these Tables it will be seen that nodules from all localities have similar composition. The most important difference between the littoral and the oceanic nodules is, that in the former the manganese is less highly oxidised than in the latter. In the oceanic nodules, when we consider the manganese alone, the peroxide is very little short of MnO_2; in the littoral ones it is very little over Mn_2O_3. In both oceanic and littoral ones the manganese is more highly oxidised in the kernels than in the rinds; this difference is particularly marked in the littoral ones. Nickel, cobalt, and copper are probably present in all, but their relative proportions are different in the littoral and in the oceanic nodules. Cobalt can be determined in both, and ranges from 0·02 per cent. in Loch Fyne nodule *M* to 0·34 per cent. in oceanic nodule V. Nickel is present in large though variable quantity in the oceanic nodules, as much as 1·46 per cent. of oxide in II, falling to 0·47 per cent. in V. In the Loch Fyne nodules the presence of nickel was doubtful. Copper was present in both classes of nodule. In the oceanic ones, however, the traces were not always very distinct, whilst in the Loch Fyne ones they were very pronounced. In testing for thallium, it was sufficient with oceanic nodules to moisten a piece of the size of a bean with HCl and place it in a platinum triangle over a Bunsen burner to get a strong thallium line in the spectroscope. Thallium could not be detected in this way in the Loch Fyne nodules.

The insoluble residues were always tested qualitatively, and consisted, besides silica, of alumina and ferric oxide, with smaller quantities of lime and magnesia. The silica, however, of the Loch Fyne nodules was largely quartz, which was not the case with the oceanic nodules. The percentage of insoluble residue varies much in both sorts, but it is always greater in the Loch Fyne ones than in the oceanic ones. The percentage of silica in the Loch Fyne nodules varies very little from 25, leaving from 2 to 6 per cent. of bases in the insoluble residue. In the oceanic nodules, excepting IV, the silica varies between 14·68 and 17·46 per cent., leaving from 2 to 3 per cent. bases in the residue. In IV the amount would be 4½. In the kernels (*K*) the moisture driven off at 100°, the carbonic acid, and the

$Fe_2O_3 + Al_2O_3$ were determined, and we have the following tabulated summary:

Sample K. Kernels of Loch Fyne Nodules.

Carbonic acid	0·83 per cent.
Moisture at 100°	8·23 ,,
Available oxygen	6·26 ,,
MnO	37·15 ,,
$Fe_2O_3 + Al_2O_3$	4·78 ,,
CuO	trace
NiO	doubtful trace.

The insoluble residue was not determined.

In order to obtain some knowledge of the other constituents of the Loch Fyne nodules, a number of nodules were pounded up and mixed, so as to form an average sample in which the determinations were to be made. They contained a certain amount of soluble matter from the sea-water, and, if it had been intended at first to make a complete analysis, the whole of this sample would have been extracted at once. This was not done, and the various portions, *X*, *Y*, and *Z*, have been treated separately. Particulars follow.

X. *General Analysis.*

A sample of air-dried Loch Fyne nodules was thoroughly extracted with boiling water. The weight taken was 5·0405 grammes, and the insoluble portion, when dried at 100°, was found to weigh 4·5458.

A. *Estimation of Moisture.* 0·1910 of the sample was dried at 100°, and found to lose 0·0010 gramme, *i.e.*, the percentage of water in it driven off at 100° is 0·53.

B. *Estimation of the Oxygen reducible by Hydrochloric Acid.* 0·1195 gramme were boiled with strong hydrochloric acid. The chlorine liberated was passed into a solution of potassium iodide, and the amount of iodine liberated found by titration with sodium hyposulphite. There was found to be 4·104 per cent. of available oxygen, representing 22·32 per cent. of MnO_2.

C. *Estimation of Weight lost on Ignition.* 0·1263 gramme were strongly heated over a Bunsen flame. The loss of weight was found to be 0·0224 gramme; that is, 17·74 per cent.

D. *Estimation of Insoluble Residue*, Fe_2O_3, Al_2O_3, *and* Co_2O_3. 2·0880 grammes were treated with strong hydrochloric acid in a covered beaker, the mixture evaporated to dryness over a water-bath, the residue moistened with a few drops of strong hydrochloric acid, taken up with water, and the solution filtered.

The insoluble residue was heated over the water-bath with a strong solution of carbonate of soda. The heating was carried on for about forty-five minutes, water being occasionally added to keep the solution of constant strength. The solution was then decanted through a filter, more carbonate of soda solution added to the insoluble residue, and the same process repeated. This was done altogether three times. What remained undissolved was washed, ignited, and weighed. It was found to weigh 0·5790 gramme, that is, 27·73 per cent., consisting of silica and silicates, which resist the action of sodium carbonate solution.

The silica which had dissolved in the sodium carbonate was precipitated from it by hydrochloric acid as usual, filtered, ignited, and weighed. It weighed 0·11928 gramme, that is, 5·71 per cent. of soluble silica.

The *filtrate* from the insoluble residue was treated with excess of carbonate of baryta to precipitate the iron and alumina. The precipitate was filtered off and redissolved in hydrochloric acid, the baryta present precipitated with sulphuric acid and filtered off, the filtrate made up to 250 c.c. volume, and two estimations of $Fe_2O_3 + Al_2O_3$ made.

100 c.c. (= 0·8352 gramme of substance) gave 0·03368 gramme, $Fe_2O_3 + Al_2O_3$, *i.e.*, 4·03 per cent.; 75 c.c. (= 0·6264 gramme of substance) gave 0·02568 gramme, $Fe_2O_3 + Al_2O_3$, *i.e.*, 4·09 per cent. The two precipitates of Fe_2O_3 and Al_2O_3 were united and fused with bisulphate of potash. An insoluble residue of silica, weighing 0·00898 gramme, was found. This represents 0·61 per cent. of silica in the nodules, which, when added to the previously found 5·71 per cent., gives altogether 6·32 per cent.

Also, subtracting 0·61 from 4·06 (the average of the iron and alumina estimations), we find 3·45 per cent. of $Fe_2O_3 + Al_2O_3$.

In 50 c.c. of the above-mentioned 250 c.c. solution 0·00676 gramme of iron were found by titration with stannous chloride. This is equivalent to 2·31 per cent. of Fe_2O_3 in nodules. Subtracting from 3·45, there remains 1·14 per cent. Al_2O_3.

In the *filtrate* from the barium carbonate precipitate the manganese was not estimated, but the cobalt was. It was separated by means of sulphide of ammonium and acetic acid, as usual, then precipitated as Fischer's salt with nitrite of potash, and this precipitate dissolved in HCl, and the cobalt reprecipitated with a few drops of carbonate of soda solution. It should be mentioned that, before treatment with sulphide of ammonium and acetic acid, sulphuric acid was added to precipitate the excess of barium; to the filtrate from $BaSO_4$ sodium carbonate was added to precipitate both manganese and cobalt, and this precipitate dissolved up and the cobalt separated. 0·00049 gramme of Co_3O_4 were found, representing 0·025 per cent. of Co_2O_3 in the nodules.

E. *Estimation of Carbonic Acid.* 0·1792 gramme of nodules were treated with solution of phosphoric acid, the carbonic acid gas liberated passed into baryta water of known strength, and the excess of baryta determined by titration with hydrochloric acid. 0·015905 gramme of Co_2 were found, representing 8·87 per cent. of carbonic acid in the nodules.

Another estimation was made; 0·1704 gramme of nodules were found to contain 0·015285 gramme of carbonic acid gas, that is, 8·91 per cent.

F. *Estimation of Phosphoric Acid.* 0·7162 gramme of nodules were converted into chlorides by treatment with HCl (the insoluble residue being removed in the usual way), and the solution made up to 250 c.c.

125 c.c. of the solution was treated with acetate of soda, and the phosphoric acid obtained as a precipitate of ferric phosphate. This precipitate was fused with bisulphate of potash and the fused mass dissolved in water. There was just a mere trace of insoluble matter. To this acid solution a very slight excess of pure sodium carbonate was added to neutralise any free sulphuric acid present, and then excess of nitric acid. The solution was then evaporated down, large excess of

ammonium molybdate added, and the solution allowed to stand in a warm place till all the phosphoric acid was precipitated. The precipitate was filtered off and dissolved in ammonia, and the phosphoric acid precipitated with magnesia mixture as usual. The precipitate was ignited and weighed. It weighed 0·00319 gramme ($Mg_2P_2O_2$), representing 0·57 per cent. P_2O_5.

Estimation of Lime and Magnesia. 120 c.c. of the abovementioned 250 c.c. solution (equivalent to 0·3438 gramme nodules) were treated with sulphide of ammonium, filtered, the ammonium sulphide in the filtrate destroyed by means of hydrochloric acid, evaporated and filtered from sulphur, and the lime precipitated twice with ammonium oxalate as usual. After the second precipitation the calcium oxalate (which seemed to contain alumina) was redissolved, and the lime precipitated a third time with ammonium oxalate. The precipitate was filtered, ignited, and dissolved in hydrochloric acid. It weighed 0·0433 gramme. The solution was just rendered alkaline with ammonia. This precipitated the trace of alumina. It was filtered and weighed, and found equal to 0·00414 gramme. Subtracting this from 0·0433 gramme, we have 0·03919 gramme of $CaCO_3$, that is, 6·38 per cent. of CaO.

The sulphide of ammonium precipitate, together with the small precipitate of alumina, was worked up for any lime that might be present in it as phosphate. It was ignited, dissolved in hydrochloric acid, the solution evaporated. The phosphoric acid, iron, and alumina present were precipitated by adding acetate of soda. In the filtrate from this the manganese was precipitated with ammonium sulphide. In the filtrate from the manganese sulphide a small additional quantity of lime was obtained on addition of ammonium oxalate; 0·09154 gramme of $CaCO_3$ were thus obtained. This is 0·45 per cent. of CaO of the nodules. Adding this to the previously found 6·38 per cent., we have altogether 6·83 per cent. of CaO in the nodules.

The filtrates from the oxalate of lime were evaporated down, and the ammonium salts driven off. The residue was taken up with hydrochloric acid and water, and filtered. Ammonia and phosphate of soda were added. A precipitate was obtained, which was filtered off, ignited, and weighed as $Mg_2P_2O_7$. It weighed 0·03366 gramme, representing 3·53 per cent. of MgO in the substance. Another estimation gave 4·08 per cent. of MgO.

G. *Estimation of Moisture at* 162° *C.* 0·6215 gramme of substance were exposed in a small bath to a temperature of 162° C. for about forty minutes. The water driven off was collected in a calcium chloride tube. It was found after heating that the substance had lost 0·0085 gramme (1·37 per cent.), while the calcium chloride tube had gained 0·0098 gramme (1·57 per cent.); mean, 1·47.

H. *Estimation of Carbon (by combustion).* 0·6215 gramme were ignited with pure lead chromate and oxide of copper in a stream of air free from CO_2. The carbonic acid given off was collected in a soda-lime tube, and found to weigh 0·0596 gramme, that is, 9·59 per cent. As 8·89 per cent. of carbonic acid as such were found in the substance, this gives us 0·70 per cent. of carbonic acid due to organic carbon, or 0·20 per cent. of organic carbon.

Y. *Estimation of* MnO *and Alkalies.*

5·3893 grammes of the same sample of air-dried Loch Fyne nodules were boiled with water as before, filtered, and the filtrate evaporated. The residue, which crystallised out when dried at 110° C., weighed 0·1240 gramme, that is, 2·30 per cent. of the nodules were found soluble in water. The insoluble part was dried at 100° C., and separated from the filter-paper. The latter was then ignited, and the ash added to the rest. The whole was then air-dried, and this air-dried mass was used for analysis.

A. *Estimation of Moisture (driven off at* 150°–160° *C.).* 0·3609 gramme were heated in a small bath, and the water driven off absorbed by a calcium chloride tube. The heating continued for over half an hour. The substance was found to have lost 0·0194 gramme (5·53 per cent.), while the calcium chloride gained 0·0183 gramme. The moisture in *X* is 1·47 per cent. and in *Y* 5·53 per cent.; the difference is 4·06. Therefore, 95·94 parts of *Y* are equivalent to 100 parts of *X*, and we must multiply the results in *Y* by 1·042 to make them comparable with those in *X*.

B. *Estimation of Alkalies.* 3·7014 grammes were dissolved in hydrochloric acid, and the insoluble part removed as usual. The solution was made up to 250 c.c.; 100 c.c. of this solution was used for the estimation of the alkalies. In the first place, however, the barium hydrate and ammonium carbonate to be used in the separation were tested for their purity. 5·518 grammes of barium hydrate were dissolved in water, and carbonic acid gas passed through the solution to saturation. The whole was then boiled and filtered. The barium carbonate precipitate was dried and roughly weighed. It weighed from 2·95 to 3 grammes, *i.e.*, about 47·6 per cent. of $Ba(OH)_2$ in the barium hydrate used, the rest being water. The filtrate from the barium carbonate was evaporated nearly to dryness in a platinum basin, a little barium carbonate which settled out was filtered off, and the filtrate evaporated to dryness. The residue was redissolved in water, refiltered, evaporated to dryness again, and weighed. It weighed 0·0017 gramme, *i.e.*, 0·031 per cent. of alkalies as chloride in the baryta hydrate. *N.B.*—This hydrate of baryta had previously been recrystallised.

The *ammonium carbonate* was then tested. It was a solution of 1 part of salt in 5 of water. 5 c.c. of it were evaporated to dryness in a platinum basin, and the residue ignited, and then moistened with hydrochloric acid and ignited again. It weighed 0·0012 gramme, therefore 1 c.c. of the ammonium carbonate solution leaves 0·00024 gramme of residue.

To the above-mentioned 100 c.c. solution (containing 1·4805 grammes of substance) 4·088 grammes of barium hydrate were added. It dissolved completely, and the whole was allowed to stand over-night. Next morning carbonic acid gas was passed through the turbid solution to saturation, and the solution boiled to get rid of excess of carbonic acid gas; it was then filtered. The *precipitate* was dried and transferred to the original flask in which precipitation took place, and the ignited filter ash added.

The *filtrate* was evaporated down to a bulk of from 25 to 50 c.c., and the barium and calcium present precipitated by the addition of 5·5 c.c. of ammonium carbonate. This precipitate was dried and added, along with its filter ash, to the flask in which were the other bases and excess of barium carbonate. The reason for doing this was, that on evaporation of the filtrate from the bases, a further small quantity of a dark-brown precipitate (MnO_2 most likely) came down.

The filtrate from the ammonium carbonate precipitate was evaporated to dryness in a platinum basin, the ammonium salts driven off, and redissolved in water. A good deal of grey powder, probably barium carbonate, remained undissolved. This was added to the precipitate of bases mentioned above. The filtrate was evaporated to dryness and ignited. It weighed 0·0279 gramme. This, then, is the weight of alkalies as chlorides. When dissolved in water it left a mere trace of insoluble matter, coloured with carbon. This was filtered off, ignited, and weighed. It weighed 0·0004 gramme.

Thus, to find the weight of alkalies as chlorides in the 1·4805 grammes of nodules used, we have to subtract from 0·0279 gramme 0·00132 gramme (residue left by 5·5 c.c. ammonium carbonate solution) + 0·00127 (which is the residue in 4·088 grammes of barium hydrate) + 0·0004 (insoluble matter). This leaves 0·02491 gramme (or 1·68 per cent.) of alkalies weighed as chlorides.

The *potassium* in the above alkaline salts was estimated by means of platinic chloride. 0·01356 gramme of the double chloride of potassium and platinum ($2KCl \cdot PtCl_4$) were found—that is, 0·01356 gramme of potassium chloride in 1·4805 grammes of nodules, or 0·91 per cent. of KCl. This is equivalent to 0·45 per cent. of K_2O, leaving 0·41 per cent. of Na_2O.

The error due to the ratio of potash to soda in the nodules being different from that in the reagents can be but trifling.

C. *Analysis of the Precipitate of Bases obtained in separating the Alkalies.* Dissolved this precipitate in hydrochloric acid, and made up the solution to 250·85 c.c.

Estimation of Manganese. Took 100 c.c. of the above solution (0·5902 gramme nodules). Evaporated to dryness to get rid of excess of acid, took up the residue with water, and, as usual, added barium carbonate to precipitate iron and alumina. In the filtrate precipitated the manganese, as usual, as MnO_2 by means of bromine, filtered off the MnO_2, dissolved it in hydrochloric acid, removed any baryta as $BaSO_4$, evaporated solution to dryness, took up with water, and precipitated the manganese with sodium carbonate as $MnCO_3$. Filtered, ignited to Mn_3O_4, and weighed. The Mn_3O_4 weighed 0·19646 gramme (or 33·29 per cent.). This is equivalent to 30·96 per cent. of MnO (including cobalt) in the nodules.

$BaCO_3$ *Precipitate.* Dried this precipitate, redissolved it in hydrochloric acid, and precipitated the baryta with sulphuric acid. Evaporated the solution over water-bath to get rid of excess of acid. Took up with water, filtered from traces of insoluble matter, diluted, and precipitated with a slight excess of ammonia, boiling till all excess was expelled. The precipitate of $Fe_2O_3 + Al_2O_3$ when ignited weighed 0·0266 gramme. Fused it with bisulphate of potash in order to see if silica was present. 0·00194 gramme of the fused mass was insoluble, leaving 0·0246 gramme of $Fe_2O_3 + Al_2O_3$, *i.e.*, 4·17 per cent.

To make the figures obtained for the alkalies and manganese comparable with those obtained in *X*, we have in both cases used substances extracted with water, and therefore free from soluble matter; but in *X* the substance was used dried at 100° C., while in *Y* it was exposed to the air after extraction, and was therefore in an air-dry condition. *X* contained 1·47 per cent. moisture, which was driven off at 162° C.; *Y* contained 5·53 per cent., the difference being 4·06. Therefore 95·94 parts of *Y* are equivalent to 100 parts of *X*, and we must multiply the results in *Y* by 1·042, and we have MnO = 32·23 per cent., K_2O = 0·47 per cent., and Na_2O = 0·43 per cent.

Some of the same air-dried samples of nodules were examined without being extracted with boiling water, or rather the extraction formed part of the analysis.

A. *Estimation of Moisture driven off at* 180° *C.* 0·8078 gramme of nodules (unextracted) were heated in an air-bath to 180° C. The heating was continued for about three hours, two hours being between 160° and 180°. The loss of weight was found to be 0·0363 gramme (4·50 per cent.).

B. *Estimation of Matter soluble in Water.* 8·8078 grammes

were (after the moisture had been driven off) extracted with hot water. The solution was evaporated to dryness, ignited, and weighed. It weighed 0·0169 gramme, that is, 2·09 per cent. of the nodules.

The dry residue was redissolved and tested for *potash*, and found to contain a little.

C. *Estimation of Insoluble Residue, Manganese, Iron and Alumina.* Treated what was insoluble in water of the above 0·8078 gramme with hydrochloric acid, and separated the silica and silicates as usual. The insoluble residue weighed 0·26106 gramme, equal to 32·32 per cent.

The *filtrate* from the insoluble residue was made up to 100 c.c. 50 c.c. of the solution (equal to 0·4039 gramme of nodules) were used for manganese estimation, and 0·1349 gramme of Mn_3O_4 found. This gives 31·11 per cent. of MnO.

The *iron and alumina* were precipitated by barium carbonate in 100 c.c. of the solution. (It was the filtrate from this precipitate that was used for the manganese estimation.) The precipitate was redissolved in hydrochloric acid, the baryta present precipitated with sulphuric acid, and iron and alumina precipitated from the filtrate from the barium sulphate. 0·0146 gramme of $Fe_2O_3 + Al_2O_3$ was found, representing 3·60 per cent. in the nodules. This was fused with bisulphate of potash, and found free from silica.

To make the figures comparable with those of *X*, we have 2·09 soluble matter and 4·05 moisture, together 6·14; from which we deduct 1·47 moisture in *X*, leaving 4·67; whence 95·33 of *Z* are equivalent to 100 of *X*. We must therefore multiply the figures obtained in *Z* by 1·049. We have then MnO = 32·62 per cent., insoluble residue = 33·90 per cent., and $Fe_2O_3 + Al_2O_3$ = 3·77 per cent. After making all reductions we have—

Insoluble Residue. In *X* 34·05 per cent., and in *Z* 33·90 per cent. As the soluble silica was determined in *X*, and the value of the total residue in *Z* agrees closely, we take the values for *X*, namely, soluble SiO_2 6·32 per cent., and remainder of residue 27·73 per cent.; together, 34·05 per cent.

Fe_2O_3 *and* Al_2O_3. As the figure found in *Z* for the sum of these constituents agrees sufficiently with that found in *X*, we take the value for the individual constituents found in it, namely, Fe_2O_3, 2·31 per cent., and Al_2O_3, 1·14 per cent.

Manganese. The results obtained in *Y* and *Z* agree well, and we take the mean, 32·43 per cent. of MnO.

Loss on Ignition. The amount of substance taken for this determination was 0·1263 gramme. Recent experiments show that it would lose about three-sevenths of its available oxygen on ignition, and it may therefore be safely concluded that all the CO_2 would be driven off, also all the water, and two-thirds of the available oxygen. We have then

Moisture removed at 162° C.	1·47	per cent.
CO_2	8·87	,,
Three sevenths available oxygen	1·76	,,
Balance	5·64	,,
Total loss found	17·74	,,

We have then the following composition:

Moisture	1·47	per cent.
Carbonic acid	8·89	,,
Available oxygen	4·10	,,
MnO	32·43	,,
Co_2O_3	0·025	,,
CuO	traces	
Fe_2O_3	2·31	per cent.
Al_2O_3	1·14	,,
CaO	6·83	,,
MgO	4·08	,,
K_2O	0·47	,,
Na_2O	0·43	,,
P_2O_5	0·57	,,
Soluble SiO_2	6·32	,,
Balance of insoluble residue	27·73	,,
Do., loss on ignition	5·64	,,
	102·43	,,

If we assign the CO_2 first to the CaO and then to the MgO, we shall have

$CaCo_3$	12·20	per cent.
$MgCO_3$	5·17	,,
MgO	2·41	,,

Analysis of the Mud in which the Nodules from Loch Fyne were found.

The mud was air-dried, and had been freed from manganese nodules and shells. The moisture, carbonic acid, available oxygen, manganous oxide, and insoluble residue were determined, with the following results:

Moisture at 162° C.	5·92 per cent.
Carbonic acid	2·55 ,,
Available oxygen	merest trace
MnO	0·71 per cent.
Insoluble residue	74·26 ,,
CuO	distinct traces.

With regard to the available oxygen, 0·5750 gramme were taken, and, after boiling with strong HCl as usual, the solution of potassium iodide remained colourless. On adding a little starch solution to it, however, a slight purple tint was produced, which the smallest drop of hyposulphite sufficed to decolourise. There is, therefore, only the smallest possible trace of MnO_2 present, and the 0·71 per cent. MnO is really present as MnO.

Analysis of Pecten Shells picked out of the Loch Fyne Mud.

Qualitative Analysis. These shells were well scrubbed with pure water, so as to free them from all adhering mud, and then air-dried. They contained—

Organic matter	A little.
CO_2	A large amount.
SO_3	Small amount.
SiO_2	Very small amount.

No metals precipitable by H_2S in HCl solution. Very small precipitate with sulphide of ammonium—

CaO	Very large amount (no BaO).
MgO	Trace.

Qualitative Estimation of MnO *and* P_2O_3. 1·5695 grammes were treated with dilute HCl. There was very little organic matter present, which shows that the shells were very old. The insoluble residue was filtered, ignited, and weighed; it weighed 0·0063 gramme = 0·4 per cent. insoluble residue. To the filtrate from the insoluble residue added Fe_2Cl_6 and precipitated all the P_2O_5 with acetate of soda. In the filtrate from this the manganese was precipitated with bromine. The following results were obtained:

Insoluble residue	0·40 per cent.
P_2O_5	0·09 ,,
MnO	1·00 ,,

Referring to Table II, page 180, it is apparent that, while the percentage of insoluble residue in the oceanic and the littoral nodules stands in the proportion 2 : 3, the percentage of silica in these residues is almost identical. This is particularly remarkable when it is borne in mind that the insoluble residue of the Loch Fyne nodules consists principally of quartz sand, and that in the nodules and muds of the great ocean the occurrence of quartz is very exceptional, and then only in districts frequented by icebergs.

No. 10. [*From The American Journal of Science, Vol.* XXI. *January*, 1906.]

ON A METHOD OF DETERMINING THE SPECIFIC GRAVITY OF SOLUBLE SALTS BY DISPLACEMENT IN THEIR OWN MOTHER-LIQUOR; AND ITS APPLICATION IN THE CASE OF THE ALKALINE HALIDES[1]

DURING the summer of 1904 I was occupied with the determination of the specific gravity of various saline solutions by the hydrometric method, which I designed for use on board the "Challenger" and have perfected in the course of years, since that date. The most important condition of success with this method is to operate always at the same temperature, and during an operation to keep that temperature perfectly constant. The temperature which I used was 19·5° C., both because a great quantity of similar work has been done at this temperature and because, in the room which it was my privilege to occupy in the Davy-Faraday Laboratory, this temperature is one which is very easily produced and kept constant, so long as the temperature of the air outside does not exceed it. This work was put a stop to by the arrival of the great anti-cyclone of the summer of 1904 which persisted over northern Europe for nearly six weeks and produced tropical conditions, which were evidenced alike by the high temperature of the air and by its insignificant diurnal variation.

In these circumstances I decided to make use of the time by putting into practice a method of determining the specific gravity of soluble salts which I have long intended to try.

The specific gravity of an insoluble substance is determined by the amount of distilled water which a known weight of it displaces. In the case of soluble salts it has been the custom

[1] Read at the meeting of the Chemical Society of London on 6th April, 1905.

to replace the water by a hydrocarbon or mineral oil. The objections to the use of this liquid are numerous, especially when the salt, the specific gravity of which it is desired to determine, is rare or costly. Moreover, to judge by the want of agreement among the values of the specific gravity of the same salt found by different chemists, there is greater uncertainty about the numerical results than there should be. One reason for this may be that the salts are not insoluble, but only sparingly soluble in the oil, and that sufficient attention has not been given to this point.

There is one liquid in which every soluble salt is quite insoluble, and that is its own mother-liquor at the temperature at which the one parted from the other. By immersing the salt in its own mother-liquor at the temperature of what we may call its birth, and by making the maintenance of this temperature a *conditio sine quâ non* of every manipulation during which the two are brought together again, errors due to uncertain solubility are eliminated, and contamination of valuable preparations is avoided. It is therefore by the immersion of each salt in its own mother-liquor that I determine its displacement; and this, combined with the weight of the salt and the specific gravity of the mother-liquor, gives the specific gravity of the salt.

It is obvious that the method is applicable only to salts which *have* a mother-liquor, such as KCl; RbBr; $CaCl_2$,-$6H_2O$; $BaCl_2 2H_2O$; it is inapplicable to salts such as $CaCl_2$; $BaCl_2$; and the like, which have no legitimate mother-liquor.

The anticyclonic meteorological conditions which prevailed during the greater part of July and August 1904 were very favourable to this class of work. The anticyclone began to give way when the work was nearly finished, and it was evident that, in the absence of artificial arrangements for the preservation of a constant laboratory temperature, this class of work cannot be carried on easily or satisfactorily except in the hottest summer weather.

It is an essential condition of success that the work be carried on in a room, for the time being, especially devoted to the purpose, and occupied by one investigator. He must have in it everything that he requires, including his balance. The window of the room must face the north, and the precautions generally to be observed are similar to those prescribed by Bunsen for the practice of his gasometric method.

The salts used in this research were the chlorides, bromides and iodides of potassium, rubidium and caesium. The rubidium and caesium preparations were from the works of Schuchardt in Goerlitz and were of the highest degree of purity. The potassium salts were also unexceptionable as regards quality and were supplied by Merck. All of these salts dissolve easily, and most of them abundantly, in water. They also crystallise with great readiness.

The first operation is to prepare a hot solution of the salt such that, after standing over night, or for such length of time as may be deemed sufficient, it shall furnish about 60 c.c. of mother-liquor and about 15 c.c. of crystals. In the case of the potassium salts there was no difficulty, as their solubility at all temperatures is well known. The solubility of the rubidium and caesium salts had to be determined, at least approximately, in each case, in order to economise the costly material. The following simple method furnished the required information easily and expeditiously. A suitable vessel, beaker or flask, is weighed empty, and then with 25 grammes of distilled water, of the temperature of the air. The salt is then gradually added and the mixture stirred with the thermometer. In the case of every one of these salts the temperature falls rapidly and by as much as from 15° to 20°. The salt is added as rapidly as it is taken up by the water. When the fall of temperature slackens, a minimum is soon reached, while some salt still remains undissolved at the bottom of the vessel. It is then continually stirred; the temperature rises slowly while the salt gradually passes into solution, until, at a certain temperature the amount of salt remaining undissolved is such that a further rise of one degree of temperature will evidently cause it to disappear. The vessel is now weighed and, as a result, we have the weight of salt dissolved in 25 grammes of water at about the last observed temperature. With a little care it is easy to arrange that this temperature shall be in the neighbourhood of that of the air. The vessel with its contents is now heated, and salt added by degrees, while the temperature rises and finally reaches the boiling point or whatever other temperature may have been determined on. Salt is added until the liquid is saturated at this temperature. The vessel is again weighed and the salt dissolved at the higher temperature is ascertained. These simple experiments, which are completed in very few minutes, furnish all the information that is required for the economical employment of the material. In the absence of more detailed information the following results obtained in the above way are worth quoting:

100 GRAMMES OF WATER DISSOLVE

Grammes	98	164	264	225	51	157	93	121	156	222
of	RbBr	RbI	RbI	CsCl	CsI	CsI	CsBr	CsBr	CsBr	CsBr
at °C.	12	20	boiling	25	12	107	7·5	24·5	50	93·5

With this information there is no difficulty in preparing the solution which shall, after allowing for unavoidable loss in preparation, give the required amount of mother-liquor and of crystals. The water is warmed and the pure salt is added while the temperature is raised to that of ebullition, or to any lower temperature that may have been selected. When the salt has all passed into solution, the liquid is poured into a flat crystallising dish and crystallisation begins immediately. The area of the dish should be such that the layer of solution shall not be more than half a centimetre thick. The mother-liquor is then everywhere in close touch with the crystals. The dish is then put away in a cupboard for the night.

In the morning, the temperature of the contents of the crystallising dish and that of the air were taken very carefully. The mother-liquor was then poured off clear into a stoppered bottle, while the crystals were collected, allowed to drain, and dried in the ordinary way. The temperature which the mixture had when separated is noted as that at which the crystals and the mother-liquor were in equilibrium; and it is exactly at this temperature that they have to be brought together again in order to determine the specific gravity of the salt. It is at this temperature also that the specific gravity bottle is weighed when filled with distilled water and with mother-liquor respectively. In fact the temperature of equilibrium and of separation is the only temperature used.

In Table I (p. 202) the experimental details are given in full in the case of one salt, namely, caesium chloride. For the other salts the results only are given, and they are collected in Table II (p. 206).

All the weights given in this paper represent the weight *in vacuo*.

The specific gravity bottle which was used was one of the common and convenient form which has a thermometer for a stopper and a lateral capillary tube for the adjustment of level. Its nominal capacity was fifty cubic centimetres. On three occasions one of 25 c.c. capacity was used for determining the displacement of the mother-liquor.

The concentration (m) of the mother-liquor is determined by titration with tenth-normal silver nitrate solution. This solution was made with the greatest care and contained exactly 17 grammes of silver nitrate in one litre, at the ordinary temperature of the laboratory at the time. The burette used was divided into tenths of a cubic centimetre and had a capacity of 50 c.c. The determination of the halogen was not made until the specific gravity had been determined, and, if the concentration was not already known within narrow limits, a preliminary titration was made, after which the volume of mother-liquor was weighed, which would certainly require 40 ± 1 c.c. for titration. The capacity of the burette from 0 to 40 c.c. was determined by weight with great care. The concentration is stated in gram-molecules salt per 1000 grammes of water.

For weighing out the salt and passing it directly into the specific gravity bottle a special and convenient form of weighing tube was used. It was made out of a stoppered specimen tube with an internal diameter of two centimetres and a length of seven or eight centimetres. The lower end of this tube was opened and a piece of thin glass tube joined to it before the blowpipe. This tube, which had a length of about three

TABLE I. *Experimental Details in the case of Caesium Chloride.*

Formula and molecular weight of salt	MR	CsCl = 168·5
Temperature	T	23·1° C.
MOTHER-LIQUOR. *Determination of Specific Gravity.*		
Weight of specific gravity bottle	w_1	38·8900 gms.
Weight of specific gravity bottle filled with distilled water	w_2	89·2399
Weight of water which fills it .... $w_2 - w_1$	w_3	50·3499
Weight of sp. gr. bottle + mother-liquor	w_4	135·0620
Weight of mother-liquor .... $w_4 - w_1$	w_5	96·1720
Specific gravity of mother-liquor .... $\frac{w_5}{w_3}$	S	1·9104
Analysis.		
Weight of mother-liquor taken	w_6	1·0334 gms.
Cb. cents. $\frac{7}{10}$ $AgNO_3$ solution used	w_7	41·21 c.c.
Weight of salt equivalent to silver used ... $\frac{w_7 + MR}{10000}$	w_8	0·6944 gms.
Weight of water in w_6 gms. mother-liquor.... $w_6 - w_8$	w_9	0·3390
Concentration of mother-liquor expressed in gm. molecules salt per thousand gms. water 0·1 .... $\frac{w_7}{w_9}$	m	12·1563
SALT IN CRYSTAL. *Determination of Specific Gravity.*		
A. Weight of first portion of salt	w_{10}	22·1229 gms.
Weight of sp. gr. bot. + salt + mother-liquor	w_{11}	146·5514
Weight of mother-liquor .... $w_{11} - (w_1 + w_{10})$	w_{12}	85·5385
Weight of water displaced by mother-liquor $\frac{w_{12}}{S}$	w_{13}	44·7828
Weight of water displaced by salt .... $w_3 - w_{13}$	w_{14}	5·5671
Specific gravity of salt .... $\frac{w_{10}}{w_{14}}$	D_1	3·9739
B. Weight of second portion of salt	w_{15}	26·6220 gms.
Sum of weights of the two portions .... $w_{10} + w_{15}$	w_{16}	48·7449
Weight of sp. gr. bot. + salt + mother-liquor	w_{17}	160·4249
Weight of mother-liquor .... $w_{17} - (w_1 + w_{16})$	w_{18}	72·7900
Weight of water displaced by mother-liquor .. $\frac{w_{18}}{S}$	w_{19}	38·1085
Weight of water displaced by salt ...... $w_3 - w_{19}$	w_{20}	12·2414
Specific gravity of salt .... $\frac{w_{16}}{w_{20}}$	D_2	3·9820
C. Weight of water displaced by salt ..... $w_{20} - w_{14}$	w_{21}	6·6743 gms.
Specific gravity of salt .... $\frac{w_{15}}{w_{21}}$	D_3	3·9890
Accepted specific gravity of salt	D	3·982

centimetres, had an external diameter such that it could just pass freely through the neck of the specific gravity bottle. The wide end was closed with a glass stopper and the narrow end with a small india rubber cork.

It was the custom to work so as to have about 15 c.c. of dry salt to be added in two charges to the specific gravity bottle. These charges were intended to be nearly, though not quite, equal. The available supply was distributed between two weighing tubes by approximate weight, after which the exact weight of each portion was determined in the usual way. The two portions of caesium chloride weighed respectively 22·1229 and 26·6220 grammes, so that in the first determination of specific gravity 22·1229 grammes and in the second 48·7449 grammes were concerned. It is not immaterial whether the first portion is charged into the empty specific gravity bottle and the mother-liquor poured over the dry powder, or is charged into the bottle which is already about half full of mother-liquor. In the former case the elimination of the entangled air is difficult and takes time, during which it is not easy to prevent the temperature getting out of hand. By the latter process very little air is carried past the surface of the liquid and very little stirring with the thermometer, which is required on other grounds, suffices to eliminate it.

Owing to the readiness with which these salts crystallise and to the slowness with which all salts dissolve in an almost saturated solution, the temperature of the mixture of salt and mother-liquor, during the adjustment of level in the specific gravity bottle, must on no account be permitted to fall below T by even 0·01°, nor should it be allowed to rise above it by more than 0·1°. The regulation of temperature was effected entirely with a standard thermometer divided into tenths of a degree, each tenth occupying a length of rather more than one millimetre on the stem. The thermometer which forms part of the specific gravity bottle is used chiefly as a stopper of convenient form. So soon as the level of the liquid has been adjusted in the bottle, it is weighed. The temperature and pressure of the air are kept account of for the reduction of all weights to the vacuum.

When the first weighing has been completed, about 20 or 25 c.c. of the clear mother-liquor are drawn off and the second charge of dry salt is added and mixed, after which the level is adjusted, and the weight determined. In the absence of experience it might be thought that it would be difficult to draw off so much of the liquid without some of the solid salt; but no matter how much they may be stirred up, these crystallised salts settle at once and completely to the bottom when immersed in their saturated solutions, and the operation presents no difficulty. It was at first intended to make a series of three determinations with each salt, but two were found to be sufficient. During all these manipulations the temperature of the air in the laboratory never differed from that of crystallisation ($T = 23{\cdot}1^\circ$ C.) by more than one or two tenths of a degree, and it is only in such conditions that operations of this kind can be carried out successfully.

Before bringing the crystals together with the mother-liquor in the specific gravity bottle, the operator must realise that their common temperature when mixed is to be as nearly as possible exactly that of crystallisation or equilibrium (T); and he must take such measures as his experience dictates to arrive at this end. Preliminary experiments on a somewhat extensive scale are absolutely necessary, and the success of an operation depends almost entirely on the competence of the operator and on the trouble that he is prepared to take.

Table II gives for each salt, MR, the temperature, T, of equilibrium between crystals and mother-liquor, and, in condensed form, the experimental *data* of the determination of S, the specific gravity at T of the mother-liquor, that of water at the same temperature being unity; of m, the concentration of the mother-liquor in gram-molecules salt per 1000 grammes water, and of D_1, D_2, D_3, the three observed values, as well as D, the finally accepted value, of the specific gravity of the salt, all at T and referred to that of water at the same temperature as unity.

The letters and suffixes have the same significance as in Table I.

The figures in line T show how uniform the temperature

was during the period over which the experiments were spread. All the experiments were made between the 12th and 22nd of July, 1904, with the exception of those on caesium bromide, which were made on August 10th. By that time the anti-cyclone had begun to break and the value of T for this salt is 21·4° C. For all the other salts, T lies between 22·8° and 24·3° C.

During the whole of the period the barometer was very steady, varying between 758 and 761 millimetres, and the relative humidity of the air varied between 40 and 50 per cent.

Of the three values D_1, D_2, D_3 for the specific gravity of the salt, D_1 is obtained directly from the first portion of the salt, D_2 from the sum of the two portions, and D_3 is derived from D_1 and D_2 by subtraction.

D_2 represents very nearly the mean of D_1 and D_3 and is the accepted value for the majority of the salts. It is expressed to three places of decimals, of which units in the second place are exact.

It will be noticed that in the case of rubidium chloride the value of D_1 is accepted. The second determination depends on the approximate weight of the second portion of salt when the tube was being filled, the exact weighing on the balance of precision having been accidentally omitted. The operation was however completed, and the calculation made with the approximate weight was used as a control. The result shows that the value of D_1 may be safely accepted. In case of potassium chloride the value of D_3 (1·951) is accepted, and the reason for this is as follows: The first portion of salt was in very coarse powder, and in mixing it with the mother-liquor numerous crystalline particles were observed which contained gaseous enclosures, easily perceptible by the naked eye. As was expected, the observed specific gravity proved to be low. The second portion was much more finely powdered and the specific gravity resulting from the two was higher (1·8872). But this result is affected to the full extent by the gaseous enclosures in the first portion. We therefore calculate the specific gravity from the second portion alone, which gives 1·9510 for the specific gravity.

Discussion. It is an advantage of the method just described

TABLE II. *Experimental Results regarding each Salt in the Ennead.*

Salt: Formula *MR* / Salt: Mol. weight	KCl.	KBr.	KI.	RbCl.	RbBr.	RbI.	CsCl.	CsBr.	CsI.
	74·6	119·1	166·1	121·0	165·5	212·5	168·5	213·0	260·0
Temperature, T	23·4°	23·4°	24·3°	22·9°	23·0°	24·3°	23·1°	21·4°	22·8°
Specific Gravity	MOTHER-LIQUOR								
Weight taken, gms. w_5	59·4068	34·3044	85·9636	74·7356	81·3282	46·2696	96·1720	42·3756	78·0087
Displacement, gms. w_3	50·3524	24·9554	49·9140	49·9188	49·9196	24·9478	50·3499	24·9744	50·3658
Specific gravity, $\frac{w_5}{w}=S$	1·1798	1·3746	1·7222	1·4971	1·6292	1·8548	1·9101	1·6968	1·5488
Concentration									
Gm.-mols. p. 1000 gm. H_2O. M	4·7619	5·7250	8·9344	7·7670	6·7229	8·2307	12·1563	5·3057	3·5454
Specific Gravity	SALT IN CRYSTAL								
A. Weight of salt, gms. w_{10}	13·3684	36·7928	27·1751	19·0112	27·0906	26·4777	22·1229	27·8926	26·3890
Displacement, gms. w_{14}	7·3271	13·7498	8·9703	7·0256	8·4700	7·7248	5·5671	6·2453	5·8545
Specific gravity, $\frac{w_{10}}{w_{14}}=D_1$	1·8245	2·672	3·0295	2·706	3·198	3·428	3·974	4·466	4·5075
B. Weight of salt, gms. w_{12}	27·4258	52·5142	52·1768	43·7750	51·5438	50·6025	48·7449	57·5390	53·3916
Displacement, gms. w_{20}	14·5322	19·6005	17·1465	15·9627	16·0568	14·7658	12·2414	12·9466	11·8423
Specific gravity, $\frac{w_{16}}{w_{20}}=D_2$	1·887	2·679	3·043	(2·74)	3·210	3·428	3·982	4·455	4·5085
C. Weight of salt, gms. w_{15}	14·0574	15·7214	25·0017	(24·76)	24·4532	24·1248	26·6220	29·6464	27·0026
Displacement, gms. $w_{20}-w_{14}=w_{21}$	7·2051	5·8501	8·1762	8·9371	7·5868	7·0410	6·6743	6·7013	5·9878
Specific gravity, $\frac{w_{15}}{w_{21}}=D_3$	1·951	2·688	3·058	(2·77)	3·223	3·426	3·989	4·424	4·509
Accepted specific gravity, D	1·951	2·679	3·043	2·706	3·210	3·428	3·982	4·455	4·508

that it furnishes more than the mere determination of the specific gravity of the salt. Thus, by ascertaining almost simultaneously the specific gravity of the mother-liquor and the displacement in it of the crystals, both being at the temperature of equilibrium, *data* are obtained for *the determination of the relation between the displacement of the salt in crystal and its apparent displacement in saturated solution at that temperature.* It has not hitherto been permissible to make exact comparisons of this kind owing to the independence of the observations on the salt and on the solution, which have been available.

In discussing the results of observation it is convenient to arrange them in a more articulate form than that of Table II so as to bring each feature forward prominently and by itself.

The group of salts which forms the subject of these experiments is one of the most remarkable in nature. The salts are nine in number and include all the possible binary combinations of the members of the electro-positive triad K, Rb, Cs with those of the electro-negative triad Cl, Br, I. The two triads of simple bodies make three triads, or one *ennead*[1] of binary compounds. The relations of the different members of the ennead are best shown in a table of the form of Table III. In it the salts of the same metal, *M*, are all in one column, and those of the same metalloid, *R*, all in one line. The symbol *MR* represents both the formula and the molecular weight of the salt.

TABLE III. *Values of MR.*

	K	Rb	Cs	
(*a*)		Formula		
	KCl	RbCl	CsCl	Cl
	KBr	RbBr	CsBr	Br
	KI	RbI	CsI	I
(*b*)		Molecular weight		
	74·6	121·0	168·5	Cl
	119·1	165·5	213·0	Br
	166·1	212·5	260·0	I

Compartment (*a*) of Table III contains the formula and compartment (*b*) the molecular weight of each salt. The

[1] From the Greek ἐννεάς, which signifies a body of nine.

latter is the fundamental attribute of a substance, on which all its properties depend. The molecular weights of the salts which occur in one column differ by the amount of the difference of the atomic weights of the metalloids which they contain, that is, by 44·5 or 47. Similarly, contiguous salts in one line have molecular weights which differ by 46·4 or 47·5. If we consider the two diagonal triads in the ennead, we see that they are characterised by the fact that both the elements in each unit are different from those in either of the other units. Further, along the diagonal KCl--CsI the molecular weights of the units differ as much as possible from each other, while the atomic weights of the components of each unit are as nearly as possible identical, being close neighbours in the atomic series. On the other diagonal, KI–CsCl, the molecular weights of the units agree with each other as nearly as possible, while the atomic weights of the constituents of the units differ from each other as much as possible.

The Crystal. Table IV contains four compartments. In the first (*a*) we have the values of *T*, the temperature at which the crystals and mother-liquor of each salt were in equilibrium, and that at which the various displacements were observed.

Under the experimental conditions, which have been minutely described above, it is impossible to fix in advance the exact temperature of equilibrium of the crystallising liquid. This is given by the meteorological conditions, modified by the structural features of the laboratory and of the apartment or enclosure where crystallisation takes place.

In the second compartment (*b*) we have the values of *D*, or the specific gravity of the salt in crystal at *T*, referred to that of distilled water of the same temperature as unity. The *data* in this compartment are in most cases for different, but always neighbouring temperatures. The differences of the values of *T* are however so small and those of *D* are so great that we may discuss the specific gravities as if they had been made at one common temperature.

On examining the values of *D*, we see that they increase with those of *MR* in Table III; but the increase is not continuous, it is remittent. It takes place *triad-wise*; and this holds whether we take the triads in column or in line. Comparing salts in the same line, we see that replacing Rb by Cs causes a rise of specific gravity which is twice as great as that caused by the substitution of Rb for K. Comparing salts in the same column, the replacement of Cl by Br causes more than double the rise caused by the substitution of I for Br. However we regard it, we see that *the specific gravity of the salts is a periodic function of their molecular weight, within the ennead.*

TABLE IV. *The Salt in Crystal.*

	K	Rb	Cs	
(*a*)		Values of T (°C)		
	23·4°	22·9°	23·1°	Cl
	23·4°	23·0°	21·4°	Br
	24·3°	24·3°	22·8°	I
(*b*)		Values of D		
	1·951	2·706	3·982	Cl
	2·679	3·210	4·455	Br
	3·043	3·428	4·508	I
(*c*)		Values of $\frac{MR}{D}$		
	38·233	44·710	42·310	Cl
	44·460	51·553	47·820	Br
	54·580	61·986	57·670	I
(*d*)		Values of $\frac{MR}{18D}$		
	2·124	2·489	2·350	Cl
	2·470	2·864	2·657	Br
	3·032	3·444	3·204	I

In the third compartment (*c*) we have the values of $\frac{MR}{D}$ or the displacement of one molecule (MR) of salt stated in grammes of water, and in compartment (*d*) the same constant is stated in gram-molecules of water $\left(\frac{MR}{18D}\right)$. In dealing with the specific gravities, we saw that, whether we follow the columns or the lines, they increase with increase of molecular weight. In the case of the molecular displacements this holds for the columns but not for the lines. In these the salts of rubidium have the greatest molecular displacement, the potassium salts have the least, and the caesium salts occupy an intermediate position.

As we shall see later, this irregularity is due to a specific peculiarity of the caesium salts. Meantime it may be noted that the *volumetric equivalent* of one gram-molecule of any of the salts of the ennead varies from 2·124 H_2O to 3·204 H_2O, the iodides having the highest and the chlorides the lowest equivalents. The average difference between the volumetric equivalents of the iodides and bromides is 0·563 H_2O, and that between the bromides and chlorides is 0·343 H_2O.

The Mother-liquor. The values of T are the same for the mother-liquor as for the crystals, and are presented in Table IV (*a*). In Table V (*a*) we have the values of m or the molecular concentration of the mother-liquor. This is expressed in gram-molecules salt per 1000 grammes water. It represents also with great exactness the solubility of the salt in water at T, and we shall consider it for a moment from this point of view.

TABLE V. *The Salt in Mother-liquor.*

	K	Rb	Cs	
(*a*)		Values of m		
	4·7619	7·7670	12·1563	Cl
	5·7250	6·7229	5·3057	Br
	8·9344	8·2307	3·5454	I
(*b*)		Values of S		
	1·1798	1·4971	1·9101	Cl
	1·3746	1·6292	1·6968	Br
	1·7222	1·8548	1·5488	I
(*c*)		Values of $\frac{v}{m}$		
	31·223	38·069	49·021	Cl
	39·038	44·131	48·137	Br
	49·506	58·575	67·907	I
(*d*)		Values of $\frac{v}{18m}$		
	1·735	2·115	2·723	Cl
	2·169	2·452	2·674	Br
	2·750	3·257	3·773	I

The least soluble of the nine is caesium iodide, which has the highest molecular weight, and potassium chloride, which has the lowest molecular weight, comes next to it. Next to caesium iodide, in molecular weight and in solubility, we have caesium bromide; and, similarly, next to potassium chloride, in molecular weight and in solubility, we have potassium bromide. In the latter case the solubility increases with the molecular weight, while in the former it decreases with it. But, if Table III be referred to, it will be observed that, as regards molecular weight, KCl and CsI occupy singular positions in the ennead. On the other hand, KBr (119·1) and RbCl (121) have almost identical molecular weights, as have also CsBr (213) and RbI (212·5), yet the solubilities in each pair respectively are very different. The lowest solubilities are on the diagonal KCl–CsI and the highest solubilities on the diagonal KI–CsCl. RbBr, which occupies the middle place on both these diagonals, is also in the middle of the middle column and of the middle line, and is the centre of the ennead. Its solubility, besides being nearly the average of the group, has a symmetrical position with respect to those of the other salts. On one diagonal the solubility of its neighbours is lower, on the other higher than its own. In its column the solubility of its neighbours is higher, in its line it is lower than its own.

In compartment (*b*) of Table V we have the values of S, the specific gravity of the mother-liquor at T, referred to that of distilled water of the same temperature as unity. These numbers cannot, as they stand, be compared with each other because they refer to solutions of such different concentrations. They enable us, however, to arrive at the average apparent displacement of one gram-molecule of salt in the saturated solution which contains 1000 grammes of water at T. Thus, taking again caesium chloride as an example, we have for the weight of salt dissolved in 1000 grammes of water

$$w = \text{m. CsCl} = 2048{\cdot}34 \text{ grammes.}$$

Adding 1000 grammes to this we have for the weight of the solution

$$W = 1000 + w = 3048{\cdot}34 \text{ grammes.}$$

The specific gravity (S) being 1·9101, the displacement of the solution is

$$V = \frac{W}{S} = 1595{\cdot}92 \text{ grammes of water,}$$

whence the gross apparent displacement of the salt in solution is

$$v = V - 1000 = 595{\cdot}92 \text{ grammes,}$$

and the mean apparent displacement per molecule is

$$\frac{v}{m} = 49{\cdot}021 \text{ grammes.}$$

In compartment (*c*) we have the value of $\frac{v}{m}$ for each member of the ennead. This expresses, in grammes of water, the average apparent displacement of one gram-molecule of salt in its saturated solution at T. In compartment (*d*) the same constant is expressed in terms of gram-molecules of water $\left(\frac{v}{18m}\right)$.

Before commenting on the numbers in the table, it is important to form a clear conception of their physical meaning. We shall best arrive at this by returning to our detailed example of chloride of caesium. As the quantity of saturated solution which contains 1000 grammes of water weighs 3048·34 grammes and displaces 1595·92 grammes of water, we may imagine it to have been prepared in the following way: 1595·92 grams of water are taken and caesium chloride is dissolved in it so that each portion, as it is added, forms a saturated solution with the exact quantity of water which it requires for this purpose and the remainder of the water remains uncontaminated. Parallel with the dissolution of the salt, pure water is removed at such a rate as to keep the displacement or bulk of the liquid always the same. When no more salt will dissolve we have a saturated solution which contains 1000 grammes of water. The weight of caesium chloride which has entered the solution is 2048·34 grammes and the weight of water which has left it is 595·92 grammes, whilst the displacement of the liquid is the same at the end of the operation as it was at the beginning. In thus describing the preparation of the saturated solution, we have described an operation of substitution. *It is therefore permissible to regard solutions as products of substitution.* If we give to the above numbers their molecular interpretation, we see that the mean apparent displacement of one molecule of caesium chloride in its saturated solution at 23·1° is equal to that of 2·723 gram-molecules of water, and therefore, that, in these conditions, *CsCl is volumetrically equivalent to* 2·723 H_2O.

If we study Table V (*d*), we see that the average molecular displacement of the salts increases with their molecular weight, whether we follow the columns or the lines. The only exception is furnished by caesium bromide, the displacement of which is very slightly lower than that of caesium chloride. The greatest molecular displacement is that of caesium iodide, which has the highest molecular weight; and the least molecular displacement is that of potassium chloride, which has the lowest molecular weight. The pair, potassium bromide and rubidium chloride, which have almost equal molecular weights, have also almost equal molecular displacements. The same is true of the pair, potassium iodide and caesium chloride, but rubidium bromide has a markedly lower displacement. Finally, the pair, rubidium iodide and caesium bromide, which have almost identical molecular weights, present no resemblance in their apparent molecular displacements.

Comparison of the Displacement of the Salt in Crystal and in Mother-liquor. The molecular displacement $\frac{MR}{D}$ of the salts in crystal is given in Table IV (*c*) in terms of grammes of water; that of the salts in mother-liquor $\frac{v}{m}$ is similarly given in Table V (*c*).

If we compare these two tables, we find the remarkable result that while in the case of the potassium and the rubidium salts the figures for the displacement in crystal are greater than those for the displacement in mother-liquor, in the case of the caesium salts the reverse is the case.

In Table VI (*a*) we have the difference $\left(\frac{MR}{D} - \frac{v}{m}\right)$ of the molecular displacement of the salt in crystal from its mean molecular displacement in mother-liquor. In compartment (*b*) we have the ratio $\left(\frac{MR}{D} \cdot \frac{m}{v}\right)$ of these quantities.

Taking the figures in compartment (*a*) we see that in the

case of the salts of potassium and rubidium crystallisation is accompanied by considerable expansion, and this is what is usually met with. In the case of the caesium salts the reverse is the case, and very decidedly so in the case of the chloride and of the iodide, much less so in the case of the bromide, which, in this, as in other particulars, maintains its singular position.

TABLE VI. *The Salt in Crystal and in Mother-liquor.*

	K	Rb	Cs	
(*a*)		Values of $\frac{MR}{D} - \frac{v}{m}$		
	7·010	6·641	− 6·711	Cl
	5·422	7·422	− 0·817	Br
	5·074	3·411	− 10·237	I
(*b*)		Values of $\frac{MR}{D} \cdot \frac{m}{v}$		
	1·225	1·175	0·863	Cl
	1·139	1·168	0·993	Br
	1·103	1·059	0·849	I

In this connection it should be noted that among the ratios $\left(\frac{MR}{D} \cdot \frac{m}{v}\right)$ given in compartment (*b*), the two which are nearest to unity are those for RbI (1·059) and for CsBr (0·993) respectively; and their molecular weights are almost identical. Further, the salts situated *co-diagonally* to them, namely RbBr and CsI, have ratios whose difference from unity are almost equal, namely + 0·168 for RbBr and − 0·151 for CsI.

Taking a general view of the figures in (*b*) which give the ratios of displacement in crystal and in mother-liquor, we see great differences. The most striking examples are, as in the case of solubility, the extreme members of the ennead KCl and CsI. The former expands by more than 25 per cent., and the latter contracts by 15 per cent. on crystallising.

These figures accentuate the peculiarity of the caesium salts, that crystallisation is accompanied by contraction. An interesting conclusion can be drawn from the behaviour of the different salts in this respect, namely, that *the crystallisation of the potassium and rubidium salts of the ennead must be hindered by increased pressure, while that of the caesium salts must be helped by the same agency.*

Conclusion. The method of determining the specific gravity of a soluble salt in its own mother-liquor, as described in the first part of the paper, involves manipulations of too delicate a character to permit it to pass into general practice in competition with other methods for the same primary purpose. When, however, the specific gravity of the salt has been ascertained in this way, the relation between its apparent displacement in the state of crystal and in that of saturated solution have been ascertained at the same time. In the second part of the paper the observations are discussed from this point of view, but owing to exigencies of space the discussion has been limited to the accentuation of the salient features. One of the most important of these is the connection which reveals itself between the molecular weight of the salts and their specific gravity and displacement in crystal and in saturated solution, in definite conditions. The authority of the periodic law makes itself as clearly felt in the limited area of the ennead as it does in the realm of the elements. It is true that the caesium salts introduce some irregularity into the periodicity, but this is not to be looked on as an exception, but as an interference, the nature of which it will be interesting to trace[1].

[1] See Contents, p. xxxiii.

No. 11. [*From the Encyclopaedia Britannica, 9th edition.*]

THE MEDITERRANEAN SEA[1]

THE southern shores of Europe are separated from the northern shores of Africa by the Mediterranean Sea. It extends in a generally east and west direction from longitude 5° 21′ W. to 36° 10′ E. Its length from Gibraltar to its eastern extremity in Syria is about 2100 nautical miles. Its breadth is very various, being 400 miles from the mouth of the Rhone to the Algerian coast, 500 miles from the Gulf of Sidra to the entrance to the Adriatic, and 250 miles from the mouth of the Nile to the south coast of Asia Minor. From the very indented nature of its coasts, the general mass of the water is much cut up into separate seas, which have long borne distinctive names, as the Adriatic, the Ægean, the Sea of Marmora, the Black Sea, etc. The area of the whole system including the Sea of Azoff is given by Admiral Smythe as 1,149,287 square miles. If we deduct that of the Black Sea and Sea of Azov, 172,506 square miles, we have for the area of the Mediterranean proper 976,781, or, roughly speaking, a million of square miles.

The Mediterranean is sharply divided into two great principal basins, the western and the eastern or Levant basin. The western possesses a comparatively smooth and unindented coast-line. It is bounded on the south by the coast of Africa and the north coast of Sicily, and it is further enclosed by the coasts of Spain, France, and Italy, which form a roughly arc-shaped coast-line. There are comparatively few small islands in this basin, though some of the more important large ones occur in it. The eastern basin is by far the larger of the two, and extends from Cape Bon to the Syrian coast, including as important branches the Adriatic and the Ægean. The latter is connected directly through the Hellespont, the Sea of Marmora, and the

[1] See Contents, p. xxxiv.

Bosphorus, with the Black Sea. The entrance to the western basin and to the sea generally from the ocean is through the Straits of Gibraltar in 36° N. latitude. If this parallel be drawn out through the sea it will be found that the western basin lies almost wholly to the northward, and the main body of the eastern one to the southward of it, the mean latitude of the western basin being about 39° 30′, and that of the eastern basin 35°. They communicate with each other by the channels separating Sicily from Italy and from Africa. The former is known as the Strait of Messina, and is of insignificant size, the latter is a wide channel apparently without any distinctive name, and generally shallow. The greatest depth on the shallowest ridge reaching from the African to the Sicilian coast is under 200 fathoms, and agrees very closely with the corresponding depth at the entrance to the Straits of Gibraltar.

Depth. So far as is at present known, the maximum depth is pretty nearly alike in the two basins, being 2040 fathoms in the western and 2150 fathoms in the eastern. Many lines of soundings have been run in the Mediterranean for telegraph purposes, and they afford a very good idea of the general configuration of the bottom. Between Marseilles and Algiers the depth ranges generally from 1200 to 1600 fathoms; between Naples and Sardinia from 1500 to 2000; between Alexandria and Rhodes from 1200 to 1600; and between Alexandria and Cyprus from 900 to 1100. The basin of the Mediterranean really begins about 50 miles to the westward of Gibraltar. It is here that the shallowest ridge stretches across from Africa to Spain; the maximum depth on it is probably not more than 180, and certainly less than 200 fathoms. From this ridge the bottom slopes quickly westward into the depths of the Atlantic, and gently eastward into the Mediterranean. The depth nowhere reaches 1000 fathoms until beyond Alboran Island, 120 miles east of Gibraltar. This is a small low island separated from the mainland on all sides by water of more than 400 fathoms; it must therefore be considered an oceanic as distinguished from a continental island[1]. Further to the

[1] Continental islands are those separated from the mainland by comparatively shallow seas, generally under 100 fathoms.

north, and off the coast of Valencia, we have the Balearic Islands,—namely, Majorca, Minorca, Iviza, and Formentera. These also must be considered oceanic islands, and indeed two groups of oceanic islands. Iviza and Formentera are isolated both from the Spanish coast and from the other two islands by water of over 300 fathoms depth; Majorca and Minorca are connected by a bank with no more than 50 fathoms of water on it. Thirty miles east of Minorca there are more than 1400 fathoms; beyond that there are no soundings between the Baleares and the large and important group of Corsica and Sardinia. These islands are continental, being connected with the Italian mainland by the bank on which Elba occurs, and which is covered by little over 50 fathoms of water. The Straits of Bonifacio, which separate Corsica from Sardinia, are also quite shallow, so that Corsica and Sardinia may be looked on as a secondary peninsula attached to the Tuscan shore of Italy by a shallow bank not more than 15 or 20 miles broad, the deep water coming close up all round it. Almost the same may be said of Sicily, including the Malta group, but excluding the Lipari group, which is purely volcanic. From Cape Passaro, in the south-east end of Sicily, a line can be drawn connecting it with the town of Tripoli, and without passing over water of more than 300 fathoms. As has already been said, the west end of Sicily is connected with the coast of Tunis by a ridge in no part covered by more than 200 fathoms of water. Between these two ridges lies a small but comparatively deep basin of 600 to 700 fathoms. At the western extremity of it lies the mountainous island of Pantellaria. The bank on which Malta is situated stretches for nearly 100 miles in a southerly direction from Cape Passaro in Sicily. Opposite, on the African shore, is a similar bank of much larger dimensions, on which are the small islands Lampion and Lampedusa, belonging to Italy. In the deep channel between them and Malta is the small but lofty island Limosa. It is entirely volcanic, with an extinct crater on its north-eastern side, and three smaller ones to the southward. It resembles the Lipari group off the north coast of Sicily, which rise abruptly out of deep water, being connected by no bank either with the African or the Sicilian coasts. Some of the Lipari group are still active, Stromboli and Vulcano being of the number. Off the south coast of Sicily, and between it and the island of Pantellaria, occurs the famous Graham's shoal, the remains of what was for a few weeks an island[1].

The deepest water of the Mediterranean is found in its widest part between Malta and Crete, and the deep water comes close up to the Italian and Greek coasts, while on the African shore the water shoals more gradually. In the Strait of Messina, close to Reggio, there are depths of over 500 fathoms, and similar depths are found inside gulfs, such as those of Taranto (nearly 1000 fathoms), of Corinth, Kalamata, and others. Also all through the Ægean in its many bights and channels very deep water is met with; in the Sea of Marmora we have 500 fathoms, and in the Black Sea over 1000 fathoms. All along the south coast of Asia Minor the water is very deep, and the large islands of Cyprus and Crete are both separated by very deep water from the mainland. If we take the eastern basin, and run along its western and southern coasts from the mouth of the Po along the shore of Italy, Sicily, and Africa to the mouth of the Nile, and even further along the Syrian shore, we do not find a single off-lying island of any importance except the Malta group, while all along the eastern and northern coasts from Trieste to Asia Minor the coast is deeply indented, and the water broken up by many large and important islands. These islands are grouped along the west coast of Turkey and Greece, and irregularly throughout the Ægean. The east coast of the Adriatic is studded with islands and inlets, and resembles in this respect the Ægean; the west coast, on the other hand, is

[1] With regard to its appearance and disappearance Admiral Smythe (*Mediterranean*, p. 111) says: "It seems that, as early as the 28th of June 1831, Captain Swinburne, in passing nearly over the spot, felt several shocks of a sea-quake, proving that the cause was then in operation; but on the 19th of the following July the crater had accumulated to a few feet above the level of the sea, and was in a great activity, emitting vast volumes of steam, ashes, and scoriæ. From that time it gradually increased in all its dimensions till towards the end of August its circumference was about 3240 feet and its height 107; then from October various changes took place, and it entirely disappeared in December." Since that time it has changed considerably. In 1863 the least water on it was 15 feet. It has two heads close together, and at the distance of about 20 yards all round there are from seven to nine fathoms of water.

low, and the water off it shallow, and there are few harbours. The Adriatic stretches in a north-westerly direction for about 460 miles from its entrance between Cape Sta Maria di Leuca and the island of Corfu to the Venetian shore in the Gulf of Trieste. Its average width is about 100 miles. A ridge with little over 400 fathoms appears to run across its entrance. Inside this the water reaches a depth of 765 fathoms, but shoals again rapidly towards Pelagosa Island, from which to the northward, including quite two-thirds of the sea, the depth is under 100 fathoms; indeed no part of the sea within 150 miles of its northern extremity is over 50 fathoms deep. There is authentic historical evidence of the encroachment of the Italian shores on the Adriatic, causing thereby a diminution of its area. As a consequence many towns which were once thriving seaports are now many miles inland; thus Adria, which was a station of the Roman fleet, is now 15 miles inland, and there are many similar examples. The large rivers Po and Adige, which bring the drainage of the southern slopes of the Alps to the sea, deliver large quantities of sediment in the course of the year. The distribution of this mud is affected, not only by its own weight tending to make it sink to the bottom, but also by the set of the currents, which, running up the eastern coast, turn to the westward and southward at the upper end of the sea, and so tend to distribute the river mud along the bottom in the neighbourhood of the Italian coasts. The fact that towns which were formerly seaports are now inland does not therefore necessitate the assumption of a general rise of the land, it is merely a reclamation by natural agencies of land from the sea at the expense of the inland mountainous country. Precisely similar phenomena are observed in the neighbourhood of the mouths of the Rhone and of the Nile.

Specific Gravity, Currents, etc. On the specific gravity Dr Carpenter reports many and interesting observations. In round numbers, that of the surface-water of the Atlantic off the Straits of Gibraltar is 1·0260 to 1·0270, that of the western basin of the Mediterranean 1·0280 to 1·0290, and that of the eastern basin 1·0290 to 1·0300, while that of the Black Sea is

1·0120 to 1·0140. It will thus be seen that the water of the Mediterranean proper is very much sal'ter than either the Atlantic on the west or the Black Sea on the east, and this great density of the water affords a useful means of recognising it when investigating the interchange of waters which takes place at the two extremities of the sea. Both the temperature and the specific gravity of the water are evidences of the local climate. The great concentration of the water shows how dry the atmosphere at the surface must be, and how insignificant the contributions of fresh water. With regard to the balance existing between the two factors, evaporation and precipitation, it would be impossible to give figures with any claim to accuracy, but a rough estimate may be formed by taking such data as Fischer has given. He puts the rainfall over the whole Mediterranean drainage area at 759·4 millimetres, or almost exactly 30 inches. If we remember that the average rainfall of the eastern slopes of Great Britain is less than 30 inches, and that therefore this may be taken as the maximum yearly supply to the North Sea, we may be sure that the Mediterranean does not receive more than 30 inches of fresh water in the year With regard to the rate of evaporation over the area of the Mediterranean there is but very meagre information, but wherever it has been observed it has been found to exceed the rainfall, even as much as three times. Thus at Madrid it is 65 inches, or more than four times the rainfall, at Rome 105 inches, and at Cairo 92 inches. It may therefore without exaggeration be assumed that the evaporation is at least twice as great as the precipitation. Putting the latter at 30 inches, we should have 60 inches for the yearly evaporation, and a balance of 30 inches evaporation over precipitation. Were there no provision for making good this deficiency, the level of the Mediterranean would sink until its surface was so far contracted as to lose no more by evaporation than would be supplied by rain and rivers. This condition would probably not be fulfilled before all the Ægean and Adriatic and the whole of the western basin west of the island of Sardinia were laid dry, and what is now the Mediterranean would be reduced to two "Dead Seas," one between Sardinia and Naples and the other between Africa and the mouth of the Adriatic. That the level and the salinity of the Mediterranean remain constant is due to the supply of water which enters at the Straits of Gibraltar. The currents in this passage have frequently engaged attention both from their scientific and their nautical interest. The most detailed investigation was that carried out by Captain Nares and Dr Carpenter in H.M.S. "Shearwater" in the year 1871[1]. From these investigations it appears that there are usually two currents in the Straits at the same time, one superposed on the other. Both are affected by tidal influence, but, after allowing for it, there is still a balance of inflow in the upper and of outflow in the under current. The waters of the two currents are sharply distinguished from each other by their salinity. Further, the upper current appears to affect by preference the middle of the channel and the African coast, while the under current appears to crop out at the surface on the Spanish coast. This distribution, however, is much modified by the state of the tide, and it must be remembered that in such places the surface separating the upper and under currents is rarely, if ever, a horizontal plane. That there is a balance of outflow over inflow at the bottom was well shown by the result of soundings as much as 200 miles north-west of the entrance of the Straits, where, in a depth of 1560 fathoms, water of decided Mediterranean origin was got from the bottom. There can be no doubt that this outflow of warm and dense Mediterranean water is largely instrumental in causing the comparatively very high bottom temperature in the eastern basin of the North Atlantic.

We have assumed that the balance of water removed by evaporation is 30 inches, or 2·5 feet. If we take the area of the Mediterranean to be 1,000,000 square miles, we have the volume of water removed:

$$v = 2{\cdot}5 \times 36 \times 10^{12} = 90 \times 10^{12} \text{ cubic feet.}$$

This quantity of water has to be supplied from the Atlantic without raising the total quantity of salt in the sea. We have seen that the only provision for the removal of the surplus

[1] *Proc. Roy. Soc.* (1872), XX. 97, 414.

salt is the outward under current in the Straits[1]. Hence the inward upper current must be sufficient to replace the water lost both by evaporation and by the outflow of the under current. We may take the Atlantic water to contain 3·6 per cent. and the Mediterranean to contain 3·9 per cent. of salt. In order that the under current may remove exactly as much salt as is brought in by the upper one, their volumes must be in the inverse ratio of their saline contents, or the volume of the upper current must be to that of the under one in the ratio 39 : 36 or 1000 : 923; so that only 7·7 per cent. of the inflow goes to replace the water removed by evaporation, while the remaining 92·3 per cent. replaces the water of the under current. We have then for the total volume of the inward current per annum:

$$v = \frac{100}{7 \cdot 7} v = 1170 \times 10^{12} \text{ cubic feet.}$$

The width of the Straits from Tarifa to Point Cires is eight miles, or 48,000 feet, and the average depth of the stream may fairly be taken as 100 fathoms; hence the sectional area is in round numbers 29,000,000 square feet.

Dividing the volume by the area, we have for the mean annual flow:

$$R = \frac{1170 \times 10^{12}}{29 \times 10^{6}} = 40 \times 10^{6} \text{ feet.}$$

Reducing this to miles per day, we find that if the above data are correct the inflowing current at the Straits of Gibraltar ought to be equivalent to a current eight miles wide, 100 fathoms deep, and running with the uniform velocity of 18·3 miles in twenty-four hours. As the currents are reversed with the tides this is the balance of inflow over outflow in the upper current. It is worthy of remark that the flood tide runs to the westward at the surface and the ebb to the eastward. The following table of tides at places inside and outside the Straits will show that the mere differences of level due to the different tidal ranges at adjacent localities are sufficient to cause strong local currents.

[1] See Contents, p. xxxv.

A similar phenomenon is witnessed at the other end of the sea. Here the fresher waters of the Black Sea rush in through the narrow channel of the Dardanelles, causing a surface flow of comparatively fresh water, while there is an outflow below of denser Mediterranean water. The dimensions of the Straits are too small to make the phenomenon of any importance for the supply of the Mediterranean. The conditions both in the Dardanelles and in the Bosphorus were examined very

Places	High water. Full and change		Springs rise		Neaps rise		Neaps range	
	h.	m.	ft.	in.	ft.	in.	ft.	in.
Chipiona	1	30	12	5	8	0	3	6
Rota	1	24	12	6	8	0	3	6
Cadiz	1	23	12	9	8	2	—	
Conil	1	18	12	0	7	5	3	3
Cape Plata	1	45	8	0	5	3	2	6
Tarifa	1	46	6	0	3	6	1	3
Algesiras	1	49	3	9	2	6	1	3
Gibraltar	1	47	4	9	2	7	1	3
Ceuta	2	6	3	7	2	5	1	3
Tetuan	2	23	2	6	1	6	0	6
Tangier	1	42	8	3	5	1	2	0
Rabat	1	46	11	0	7	1	3	3
Mogador	1	18	12	4	8	0	3	6

carefully in the year 1872 by Captain Wharton, R.N., of H.M.S. "Shearwater," and his results are published in an interesting report to the Admiralty of that date. It is remarkable that the comparatively fresh water of the Black Sea persists without sensible mixture through the Sea of Marmora and into the Dardanelles, while there is constantly a current of Mediterranean water running underneath, and the depth in the two channels is only from 30 to 50 fathoms. There can be little doubt that the saltness of the Black Sea is due wholly to the return current of Mediterranean water entering through the Bosphorus. Were the exit of the Black Sea a channel with sufficient fall to bring the surface of the Sea of Marmora below the level of the highest part of its bottom, so that no return current could take place, the waters of the Black Sea would be fresh.

In the body of the sea the rise and fall are much less than at any of the places in the above table. At Algiers a self-recording tide gauge was set up by Aimé in 1848, and from its records he deduced a rise and fall of 88 millim. (say $3\frac{1}{2}$ in.) at springs and half that amount at neaps, a fluctuation which would escape ordinary observation, as it would be masked by the effects of atmospheric disturbances. At Venice and in the upper reaches of the Adriatic, the true lunisolar tide seems to be more accentuated than in other parts; but here also its effects are subordinate to those of the wind. In summer the Mediterranean is within the northern limit of the north-east trade wind; consequently, throughout a great part of the year, the winds are tolerably constant in direction; and, blowing as they do over large areas of water, they are instrumental in moving large masses of it from one point to another, and so producing streams and currents.

The effect of wind on a surface of water is twofold: it produces the rhythmic motion of waves and the motion of translation of currents. Besides the motion produced by the direct action of the wind on the surface water there are currents due, in the first instance, to the accumulation of water produced by a wind which has been blowing constantly in one direction. The phenomenon of an abnormally high tidal rise with a gale of wind blowing on shore is one with which inhabitants of the British Islands are familiar. It is also a matter of frequent observation that, for instance, a south-west gale which exaggerates the height of high water on the western coasts of Britain reduces it on the east coasts. It blows the water on the west coast and off the east coast, so that the difference in the high-water levels on the two coasts is very pronounced. Supposing free communication were quickly made between the two coasts, a current would be the result, and its violence would be much greater than would be due to the local action of the wind on its surface. In the Mediterranean the winds blow during a great portion of the year very constantly from one direction or another and generally from north and east. The extent of the sea is so great that the slope produced by the transference of the surface water constantly in one direction might have a sine or arc capable of being measured in feet and inches when the radius is as much as 200 miles long. Thus at Port Mahon, in the island of Minorca, according to the *Admiralty Sailing Directions*, the water rises and falls according to the direction of the wind. With wind from south-east or south-west the water rises, but from north-west or north-east it falls. When northerly or north-westerly winds prevail, and this is the case for two-thirds of the year, a strong current sets to the south-west off Ayre Island, which is reversed in seasons when south-westerly winds prevail. This current is due to the water escaping round the end of Minorca having been driven southward so as to raise a head on the north coasts of the island. Similarly, in the Faro or Strait of Messina the currents, of which the famous Scylla and Charybdis are swirls or eddies, are the evidence of a tendency towards equalizing the levels of the eastern extremity of the western basin and of the western extremity of the eastern basin. In addition to this peculiarity of position with reference to the two basins, it has been found that there is a very strong purely tidal influence at work which alone produces an alteration in the direction of the currents, and thus adds to the confusion of the waters. At Capo di Faro the rise is scarcely perceptible, at Messina it may attain a maximum of 10 to 13 inches. In the Straits of Bonifacio, between Corsica and Sardinia, the currents follow entirely the direction of the prevailing winds, and are at times very rapid. In the channel between Sicily and the African coast the currents also follow the winds. In long periods of calm weather a steady easterly set is observed, no doubt a prolongation or reproduction of the Gibraltar current.

Temperature. Nothing whatever was known of the temperature of the deep water of the Mediterranean until Saussure extended to it his classical investigation into that of the Swiss lakes. In October 1780 he sank his thermometer to a depth of 160 fathoms off Genoa and of 320 fathoms off Nice, and at both depths he found the temperature of the water to be 55·8° F. These observations have a special value, for, owing to Saussure's method of experimenting, his results were not affected by the pressure obtaining at great depths in the sea. Fifty years elapsed before any similar experiments were made, when D'Urville, in the "Astrolabe," made a few observations at the beginning and the end of his famous expedition. There is some uncertainty about his observations in 1826 and 1829, and also about the later ones of Bérard in 1831, as we are not informed whether the self-registering instruments used were protected from pressure or not. Mr Prestwich[1], however, who has collected and critically discussed all the older deep-sea temperature observations, concludes from a comparison of their results with those obtained by Aimé with protected instruments, that they were so protected, and admits their results into his tables without correction. In the deep water to the northward of the Balearic Islands, D'Urville found in April 1826 54·5° F. in 270 fathoms, and in March 1829 54·7° at the same depth, and the same temperature (54·7°) in 530 fathoms. Bérard, experimenting in the sea between the Balearic Islands and Algeria, found the temperature of the deep water nearly a degree higher, namely 55·4° F., in depths of 500 to 1000 fathoms. Aimé[2] relates his own careful experiments on the temperature of both surface and deeper water in the neighbourhood of Algiers, and discusses them in connection with those of other observers with very great ability. He concludes from his own observations and those of Bérard that the uniform temperature at great depths is 54·86° F. From a consideration of the general climate of the Mediterranean, he comes to the conclusion that the temperature in the deeper layers of the sea ought to be lower than the annual mean of the surface, and that it ought to be not very different from the mean surface temperature in the winter months. From observations at Toulon and Algiers, he finds that at neither place does the surface temperature fall below 50° F., and that the mean surface temperatures in the months December, January, February, March, and April is at Toulon 53·06° F. and at Algiers 56·84° F. The mean of these two temperatures is 54·9° F., which is almost exactly what he finds to be the mean

[1] *Phil. Trans.* 1875, part ii. p. 601.
[2] *Ann. Chem. et Phys.*, 145, XV. p. 5.

annual temperature of the deepest water of the western basin. During the forty years which have elapsed since Aimé made his experiments and speculations, further observations have only tended to confirm his theory. It is true that the temperatures observed in the many soundings which have been made of late years have not shown absolute identity of temperature, and it is probable that the greater the refinement in the instruments used the more decided will the local differences appear. Especially it will be apparent that the bottom temperature varies with the climate of the preceding winter, and the distribution of temperature varies much with the prevalence of the winds. At the few stations where the temperature of the sea-water and that of the air are regularly examined, it appears that the water is generally for the greater part of the year warmer than the air, and in winter considerably so. The existing observations, however, are too few to justify any very definite statement on the subject. At Palermo the sea is warmer than the air throughout the whole year with the exception of the months May and June. In Algiers Aimé found but little difference; in autumn and winter the water was slightly warmer, in spring and summer slightly colder, than the air. In the eastern basin we have first Admiral Spratt's observations in July 1845 in Ægina Gulf. In all his experiments made previous to the year 1860 he determined the temperature of the bottom water by taking that of the mud brought up in the dredge. This is a very excellent method; in fact, it is probably the best of all methods if a sufficient quantity of mud be obtained. From 1860 he used self-registering unprotected thermometers, which gave results necessarily too high, and it is impossible to apply any reliable correction to them without experimentally determining it on each thermometer which was used. By the first method Admiral Spratt found 55·5° F. at depths between 100 and 200 fathoms.

From these observations it seemed reasonable to conclude, as Aimé had done, that all over the Mediterranean a practically uniform temperature is found at all depths greater than 100 or 200 fathoms, and that this temperature is 54° to 56° F. In order thoroughly to investigate this matter, as well as the biological conditions of the deep water of the Mediterranean, H.M.S. "Porcupine," Captain Calver, with Messrs Carpenter and Gwyn Jeffreys, visited the western basin of the Mediterranean in the autumn of 1870. A large number of temperature observations were made in the western basin near its southern coasts, and one sounding with temperature observation in the eastern basin a short distance from the Sicilian coast, the result of which was to confirm the conclusion arrived at from earlier observations, that, however high the temperature of the surface may be (and it may reach 90° F.), the water becomes rapidly cooler as we go below the surface until we reach a depth of about 100 fathoms, where a temperature of 54° to 56° F. is found, and persists without sensible variation to the greatest depths. The average of all the bottom temperatures in the western basin was 54·88° F. Three soundings were made in the intermediate basin to the eastward of Pantellaria in depths of 266, 390, and 445 fathoms, and in each case the bottom temperature was found to be 56·5° F., or about a degree and a half warmer than in the deeper western basin. This is precisely what might have been expected from what we know of inland seas divided into several basins. In summer the shallower basin has usually a higher temperature at the bottom than is found at the same depth in the deeper one. Only one observation was made in the eastern basin, namely off Cape Passaro in 1743 fathoms, with a bottom temperature of 56·0° F. That the temperature in this basin should be lower than in the Pantellaria basin is due to its greater depth, and that it should be higher than is found in the western basin is due to its lower latitude. These researches were further prosecuted in the autumn of 1871 in the "Shearwater," Captain Nares, accompanied by Dr Carpenter. At two stations in the eastern basin "serial temperatures" were taken. At the first, 35° 54′ N. latitude, 16° 23′ E. longitude, depth 1650 fathoms, the bottom temperature was 56°, or the same as had been observed the year before in 1743 fathoms; at the second, 32° 17½′ N. latitude, 26° 44′ E. longitude, depth 1870 fathoms, the bottom temperature was 56·7°, and the temperature at all intermediate depths was much higher than at the first station The mean temperature of the water from the surface to a depth of 200 fathoms was, at the first station 63·75° F., and at the second 66·78° F., or three degrees higher. At the first station all the temperatures down to 100 fathoms are higher than were observed in 1870 in the western basin, but it must be remembered that temperature observations made in different years cannot with justice be closely compared, as the climates of the two years are sure to differ considerably, and in the present case the difference in climate between the summers of 1870 and 1871 appears to have been very considerable.

In the autumn of 1881 a very interesting series of observations were made by Captain Magnaghi, hydrographer of the Italian navy, and Professor Giglioli, on board the Italian surveying ship "Washington," in that part of the western basin which is enclosed between the islands Corsica and Sardinia on the one side and the Italian coasts on the other. It is here that the deepest water of the western basin was found; and, apart from the great interest attaching to the physical results obtained, the collections made with the dredge in the comparatively lifeless waters were of the very highest importance, showing, as they did, a practical identity in the abyssal fauna with that of the open ocean. This is the more remarkable as we have hitherto been accustomed to consider the similarity in the fauna of portions of the ocean remotely distant from each other as being due to the likeness of their temperatures. In the Mediterranean, however, the bottom temperature is quite 20° F. higher than is found in great depths anywhere in the open ocean.

For determining the temperature of the deep water Captain Magnaghi used the half-turn reversing thermometer of Negretti and Zambra, which in itself is a very beautiful instrument. The mechanical arrangement, however, for reversing, even as improved by Magnaghi, was not so satisfactory, and from certain irregularities in the temperature observations reported the writer is inclined to think that some of the remarkable results obtained, for instance on the 11th August, are due to this instrumental imperfection. On that day the water at

70 metres was found to have a temperature of 25·1° C., while that at 50 metres was 20·1° C., and that at 90 metres was 16·7° C. The results obtained in the deep water are no doubt quite reliable, for the temperature is so uniform that a few fathoms more or less in the depth at which it turned would make no difference in the temperature registered. In the more northern parts of this portion of the western basin, off the coast of Corsica, we find a practically uniform temperature from 250 metres down to the bottom in 2800 metres, the mean bottom temperature being 55·96° F. Further to the south the temperature of the abyssal water appears to be distinctly higher. Thus between the south end of Sardinia and the Bay of Naples, in the deepest water, the practical uniformity of temperature is not reached until a depth of 1000 metres has been passed, and it is there 56·7° F. It is unfortunate that we do not know what the bottom temperature in other parts of the Mediterranean was. In this summer of 1881 it was quite one degree higher than that observed by Dr Carpenter in 1870.

The great value of such a volume of water as an equalizer of temperature on its shores must be apparent, though in this respect it is inferior to the Atlantic Ocean in its immediate neighbourhood. Places on the west coast of Spain and Portugal have a much higher winter temperature and lower summer temperature than places in the same latitude in Italy. The reason of this is simple: on the Atlantic coast the principal winds in winter are from the south-west, and have a warming effect, while in summer the source of the north-east trade wind is pushed back into the Bay of Biscay, causing in this season constant northerly winds along the coast of Portugal. The winds of the Mediterranean have no seas of remote latitudes to draw on either for heating or cooling purposes, though the sandy deserts of Africa which bound its southern coasts have at certain seasons a very decided influence on the climate. The tempering action of the sea does not extend very far inland, as is evident from the climate of inland towns in Italy. As the Mediterranean shores have so much importance as health-resorts, the data presented in the following table are of interest. They are taken chiefly from Theobald Fischer's *Studien über das Klima der Mittelmeerlander.*

TABLE *of mean January temperature (J.), of mean temperature of three winter months, December, January, and February (W.), also Rainfall (R.) in the same three months, for places on the Mediterranean, with those for some others for comparison.*

Place	J.	W.	R.	Place	J.	W.	R.
	° F.	° F.	In.		° F.	° F.	In.
Bilbao	46·4	48·1	14·2	Ajaccio	50·45	52·2	8·0
Oporto	49·5	50·54	23·1	Trieste	39·95	41·3	7·8
Lisbon	50·54	50·9	11·3	Corfu	50·45	51·28	22·5
Tarifa	52·88	53·6	9·6	Athens	47·57	49·2	5·7
Gibraltar	54·0	54·5	12·4	Constantinople	40·28	41·6	10·1
Malaga	54·0	57·38		Jerusalem	48·74	49·1	12·7
Valencia	50·72	52·52	4·3	Port Said	57·38	57·2	
Mahon	51·62	52·52		Cairo	56·84	58·1	
Barcelona	48·02	49·64	4·1	Alexandria	60·0	60·1	5·6
Montpellier	42·08	43·16	9·3	Suez	56·3	56·58	
Marseilles	43·52	46·04	5·0	Tunis	53·06	55·76	
Nice	46·94	48·92	8·5	Algiers	59·18	59·65	14·4
Mentone	48·2	48·2		Oran	51·8	52·76	9·0
San Remo	47·48	48·38	8·0				
Genoa	46·4	47·66	13·0	S. Cruz (Teneriffe)	63·84	64·65	
Turin	32·0	35·06	4·6	Funchal (Madeira)	60·40	61·60	
Milan	32·9	35·42	8·0				
Venice	36·86	39·4		Valentia (Ireland)	45·00	44·60	18·3
Florence	41·54	43·16	12·0	Scilly	44·90	45·76	12·1
Rome	45·7	46·6	9·9	Jersey	41·70	42·97	10·1
Naples	48·2	49·3	10·5	Ventnor	41·80	42·60	
Catania	51·62	52·7	7·5	Pembroke	41·10	41·97	12·2
Palermo	51·62	52·7	8·8	Monach (Hebrides)	42·90	43·27	15·9
Malta	54·5	56·0	17·5	St Kilda	44·70	44·50	

Nature of the Bottom. In the western basin the bottom consists chiefly of clay of a grey to brownish colour. Without doubt, when freshly collected, the surface layer is reddish-brown and the lower ones dark grey. There is always some carbonate of lime, chiefly due to *Foraminifera.* The mud very much resembles that obtained from similar depths in those parts of the open ocean whose bottom waters are shut off from free communication by ridges which may not approach within 2000 or 1500 fathoms of the surface, and with the exception of the *Foraminifera* it much resembles the mud from enclosed and comparatively shallow basins off the west coast of Scotland. In the following table the analyses are given of a few samples on the line of the submarine cable connecting Marseilles with Algiers.

Locality			Composition per cent.					
			Insoluble in HCl		Soluble in HCl			
Latitude N.	Longitude E.	Depth in fathoms	Residue	Per cent. SiO_2 in residue	$CaCo_3$	Fe_2O_3	FeO	Al_2O_3
37° 39′	3° 23′	1343	66·13	63·98	19·79	3·09	0·39	3·46
38° 11′	4° 6′	1469	39·16	79·98	38·25	2·44	0·25	10·93
39° 26′	4° 36′	782	28·13	78·98	47·50	2·21	0·20	2·54
42° 47′	5° 11′	780	48·63	70·16	31·52	2·09	0·33	4·26
43° 1′	5° 15′	265	48·04	78·60	30·80	2·40	0·36	4·58

To the student of the physical conditions of the sea the Mediterranean possesses a very high interest; its size is such as to entitle it to rank among oceans, while it is so completely cut off from the remaining world of water that it presents us with a type which is purely local, and one might almost say provincial.

No. 12. [*From Proc. Roy. Soc. Edin.* 1891, *Vol.* XVIII. *p.* 131.]

ON THE COMPOSITION OF SOME DEEP-SEA DEPOSITS FROM THE MEDITERRANEAN

THE muds, the analyses of which are reported in this paper, were collected in September 1879, during the laying of a cable between Marseilles and Algiers by the India Rubber, Gutta-Percha, and Telegraph Works Company, Limited, of Silvertown, the ship employed being the s.s. "Dacia." The numbers of the samples are those which were affixed to them on board ship. Nos. 31 to 43 are all from localities lying near the African coast; Nos. 45 and 46 are from positions between the African coast and the Balearic Bank. Nos. 64 and 65 are from the Balearic Bank, and Nos. 86 to 89 are from the Gulf of Lyons. The positions and depths are collected in Table I.

TABLE I. *Giving the Position of the Ship where each Sample of Mud was Collected, and the Depth of Water there.*

No. of sample	Position		Nature of bottom	Depth (fathoms)
	Latitude N.	Longitude E.		
31	36° 57½′	3° 21′	Soft mud	1080
32	37° 3′	3° 17′	,,	1238
35	37° 5′	3° 26′	Clayey mud	1258
36	37° 9½′	3° 23′	Mud	1343
39	37° 12′	3° 31′	,,	1454
41	37° 21′	3° 38′	Soft mud	1502
43	37° 39′	3° 53′	,,	1536
45	37° 56′	4° 6′	,,	1494
46	38° 11′	4° 6′	,,	1469
64	39° 26′	4° 36′	,,	782
65	39° 35′	4° 40′	,,	646
86	42° 47′	5° 11½′	Grey ooze	780
87	42° 53′	5° 18½′	Clayey mud	542
88	43° 3′	5° 12′	Mud and ooze	530
89	43° 1½′	5° 15′	Mud	265

The samples, as received, were in the condition in which they had been collected, having been transferred from the sounding-tube to the bottle without any form of preparation or drying. Some were therefore much wetter than others, and the diversity in their condition in this respect is well shown in the percentage column of Table II. The actual state of the mud when put up in the sample bottle on board depends on so many fortuitous circumstances that no physical importance must be attached to the figures in this Table. In order to bring all the muds, as far as possible, into a similar condition, they were heated in the water-bath until they ceased to lose weight. It was necessary, therefore, to determine their weights, and they have accordingly been tabulated, and will give roughly an idea of the difference between a wet mud and a dry one.

TABLE II. *Preparation of Samples for Analysis, by Drying on the Water-Bath.*

No. of sample	Weight of mud taken (grammes)	Weight of mud dry (grammes)	Loss (grammes)	Per cent. of loss
	a	b	$c=a-b$	$d=100\frac{c}{a}$
31	28·4	19·3	9·1	32·04
32	23·8	18·15	5·65	23·71
35	22·9	15·3	7·6	33·18
36	23·0	16·4	6·6	28·69
39	14·7	11·7	3·0	20·40
41	26·3	16·9	9·4	35·73
43	29·5	20·0	9·5	32·20
45	25·1	20·2	4·9	19·52
46	19·9	16·1	3·8	19·09
64	24·6	19·7	4·9	19·91
65	38·5	27·3	11·2	29·09
86	25·4	19·2	6·2	24·41
87	25·4	20·7	4·7	18·50
88	29·5	19·8	9·7	33·25
89	22·0	16·9	5·1	23·18

Table II. *Preparation of Substance for Analysis.* About half of the sample was placed in a tared porcelain basin and dried in the water-bath till it was in a fit form for handling. It was then weighed, and the loss of weight called water.

The dried portion was broken up in an agate mortar, and preserved in a well-stoppered bottle. Sufficient quantities of each sample were thus prepared in a uniform manner, and the bottles in which they were preserved were carefully weighed and kept under a bell-jar. In this way any alteration in the substance is at once detected. If this precaution be not taken it is necessary, in dealing with substances which are more or less hygroscopic, to weigh out at once all the portions of any one sample which will be required for the various determinations which are to be made, in order to be certain that a uniform material is used for each. This is attended with much inconvenience, which is obviated by preserving the sample in such a way that it will be unlikely to alter, and by keeping strict account of its weight, so as at once to detect any alteration which may occur.

TABLE III. *Determination of Loss on Ignition, and of the Water and Carbonic Acid expelled thereby.*

No. of sample	Weight of sample		Loss (grams)	Per cent. of loss	Weight of water absorbed by $CaCl_2$ (grams)	Per cent. of H_2O	Weight of CO_2 absorbed by soda-lime (grams)	Per cent. of CO_2	$i=f+h-d$
	Before heating (grams)	After heating (grams)							
	a	b	$c=a-b$	$d=100\frac{c}{a}$	e	$f=100\frac{e}{a}$	g	$h=100\frac{g}{a}$	i
31	0·8180	0·7905	0·0275	3·36	0·0327	3·99	0·0086	1·05	1·68
32	0·7114	0·6873	0·0241	3·38	0·0200	2·81	0·0125	1·75	1·18
35	0·8978	0·8670	0·0308	3·38	0·0200	2·22	0·0240	2·67	1·51
36	0·7337	0·7068	0·0269	3·66	0·0242	3·29	0·0120	1·63	1·26
39	0·6940	0·6702	0·0238	3·42	0·0205	2·75	0·0007	0·10	−0·57
41	1·0103	0·9724	0·0379	3·75	0·0322	3·18	0·0000	0·00	−0·57
43	0·7640	0·7363	0·0277	3·62	0·0252	3·29	0·0041	0·53	0·20
45	0·4669	0·4539	0·0130	2·90	0·0168	3·59	0·0083	1·77	2·46
46	0·5529	0·5317	0·0212	3·83	0·0156	3·00	0·0088	1·59	0·76
64	0·5160	0·4598	0·0562	10·89	0·0451	8·74	0·0084	1·62	−0·53
65	0·6972	0·6551	0·0421	6·03	0·0364	5·22	0·0090	1·29	0·48
86	0·7951	0·7500	0·0451	5·67	0·0301	5·68	0·0182	2·29	2·30
87	0·7999	0·7442	0·0557	6·96	0·0236	2·95	0·0300	3·75	−0·26
88	0·6339	0·6071	0·0268	4·22	0·0207	3·26	0·0137	2·16	1·20
89	0·6010	0·5663	0·0347	5·77	0·0220	3·66	0·0065	1·08	−1·03

Table III. *Determination of the Moisture, Carbonic Acid, and Total Loss.* A quantity of the substance was weighed into a porcelain boat, placed in a combustion tube, and heated strongly in a current of air freed from moisture and carbonic acid by passing it through a tube filled with soda-lime and another filled with calcium chloride. The water was collected in a calcium chloride tube, and the carbonic acid in a soda-lime tube, and weighed. The boat was again weighed after the heating, and the difference in weight is called total loss. In every case the mud was of a reddish colour after heating. It will be observed that in all cases the loss of weight of the substance is different from the gain of weight of the tubes; as a

TABLE IV. *Determination of Amount of Carbonic Acid.*

No. of sample	Weight of sample taken (grms.)	Weight of CO_2 absorbed by 1st soda-lime tube (grammes)	Weight of CO_2 absorbed by 2nd soda-lime tube (grammes)	Total weight of CO_2 (grammes)	Per cent. of CO_2	Per cent. of $CaCO_3$
	a	b	c	$d=b+c$	$e=100\frac{d}{a}$	f
31	1·7395	0·1396	—	0·1396	8·037	18·3
32	1·1346	0·0987	—	0·0987	8·10	18·4
35	1·5251	0·1268	0·001	0·1278	8·38	19·1
36	1·5970	0·1385	0·0006	0·1391	8·71	20·0
39	1·0160	0·1083	0·0017	0·1100	10·83	24·5
41	1·3468	0·1238	—	0·1238	9·19	21·0
43	1·5683	0·1600	—	0·1600	10·10	23·6
45	1·3200	0·1853	0·0002	0·1855	14·05	32·4
46	1·2864	0·2165	0·0009	0·2174	16·90	38·2
64	1·1379	0·2378	—	0·2378	20·90	47·1
65	1·8921	0·3129	—	0·3129	16·08	36·6
86	1·2456	0·1767	—	0·1767	14·19	32·5
87	1·4305	0·1909	—	0·1909	13·34	30·8
88	1·5087	0·2131	—	0·2131	14·12	32·5
89	1·4595	0·2024	—	0·2024	13·87	31·8

rule, it is decidedly less. In so complex a substance as a deep-sea mud it is impossible to account for this in detail; but organic matter, which is never absent from such muds, would, by its oxidation, increase the weight of the tubes at the expense of the air, while the oxidation of the fixed components, such as ferrous oxide, would have a like effect on the ignited mud. The figures, therefore, in the Table give a complex result, from which it is impossible to isolate the separate items. It is evident, from the general agreement of the figures, that the drying process (Table II) has brought the various samples into a very fairly comparable condition.

Table IV. *Determination of the Carbonic Acid.* A weighed quantity of the substance was placed in a flask and sulphuric acid added. The carbonic acid was collected in soda-lime tubes, being first dried by passing it through a U tube filled with pumice, moistened with concentrated sulphuric acid, and a tube filled with calcium chloride.

TABLE V. *Determination of the Residue Insoluble in 20 per cent.* HCl, *and of the Total Silica.*

No. of sample	Weight of sample taken (grams)	Weight of insoluble residue (grams)	Per cent. of insoluble residue	Per cent. soluble in acid	Weight of SiO_2 in insoluble residue (grams)	Per cent. of SiO_2 in sample	Per cent. of SiO_2 in insoluble residue
	a	b	$c=100\frac{b}{a}$	$100-c$	d	$e=100\frac{d}{a}$	$f=100\frac{d}{b}$
31	1·4796	0·8836	59·72	40·28	0·6456	43·64	73·06
32	1·8465	1·0952	59·31	40·69	0·8071	43·71	73·69
35	1·1713	0·4136	35·31	64·69	0·3213	27·43	78·08
36	0·8916	0·5896	66·13	33·87	0·3772	42·31	63·98
39	1·1890	0·6561	55·18	44·82	0·5276	44·37	80·41
41	1·7195	0·9598	55·81	44·19	0·7337	33·89	76·44
43	1·9403	1·0937	56·37	43·63	0·7219	32·41	66·00
45	1·3528	0·5774	42·68	57·32	0·4433	32·77	76·77
46	1·5053	0·5895	39·16	60·84	0·4715	31·32	79·98
64	1·3351	0·3755	28·13	71·87	0·2966	22·21	78·98
65	1·4602	0·4244	29·06	70·94	0·2160	14·79	50·89
86	1·7149	0·8339	48·63	51·37	0·5851	34·12	70·16
87	1·4539	0·6300	43·23	56·77	0·5349	36·79	85·10
88	1·5478	0·7622	49·25	50·75	0·5820	37·60	76·36
89	1·6682	0·8015	48·04	51·96	0·6300	37·68	78·60

The flask was boiled to expel the gas, and a current of air, free from carbonic acid, was drawn through the apparatus to sweep out all the carbonic acid.

Near the African coast the amount of CO_2 varies between eight and ten per cent. It increases with distance from the land, being no doubt less masked by land *débris*. The maximum 20·97 of CO_2 (47·17 $CaCO_3$) is found on the Balearic Bank. The depth here was only 782 fathoms, and the land drainage is insignificant.

TABLE V. *Treatment with Hydrochloric Acid.* (*a*) A weighed quantity of the substance was placed in a porcelain basin and 100 c.c. of 20 per cent. hydrochloric acid added. The basin was placed on a water-bath and evaporated to dryness. It was then placed on an air-bath and heated so as to convert any soluble silica into the insoluble form. Then it was treated with hydrochloric acid and filtered. The precipitate was weighed, and called the "residue."

(*b*) The "residue" was fused with potassium-sodium carbonate, and the silica determined in the usual way.

TABLE VA. *Determination of* Fe_2O_3 *and* Al_2O_3 *in Hydrochloric Acid Solution of Table V.*

No. of sample	Weight of sample taken (grammes)	Weight of Fe_2O_3 (grammes)	Per cent. of Fe_2O_3	Weight of Al_2O_3 (grammes)	Per cent. of Al_2O_3
	a	b	$c=100\frac{b}{a}$	d	$e=100\frac{d}{a}$
31	1·4796	0·0826	5·58	0·0203	1·37
32	1·8465	0·0991	5·37	0·0415	2·45
35	1·1713	0·0452	3·86	0·1441	12·30
36	1·5547	0·0956	6·17	0·0538	3·46
39	1·1890	0·0788	6·54	0·0460	3·87
41	1·7195	0·1060	6·64	0·0224	1·30
43	1·9403	0·1235	6·22	0·1864	9·60
45	1·3528	0·0720	4·23	0·1389	10·27
46	1·5053	0·0732	4·86	0·1645	10·93
64	1·1908	0·0505	4·24	0·0302	2·54
65	1·3671	0·0472	3·45	0·0292	2·04
86	1·7149	0·1073	6·26	0·0920	4·26
87	1·4539	0·0783	5·39	0·1212	8·34
88	1·7470	0·0763	4·37	0·0680	3·89
89	1·6682	0·0726	4·42	0·0765	4·58

Table VA. *Estimation of Iron and Alumina.* The hydrochloric acid solution (the filtrate from the insoluble residue) was peroxidised with potassium chlorate, and ammonia was added till the precipitate locally formed was very slow in dissolving.

Acetate of ammonium was then added, and the mixture boiled and filtered, and the precipitate washed. The precipitate was dissolved in hydrochloric acid, and ferric hydrate precipitated with pure caustic potash in a platinum basin. It was then diluted and filtered, ignited and weighed. To the filtrate ammonium chloride was added, and the solution boiled till ammonia ceased to come off. The precipitate was filtered, ignited to Al_2O_3, and weighed.

In the Fe_2O_3 the variations are not great. The minimum, 3·45 per cent., is on the Balearic Bank, and the maximum, 6·64, in the deep water near the African shore. In this neighbourhood all the muds have large amounts of Fe_2O_3 and small amounts of Al_2O_3, except No. 35, where the amount of Fe_2O_3 is small and that of Al_2O_3 very large—in fact, the maximum (12·3 per cent.). It is remarkable that at No. 41 we find the maximum of Fe_2O_3 (6·64 per cent.), and the minimum of Al_2O_3 (1·3 per cent.). The deep water between Africa and the Balearic Islands covers muds comparatively rich both in Al_2O_3 and Fe_2O_3; on the Balearic Bank the amounts are small, and in the Gulf of Lyons moderate.

Table VI. *Estimation of* FeO *and* Fe_2O_3. A weighed quantity of the substance was placed in a 200 c.c. flask and the flask filled with carbonic acid gas. 20 c.c. of strong hydrochloric acid were also added, and the flask fitted with a cork pierced by a tube with an india-rubber valve. It was then heated on the water-bath for half-an-hour, filled up with boiling distilled water, corked, and allowed to cool. 50 c.c. were taken for analysis, and titrated first with potassium permanganate of $\frac{KMnO_4}{50}$ grammes per litre; and then with stannous chloride.

The stannous chloride used in the case of samples 31 to 46 was of such a strength that 1 c.c. = 0·0047 grms. Fe_2O_3, and in the case of samples 64 to 89, 1 c.c. = 0·0042 grms. Fe_2O_3. From the figures in Table VI it will be seen that the bulk of the iron extracted in this way is in the ferrous state; while from column *m* the total amount of iron, expressed as Fe_2O_3, extracted in this way is only from 40 to 50 per cent. of the amount extracted by prolonged digestion.

TABLE VI. *Determination of the State of Oxidation of the Iron extracted by Hydrochloric Acid.*

No. of sample	Weight of sample taken (grams)	Volume of $KMnO_4$ $\frac{}{50}$ (c.c.) (mean of two estimations)	Equivalent weight of FeO (grams)	Per cent. of FeO	Equivalent weight of Fe_2O_3 (grams)	Volume of $SnCl_2$ (c.c.) (mean of two estimations)	Equivalent weight of Fe_2O_3 (grams)	Weight of Fe_2O_3 present in solution as Fe_2O_3 (grams)	Per cent. of Fe_2O_3	Per cent. of total Fe extracted (expressed as Fe_2O_3)	Per cent. of Fe_2O_3 (from Table V A)	
	a	b	$c = \cdot 0072\, b$	$d = 400\frac{c}{a}$	$e = \frac{10}{9}c$	f	g	$h = g - e$	$i = 400\frac{h}{a}$	$k = 400\frac{g}{a}$	l	$m = 100\frac{k}{l}$
31	1·6273	1·3	0·00936	2·30	0·0104	2·30	0·01081	0·00041	0·11	2·66	5·58	47·67
32	1·8723	1·4	0·01008	2·15	0·0112	2·65	0·01245	0·00125	0·27	2·66	5·37	49·53
35	1·5134	1·2	0·00864	2·28	0·0096	2·20	0·01034	0·00074	0·19	2·73	3·86	70·73
36	1·9895	1·3	0·00936	1·88	0·0104	2·65	0·01245	0·00205	0·41	2·55	6·17	41·33
38	1·3194	0·8	0·00576	1·75	0·0064	2·12	0·00996	0·00356	1·08	3·02	8·72	34·63
39	1·5296	1·02	0·00734	1·92	0·00815	2·1	0·00987	0·00172	0·45	2·58	6·54	39·45
41	1·6770	0·9	0·00648	1·55	0·0072	1·82	0·00855	0·00135	0·32	2·04	6·64	30·72
43	1·5664	0·85	0·00612	1·56	0·0068	2·22	0·01043	0·00363	0·93	2·66	6·22	42·77
45	1·6763	0·8	0·00576	1·37	0·0064	1·75	0·00822	0·00172	0·36	1·96	4·23	46·34
46	1·6676	0·7	0·00504	1·21	0·0056	1·72	0·00808	0·00248	0·59	1·94	4·86	39·92
64	1·6684	0·52	0·00374	0·89	0·00415	1·86	0·00781	0·00366	0·88	1·87	4·24	44·10
65	2·0729	0·57	0·00410	0·79	0·00455	2·05	0·00861	0·00406	0·78	1·66	3·45	48·11
86	1·9550	1·05	0·00756	1·55	0·0084	2·95	0·01239	0·00399	0·82	2·53	6·26	40·41
87	1·9106	1·50	0·01080	2·26	0·0120	3·20	0·01344	0·00144	0·30	2·81	5·39	52·13
88	1·7496	1·20	0·00864	1·98	0·0096	2·57	0·01079	0·00119	0·27	2·46	4·37	56·29
89	1·8807	1·05	0·00756	1·61	0·0084	2·87	0·01205	0·00365	0·78	2·56	4·42	57·92

TABLE VII. *Summary, Percentage Composition of Muds.*

No.	SiO_2	Balance insoluble in HCl undetermined	Total insoluble residue	Fe_2O_3	Al_2O_3	$CaCO_3$	Loss on ignition	Balance soluble in HCl undetermined	Total soluble in HCl
31	43·64	16·08	59·72	5·58	1·37	18·3	3·36	11·67	40·28
32	43·71	15·60	59·31	5·37	2·45	18·4	3·38	11·09	40·69
35	27·43	7·88	35·31	3·86	12·30	19·1	3·38	26·05	64·69
36	42·31	23·82	66·13	6·17	3·46	20·0	3·66	0·58	33·87
39	44·37	11·45	55·18	6·54	3·87	24·5	3·42	6·49	44·82
41	33·89	21·92	55·81	6·64	1·30	21·0	3·75	11·50	44·19
43	32·41	23·96	56·37	6·22	9·60	23·6	3·62	0·59	43·63
45	32·77	9·91	42·68	4·23	10·27	32·4	2·90	7·52	57·32
46	31·32	7·84	39·16	4·86	10·93	38·2	3·83	3·02	60·84
64	22·21	5·92	28·13	4·24	2·54	47·1	10·89	7·10	71·87
65	14·79	14·27	29·06	3·45	2·04	36·6	6·03	22·82	70·94
86	34·12	14·51	48·63	6·26	4·26	32·5	5·67	2·68	51·37
87	36·79	5·44	43·23	5·39	8·34	30·8	6·96	5·28	56·77
88	37·60	11·65	49·25	4·37	3·89	32·5	4·22	5·77	50·75
89	37·68	10·36	48·04	4·42	4·58	31·8	5·77	5·39	51·96

In Table VII the results of the foregoing tables are collected so as to facilitate comparison of the general composition of the different muds.

No. 13. [*From Proc. Roy. Soc. Edin.* 1881, *Vol.* XI. *p.* 191.]

ON THE OXIDATION OF FERROUS SALTS[1]

IN this paper the action of permanganate of potassium on ferrous salts in hydrochloric acid is studied, and a satisfactory method of titrating such solutions is indicated. Further, a short study is made of the action of chlorates, perchlorates, and nitrates on acid ferrous salts under various conditions. The results have a certain interest for the analytical chemist.

The following solutions were used:

Ferrous Sulphate. Ordinary recrystallised green vitriol was used.

(*a*) 5·857 grammes $FeSO_4 + 7H_2O$ dissolved in water and diluted till the solution weighed 290 grammes. This solution contains 0·0407 gramme iron in 10 c.c.

(*b*) 4·4600 grammes $FeSO_4 + 7H_2O$ (= 0·8977 gramme iron) were dissolved in water and made up to 145·5 grammes. The solution therefore contains 0·062 gramme iron in 10 c.c.

(*c*) 25·222 grammes $FeSO_4 + 7H_2O$ were dissolved in water and made up to 508 c.c. When filled up to the mark, and before adding the extra 5 c.c. water, the 500 c.c. solution weighed 514 grammes, therefore the specific gravity of the solution is 1·028 at 15° C. This solution when fresh contains 0·1 gramme iron in 10 c.c. It was used for most of the observations, and extending over nine days, during which time it gradually deposited a little ochreous matter. Compared with permanganate 10 c.c. required

March 11,	17·85 c.c. permanganate.	
March 19,	17·6	,,

Hence the alteration had not been very great, considering that during these nine days the solution had been consumed in portions of 10 c.c. at a time.

Chlorate of Potassium. 3·654 grammes recrystallised chlorate of potassium were dissolved in water and made up to 1 litre.

[1] See Contents, p. xxxvii.

10 c.c. of this solution oxidise 0·1 gramme iron from ferrous to ferric salt.

Perchlorate of Potassium. The perchlorate was obtained from the University laboratory, and was very pure. On analysis the following results were obtained:

Weight of salt	0·6208 gramme
Weight of salt after heating (KCl) ..	0·3330 ,,
Difference oxygen	0·2878 ,,
Chlorine in residue	0·1581 ,,

Whence we have the following composition per cent.:

	Found in salt	Calculated in $KClO_4$
Potassium	28·18	28·22
Chlorine	25·46	25·61
Oxygen	46·36	46·17
	100·00	100·00

3·0938 grammes $KClO_4$ oxidise 10 grammes iron from the ferrous to the ferric state. 3·0940 grammes were dissolved in warm water and made up to a litre at 9° C.

Nitrate of Potassium. 6·018 grammes KNO_3 oxidise 10 grammes iron from the ferrous to the ferric state. 6·019 grammes were dissolved in water and made up to a litre. The nitrate of potassium was purified by recrystallisation.

The *sulphuric acid* used to acidify the solutions was made by diluting 1 part by weight of pure oil of vitriol with 9 parts by weight of water. It contains, therefore, very closely one gram-molecule per litre (H_2SO_4).

Permanganate of Potassium. This solution was made by dissolving 3·163 grammes crystallised salt in water and making up to a litre. 3·162 grammes $= \frac{KMnO_4}{50} = 0{\cdot}02 KMnO_4$. Tested with the double sulphate of iron and ammonia, 17·9 c.c. were found to be required to oxidise 0·1 gramme iron from the ferrous to the ferric state. 3·163 grammes $KMnO_4$ oxidise 5·602 grammes iron from ferrous to ferric, therefore 17·9 c.c. of the above solution ought to oxidise 0·1003 gramme iron.

It is well known that the use of permanganate of potassium for titrating ferrous salts in hydrochloric acid is discouraged, and indeed prohibited by the highest authorities in analytical chemistry. As a matter of fact, however, there is very little inaccuracy attending its use in hydrochloric acid solution, which is as dilute as the sulphuric acid solution is when it is commonly titrated. But the chief want of the analyst is the power to titrate strongly acid hydrochloric solutions of iron with permanganate, and I find that this can be done without difficulty. As long as there is ferrous salt in the solution the permanganate devotes itself exclusively to it; but as soon as it is all transformed into ferric salt the permanganate at once attacks the hydrochloric acid, and the characteristic odour of euchlorine is at once perceived. If the permanganate is added with sufficient care to avoid local supersaturation, the appearance of this odour indicates with very considerable sharpness the moment when all the iron has been oxidised. It can, however, also be indicated clearly to the eye. If one or two drops of a dilute solution of pure ferricyanide of potassium be added to the ferrous solution so as to colour it blue without producing a precipitate, then on titrating with permanganate the disappearance of the ferrous salt is indicated by the simultaneous disappearance of the blue colour. This is illustrated in the following table. In it the concentration of the solutions

TABLE I.

	Number of experiment						
	1	2	3	4	5	6	7
Ferrous sulphate (*a*) c.c.	10	10	10	10	10	10	10
Sulphuric acid ($0{\cdot}5H_2SO_4$) ,,	—	—	20	—	—	—	—
Hydrochloric acid ($12{\cdot}6HCl$) ,,	10	10	—	25	25	25	25
Water .... ,,	30	30	30	25	25	25	25
Permanganate .... ,,	7·55	7·5	7·5	7·6	7·45	7·7	7·6

is expressed in gram-molecules per litre: as $0{\cdot}5H_2SO_4$ and $12{\cdot}6HCl$. For each experiment 10 c.c. of ferrous sulphate solution (*a*), containing 0·0407 gramme iron, were acidified either with normal sulphuric acid ($0{\cdot}5H_2SO_4$) or with fuming hydrochloric acid ($12{\cdot}6HCl$), diluted with water so as to bring the volume to about 60 c.c., then titrated with permanganate in hydrochloric acid solution, with the addition of enough ferricyanide to colour the solution blue.

From this table it will be seen that with 10 c.c. fuming hydrochloric acid the results are quite as good as with sulphuric acid. With 25 c.c. hydrochloric acid the method becomes a little strained; but it can be perfectly well carried out if care is taken to add the permanganate slowly so as to avoid a local excess, which would involve oxidation of the hydrochloric acid. This happened in experiment No. 6. In No. 5, which was also otherwise unsatisfactory, the ferricyanide had produced a precipitate which should always be avoided.

In Table II will be found the results of experiments on the action of chlorate of potassium and of perchlorate of potassium solutions on ferrous sulphate in sulphuric acid and in hydrochloric acid solution when left together for a short time.

TABLE II.

	Number of experiment						
	8	9	10	11	12	13	14
Ferrous sulphate (*b*) c.c.	10	10	10	10	10	10	10
Sulphuric acid (H_2SO_4) ,,	20	20	20	20	—	—	20
Hydrochloric acid (12·6HCl) ,,	—	—	—	—	10	10	—
Water .... ,,	30	30	30	30	40	40	30
Chlorate of potassium solution.. ,,	—	—	—	8·8	—	8·8	—
Perchlorate of potassium ,,	—	—	6·2	—	6·2	—	—
Duration of action...min.	—	—	1·5	1·5	2	2	—
Temp. of mixture.... °C.	—	—	17	17	16·5	16·5	—
Permanganate ..... c.c.	11·2	11·3	11·3	11·0	11·3	11·2	11·2

In experiments 11 and 13 the chlorate of potassium solution used was not that described above, which is used for all the following experiments, but a somewhat weaker one, of which 14·2 c.c. were required to oxidise 0·1 gramme of iron. It will be seen that the only case which shows any decomposition is No. 11, where 11·0 instead of 11·2 c.c. permanganate have been used. Hence at ordinary temperatures the action of chlorate of potassium in dilute acid solutions does not necessarily begin instantaneously.

In Table III will be found the results of experiments on the oxidising effects of solutions of perchlorate, chlorate, and nitrate of potassium acting on ferrous sulphate solution (*c*) for different lengths of time at ordinary temperatures. The mixture consisted in each case of 10 c.c. ferrous sulphate solution (*c*), 20 c.c. H_2SO_4, 30 c.c. water, and 10 c.c. of one of the oxidising solutions.

TABLE III.

Duration	Temperature	Cub. cents. permanganate				Chlorate decomposed per cent.
		Perchlorate	Chlorate	Nitrate	Blank	
hours	° C.					
1	18 to 19	17·8	12·5	18·0	18·0	30·2
23½	14 to 19	17·9	2·25	17·8	17·8	87·4
51	9 to 16	17·7	1·2	17·8	17·8	93·2
168	9 to 18	17·7	0·3	17·75	17·75	98·3

From these results it will be seen that ferrous sulphate in sulphuric acid solution, containing 0·1 gramme iron in 60 c.c., is stable at ordinary temperatures in presence of the quantity of nitrate or perchlorate of potassium necessary for the complete oxidation of the iron. In the case of chlorate the oxidation goes on at such a rate that in one hour 30·2 per cent., and in seven days 98·3 per cent. of the iron has been oxidised. In connection with these experiments it was observed that a clear solution of bleaching powder saturated in the cold, oxidises ferrous sulphate at once and completely in the cold; and that bromine water resembles chlorate of potassium in requiring time, and being accelerated by a high temperature.

The observations in Table IV were made with a view to gain a closer insight into the behaviour of chlorate of potassium solution at ordinary temperatures. The mixture exposed was in all cases 10 c.c. ferrous sulphate (*c*), 10 c.c. chlorate solution, 20 c.c. H_2SO_4, and 30 c.c. water. In three blank experiments without chlorate 17·6 c.c. permanganate were used.

TABLE IV.

Temperature of solution, °C.	13·5	13·8	14·0	14·2	14·9	15·0	15·1	15·2
Duration of action, min. ...	16·5	33	49·5	66	99	132	165	198
Permanganate used, c.c.	16·35	15·2	14·25	13·35	11·5	10·4	9·35	8·5
Per cent. chlorate reduced	7·10	13·64	19·03	24·15	34·66	40·91	46·88	51·70

It will be seen that during this series of observations the temperature rose from 13·5° C. to 15·2° C., in consequence the decomposition has gone on during the short intervals more slowly, and during the long intervals more quickly than it would have done at a mean constant temperature.

In Table V are given the results of observations on the action of nitrate and perchlorate solutions at higher temperatures. The composition of the solutions was as before, 10 c.c. ferrous sulphate (*c*), 20 c.c. H_2SO_4, 30 c.c. water, and 10 c.c. of perchlorate or nitrate of potassium. They were heated for five or six minutes to two intermediate temperatures, and also boiled for a like time. In three blank experiments made during the series 17·55, 17·50, and 17·55 c.c. permanganate were used.

TABLE V.

Oxidising agent	Perchlorate			Nitrate		
Temperature, °C.	46	81	boiling	56	84	boiling
Duration in minutes...	5	5	6	5	5	5
Permanganate used, c.c.	17·5	17·6	17·55	17·45	17·6	17·55

From these results it will be seen, and in the case of nitrate at least with surprise, that even boiling for five minutes has had no effect in bringing about a reaction between the oxidising agent and the ferrous salt. In the case of chlorate it is otherwise. In the following three experiments the chlorate was allowed to act for five minutes. In a blank experiment 17·6 c.c. permanganate were used.

TABLE VI.

Temperature, ° C.	28	47	70
Permanganate used, c.c.	15·95	11·9	3·8
Chlorate used, per cent.	9·2	33·5	77·1

The effect of dilution on the rate of action will be apparent from Table VII. For each experiment 10 c.c. of ferrous sulphate (*a*) were allowed to react on 10 c.c. chlorate of potassium along with 10 c.c. H_2SO_4, and different quantities of water, which will be apparent from the figures giving the total volume of the solution. After standing exactly half an hour they were titrated with permanganate. In the last four experiments in this table the sulphuric acid used was formed by diluting 1 part of oil of vitriol with 4 parts of water (both by weight); it may therefore be written $2H_2SO_4$, as there are two gram-molecules in the litre. Three blank experiments gave 17·6 as the permanganate equivalent to the 10 c.c. ferrous sulphate.

TABLE VII.

Temperature, ° C.	18	17	16·7	12·8	12·7	12·7	12·8
Volume of solution, c.c.	40	80	120	30	60	90	120
Permanganate used, c.c.	10·95	15·35	16·4	9·2	14·5	16·15	16·8
Chlorate used, per cent.	37·7	12·7	6·8	47·8	17·8	8·5	4·8

It is shown by these experiments that the energy of the oxidising action of the oxides of Chlorine and of Nitrogen is in inverse proportion to the amount of oxygen which they contain; and that the resistance to reduction is greatest in the case of the saturated bodies, H_2O, Cl_2O_7 and H_2O, N_2O_5.

No. 14. [*From The Times, December* 25, 1883.]

THE SUNSETS OF AUTUMN, 1883. OBSERVED AT SEA IN THE CABLE-SHIP "DACIA"[1]

So many letters have appeared in *The Times* describing recent sunsets that they now form a valuable repertory of observations, among which I hope the following may be fortunate enough to find a place.

During the months of October and November I was cruising among the Canary Islands and in the waters between them and the south-west coast of Spain. These months are considered to be the rainy season of the islands, and this year it was a reality. Consequently on many days the sunset was obscured by clouds, and on many more by the high land which lies to the westward of nearly all the anchorages in the group. Occasionally, however, it was unobscured by either cloud or land, and then its brilliancy was such as to defy description. On the 27th October especially, when the ship was approaching the island of Lanzarote from the westward, the sun set almost immediately behind the Peak of Teneriffe, which, though 115 miles distant, was as clearly and sharply defined as if it had been only ten miles off. I will not attempt to describe the colours of the sky, except to say that they graduated from a delicate violet near the zenith to a deep and intense scarlet on the horizon. Looking towards the west, and therefore towards the shaded sides of the waves, the sea looked dark against the illuminated sky; turning towards the east, the reverse was the case—the sea was blood red and the sky pale.

Leaving the islands and approaching the coast of Africa, the weather was bright and clear, and every successive evening showed a more brilliant sunset. It is well known that the prevailing wind in these regions—the north-east trade wind—blowing over the sandy deserts of Africa, carries much of the dust out to sea, and there are many instances of it having

[1] See Contents, p. xxxviii.

fallen in showers upon ships hundreds of miles from land. In Cadiz the sunset was as brilliant as close to the coast of Africa. In Seville the colours were distinctly less deep than at sea or on the coast, but when reflected from the roofs and towers, and especially from the Cathedral and the Giralda, the scene was one which it would be impossible to forget. Going further inland, the sunsets at Madrid could not in any way be compared with even those of Seville in brilliancy and depth of colour. It must be mentioned that during the fortnight over which these observations on land extended the same bitterly cold, but clear and cloudless weather, with north-east wind, prevailed all over Spain, from Madrid to Cadiz.

With regard to the sunrises, I can speak only of those occurring in the genial climate of the Canaries. For about ten days I was able to observe, from the island of La Palma, or the anchorage in front of its chief town, Santa Cruz, the sun rise every morning behind the Peak of Teneriffe, or, more correctly, behind its western shoulder, formed by the secondary crater of Chahorra. As soon as the light became sufficiently strong a bank of clouds was seen to stretch along the eastern horizon, and as the sky behind became better illuminated the cloudy upper edge of this bank was seen to be interrupted by the sharp lines of the familiar profile of the Peak, which stood out as a cold, grey screen against the more distant sky, already illuminated by the direct rays of the sun. As the moment of sunrise approached the Peak stood out more and more grandly, until the sun actually rose, when its outline became dim and it rapidly faded from view, its place being taken by the now illuminated particles of dust and spray in the intervening 70 miles of air. The colours attending the sunrise were in no way remarkable either for variety or brilliancy. Almost exactly the same view repeated itself every morning, the Peak being visible daily from dawn till about half an hour after sunrise.

From the fact that the brilliancy of the colours certainly increased on approaching the African coast from the sea, I conclude that much of the effect was due to dust floating in the air. On the other hand, looking to the fact that at sunrise there were no colour effects comparable with those of sunset, and that as the position of observation was shifted inland and to a higher level above the sea the sunset colours became paler and less brilliant, I conclude that water, whether as one of the gaseous constituents of the atmosphere or as floating particles, solid or liquid, was an important factor of the phenomenon.

(1918.) The study of the *Report of the Krakatoa Committee of the Royal Society*, published in the year 1888, showed that the atmospheric effects witnessed after the eruption fall into two periods. The earlier period covers the effects produced by the dust which remained visibly suspended for some weeks after the eruption and gave rise to the phenomenon usually recorded as "blue suns." Most of the observations during this period come from the Indian Ocean and its confines. This, the blue-sun period, closed before the end of September, 1883, and during that time abnormal sunrises were recorded about as often as abnormal sunsets.

The second period may be taken as beginning on 1st October, 1883, after which no blue suns are recorded. Most of the observations during this period come from the Atlantic Ocean and its confines. Between 27th September and 31st October twenty-two observations of remarkable sunsets are reported from ships at sea, and only two reports of sunrises. One of these from the s.s. "Olbers" reports—"Sunrises very pale," and the other from the s.s. "Orissa" reports—"At sunrise ugly looking sky, brick red under sun to 20° altitude." This report was from the Indian Ocean.

My letter to the *Times* is quoted at page 293: "Lanzarote, about 29° N. 14° W., Mr Buchanan, *Times*, date 27 October:—Splendid glows." From this most readers would conclude that I had observed "glows" at both sunrise and sunset; whereas the important point in the letter is that I established by direct observation the *absence* of any remarkable colouration of the sky at sunrise, when, on every day, the corresponding sunsets were at the very acme of their splendour.

It was not till long after the publication of my letter that a causal connexion was claimed between the remarkable sunsets of the Atlantic and the antecedent eruption of Krakatoa.

Admitting the existence of the cloud of fine volcanic dust in the higher regions of the atmosphere, postulated by the authors of this theory, it is difficult to account for the nature of the influence which made the rays of the rising sun differ so markedly from those of the setting sun in their behaviour when they encountered it.

No. 15. [*From Comptes Rendus*, 1893, *Vol.* CXVI. *p.* 1321.]

SUR LA DENSITÉ ET L'ALCALINITÉ DES EAUX DE L'ATLANTIQUE ET DE LA MÉDITERRANÉE[1]

Les observations rapportées dans le Tableau suivant ont été faites à bord du yacht "Princesse Alice," commandé par S.A.S. le Prince de Monaco qui avait bien voulu m'inviter à faire le voyage de Dartmouth à Gênes pendant les mois d'août et de septembre 1892. J'ai cherché à déterminer si le rapport entre la densité et l'alcalinité varie ou reste constant en passant de l'Atlantique dans la Méditerranée.

La détermination de la densité ($S_{t'}$) a été faite avec un aréomètre du type qui a servi à bord du "Challenger," mais avec cette amélioration que l'on peut ajouter autant de poids que l'on veut, et faire autant d'observations indépendantes de la densité de la même eau. L'instrument (n° 12) pesait $152^{gr},4075$ réduit au vide. La tige était divisée en millimètres sur une longueur de 100^{mm}. Elle avait un diamètre presque uniforme et les 100^{mm} déplaçaient $0^{gr},814$ d'eau distillée à 4°. Le volume du corps cylindrique de l'instrument a été déterminé à diverses températures en l'immergeant dans de l'eau distillée dont la température était déterminée par un thermomètre normal divisé en dixièmes de degré. On a, de plus, déterminé de temps en temps et parallèlement à celle de l'eau de mer la densité d'eau distillée préparée immédiatement avant le départ. On a pu observer ainsi directement les poids d'eau de mer et d'eau distillée avec le même instrument, à des températures très rapprochées et souvent identiques. On a toujours fait au moins trois observations (quelquefois huit et même dix) sur chaque échantillon. La moyenne des trois observations est exacte à trois unités près de la cinquième décimale, pour la température de l'eau au moment de l'expérience. Pour

[1] See Contents, p. xxxix.

profiter de tous les avantages de cette méthode, il faut que l'eau soit à la température de l'atmosphère et que celle-ci soit constante. Toutefois la détermination exacte de la température du liquide en observation est de la première importance pour la connaissance de sa densité, et c'est ici un grand avantage de l'aréomètre sur le pycnomètre. Avec celui-là le liquide est hors de l'instrument et sa température peut être déterminée exactement; avec celui-ci, il faut juger de la température de l'intérieur de l'instrument par celle du milieu dans lequel il se trouve. Il y a ainsi perte de temps et incertitude dans le résultat.

En considérant les densités, la température restant constante, on observe que la densité de l'eau est la même que celle de l'Atlantique tout le long de la côte méridionale de l'Espagne jusqu'au cap de Gata. Cela est confirmé par la présence d'un fort courant vers l'est, que l'on constate en même temps. Au delà du point indiqué, on n'a que l'eau plus dense de la Méditerranée.

Pour établir le rapport de la densité avec l'alcalinité, on a calculé le poids de 1^{lit} d'eau de mer à 23° C., température très voisine de la température moyenne 22°,96 C., des échantillons à l'époque de la détermination de la densité. On évite ainsi la plupart des erreurs de réduction qui dépendent de la dilatation de l'eau de mer. On considère comme proportionnelle à la quantité de sel dissous dans 1^{lit} d'eau, la différence D entre le poids de 1^{lit} d'eau de mer à 23° C. et $997^{gr},655$, poids du même volume d'eau distillée à la même température.

Pour déterminer l'alcalinité, on acidifie 250^{cc} d'eau dans une capsule de porcelaine avec une solution d'acide sulfurique dont 1^{cc} équivaut à 1^{mgr} d'acide carbonique CO^2; on ajoute quelques gouttes d'une solution d'aurine (rosolate de potasse) et l'on fait bouillir pendant cinq minutes pour éliminer tout l'acide carbonique. On neutralise alors l'eau, toujours bouillante, avec une solution de soude caustique équivalente par volume à l'acide sulfurique. On détermine le point de neutralisation à plusieurs reprises. On a ainsi le volume d'acide sulfurique neutralisé par 250^{cc} d'eau de mer. Le chiffre obtenu, multiplié par 4, donne en milligrammes de CO^2 par litre d'eau l'alcalinité A de l'eau de mer.

Density and Alkalinity of Sea-Water

Numéro de l'échantillon	Jour et heure de la prise de l'échantillon	Température de l'eau de mer		Densité de l'eau de mer[1]		Alcalinité de l'eau de mer en milligr. de CO_2 par litre A	Poids d'un litre d'eau de mer à 23° C. W	Différence[2] entre W et 997,655 D	Coefficient d'alcalinité $D_A = \frac{D}{A}$	Point d'origine de l'échantillon[3]	
		in situ t	au moment de l'observation de la densité t'	observée à t' $S_{t'}$	réduite à 15°·56 C. $S_{15°·56}$					latitude N.	longitude de Greenwich
	h m	°	°							° ′ ″	° ′ ″
1	22 août 9.30 *a*	19,10	20,05	1,025400	1,026626	54,24	1024,706	27,051	0,4987	45. 3.	6.27. 5 W.
2	22 ,, 9.30 *a*	19,50	20,20	5389	6539	54,88	4,619	26,964	0,4932	45.28. 5	6.54. 7
3	23 ,, 9.15 *a*	19,40	20,85	5283	6613	54,00	4,693	27,038	0,5007	44.29	7.47
4	24 ,, 7. *a*	17,10	20,60	5291	6551	54,32	4,631	26,976	0,4966	42.38	9.34
5	24 ,, 1.15 *p*	17,40	20,60	5281	6541	54,68	4,621	26,966	0,4932	42. 9	8.53. 5
6	24 ,, 2.55 *p*	18,80	20,70	5214	6504	54,44	4,584	26,929	0,4946	42.13.25	8.43. 5
7	25 ,, 5.25 *p*	17,15	20,95	5234	6589	54,24	4,653	26,998	0,4977	41.34	9. 7
8	26 ,, 7.30 *a*	18,90	20,55	5460	6705	53,36	4,795	27,140	0,5086	39.57. 5	9.30
8	26 ,, 7.30 *a*	18,90	20,50	5485	6715	53,36	4,795	27,140	0,5086	39.57. 5	9.30
9	26 ,, 10. 0 *a*	18,25	21,05	5332	6717	53,78	4,797	27,142	0,5047	39.25	9.38
9	26 ,, 10. 0 *a*	18,25	20,65	5469	6744	53,78	4,797	27,142	0,5047	39.25	9.38
10	26 ,, 2.30 *p*	16,15	20,90	5392	6732	54,36	4,812	27,157	0,4996	38.47.25	9.32
11	26 ,, 3.40 *p*	16,50	21,10	5302	6702	53,68	4,782	27,127	0,5053	38.33. 5	9.25
12	27 ,, 8. 0 *a*	19,40	20,50	5673	6903	54,52	4,983	27,328	0,5012	36.50. 5	8.36
13	27 ,, 11.30 *a*	21,75	22,32	5507	7233	54,84	5,313	27,658	0,5043	36.43	8.13
14	27 ,, 4. 0 *p*	22,90	22,85	5345	7220	54,72	5,300	27,645	0,5052	36.36	7.37.25
15	27 ,, 6.40 *p*	22,40	23,30	5201	7201	55,20	5,281	27,626	0,5005	36.30	7.20. 5
16	28 ,, 10. 0 *a*	21,15	23,60	5047	7137	55,04	5,217	27,562	0,5008	36. 3	5.48. 5
17	28 ,, 1.30 *p*	18,10	23,08	5225	7169	55,04	5,249	27,594	0,5013	36. 1	5.31.25

18	28 août	3. 0 *p*	17,40	23,57	5175	7366	55,96	5,446	27,791	0,4966	36. 4	5.22. 7
19	29 ,,	9.15 *a*	17,40	23,10	5115	7065	54,66	5,145	27,490	0,5029	36. 9	5. 5. 5
20	29 ,,	6.45 *p*	22,40	24,55	4759	7119	55,72	5,199	27,544	0,4943	36.28	3.52
21	30 ,,	6. 0 *a*	23,40	24,32	5511	7806	57,48	5,886	28,231	0,4911	36.39	2. 8
21	30 ,,	6. 0 *a*	23,40	25,58	5551	7814	57,48	5,886	28,231	0,4911	36.39. 5	2. 8
22	30 ,,	12.45 *p*	23,10	25,20	5191	7451	56,60	5,532	27,876	0,4925	37. 4. 5	1.24
23	30 ,,	6.20 *p*	24,70	24,75	5492	7907	58,32	5,987	28,332	0,4858	38.28	0.45 E.
24	31 ,,	6. 0 *a*	24,90	24,70	5663	8084	58,28	6,164	28,509	0,4892	38.40	0.21
25	31 ,,	11.15 *a*	25,00	25,04	5581	8203	59,24	6,283	28,628	0,4815	39.11	0.43
26	31 ,,	3. 0 *p*	25,40	25 85	5494	8239	58,60	6,319	28,664	0,4891	39.35	0.59
26	31 ,,	3. 0 *p*	25,40	23,72	6168	8292	58,60	6,319	28,664	0,4891	39.35	0.59
26	31 ,,	3. 0 *p*	25,40	21,30	6773	8217	58,60	6,319	28,664	0,4891	39.35	0.59
27	31 ,,	6.15 *p*	24,90	23,67	6191	8302	59,36	6,382	28,727	0,4839	39.51	1.10.25
28	1 sept.	9.30 *a*	24,15	23,72	5971	8105	58,08	6,185	28,530	0,4912	40.47	1.51
28	1 ,,	9.30 *a*	24,15	20,90	6754	8094	58,08	6,185	28,530	0,4912	40.47	1.51
29	2 ,,	9.30 *a*	21,40	23,50	6227	8287	58,44	6,367	28,712	0,4913	42.11	4.10
29	2 ,,	9.30 *a*	21,40	23,90	6108	8278	58,44	6,367	28,712	0,4913	42.11	4.10
30	2 ,,	6.30 *p*	22,40	21,20	7060	8480	58,44	6,367	28,712	0,4913	42.38	5.31
31	3 ,,	9. 0 *a*	22,90	21,20	7011	8434	58,44	6,367	28,712	0,4913	43.25	7.10
32	25 ,,	6. 0 *p*	21,30	24,25	5041	7316	58,44	6,367	28,712	0,4913	42.56. 3	8.58
33	25 ,,	6. 0 *p*	13,10	24,35	6054	8368	58,44	6,367	28,712	0,4913	42.56. 3	8.58
34	25 ,,	6. 0 *p*	13,40	24,60	6317	8667	59,92	6,747	29,092	0,4855	42.56. 3	8.58
35	25 ,,	6. 0 *p*	13,10	24,70	6235	8635	60,20	6,715	29,060	0,4827	42.56. 3	8.58
36	26 ,,	6. 0 *p*	13,10	24,25	6338	8613	59,84	6,693	29,038	0,4853	42.56. 3	8.58
37	26 ,,	6.30 *p*	21,90	24,80	5986	8416	59,84	6,693	29,038	0,4853	43.25	6.50
38	26 ,,	6.30 *p*	17,70	24,85	5898	8343	59,84	6,693	29,038	0,4853	43.25	6.50
39	26 ,,	6.30 *p*	14,00	24,75	5821	8228	58,73	6,308	28,653	0,4879	43.25	6.50

[1] La densité de l'eau distillée à 4° C. est prise pour unité.

[2] 997^{gr},655 est le poids d'un litre d'eau distillée à 23° C.

[3] Tous les échantillons d'eau analysés ont été pris à la surface, sauf les échantillons suivants: n° 33 pris à 100^{m}; n° 34 à 500^{m}; n° 35 à 1000^{m}; n° 36 à 150^{m}o; n° 38 à 30^{m}; et n° 39 à 50^{m}.

Considérant que les données de la colonne *D* sont proportionnelles à la salinité de l'eau, en divisant ces chiffres par ceux de la colonne *A*, on a le coefficient D_A qui exprime le rapport entre la salinité et l'alcalinité. On voit que ce coefficient est plus grand dans l'Atlantique que dans la Méditerranée. La différence n'est pas grande, mais elle est fort nette. Le coefficient moyen des n°s 1 à 20 (Atlantique) est 0,5000, le maximum étant 0,5086 et le minimum 0,4932. Pour la Méditerranée (n°s 21 à 39), le coefficient moyen est 0,4875, le maximum étant 0,4925 et le minimum 0,4815.

Puisque le coefficient maximum des 13 échantillons de la Méditerranée est inférieur au minimum des 20 échantillons de l'Atlantique, il est certain que nous avons ici une véritable différence entre les deux eaux. Les grandes différences observées par M. Gibson entre les eaux de la Baltique et celles de l'Atlantique peuvent être imputées aux gelées d'hiver. Dans la Méditerranée, les faibles différences sont peut-être dues à l'abondance des roches calcaires sur ses côtes.

No. 16. [*From The Times, June* 15, 1896.]

MONACO A WHALING STATION

THE enclosed extract from a letter which I have just received from H.S.H. the Prince of Monaco requires no comment. It will be read with equal interest by lovers of science and lovers of sport.

Last year, whilst pursuing deep sea research in his yacht "Princesse Alice" in the archipelago of the Azores, a native crew killed a sperm whale which died under the bottom of the yacht, having charged it in its death agony as an enemy. At the same time it rendered the remains of its last meal, which proved to be morsels of gigantic cuttle-fish hitherto absolutely unknown to science. They have already been described in communications to the Academy of Sciences. So soon as the yacht returned from the Azores the Prince set about equipping her for the whale fishery and engaged Mr Wedderburn, a Dundee whaler, as his mate. The results recorded in the letter show that the choice was a happy one and that his crew of well-tried and seasoned Bretons were quick to learn their lesson. Last year's cruise added a new family to natural history, in which the popular if somewhat legendary sea-serpent finds a place which fits it. This season's cruise bids fair to do as much for the great family of the cetaceans, which are generally supposed to be well known, but only specialists are fully aware of the extent of our ignorance.

From letter, dated June 7, 1896, from H.S.H. Albert Prince of Monaco to Mr J. Y. Buchanan, F.R.S.

"We are now in Marseilles ready to leave for the cruise after a couple of days.

The trial of our whaling business has given splendid results as soon as the material was complete and in order; in 24 hours we harpooned and secured three big cetaceans and lost a whale. Each of these cases was very dramatic; the whale, a piece of about 20 mètres long, was one of those who dive very deep and straight towards the bottom. She pulled out the 400 mètres of line that we had, in three minutes or less, with such a powerful speed that the fore part of the boat took fire. We had to cut just when a few fathoms were left, and then our boat was full of water. Then the animal reappeared on the surface, about half an hour later and at a distance of three miles, we steamed after it and the run lasted the whole day without loss or gain, but after all, without the possibility for us to shoot the rocket to cause an end, the whale having got the harpoon in some part which was not deadly and losing no blood at all. At night I had, of course, to abandon the pursuit.

The next day I came into a school of the good-sized cetaceans, which we later recognized as *Orca gladiator*, almost white, of which we met one individual about Cape St Vincent, if you remember[1]. Their aspect was *unheimlich*, and they did not run away from the boat; therefore, one was shot at once and killed by the harpoon, which went right through and wounded another, which was quite near him. Then the two other ones came alongside the boat and worked so as to squeeze it between them, which did not succeed, because the dead one, having been hauled up, served as protection on one side, and also because the round shape of the boat and of the animals produced each time on the boat the effect of lifting it out of the water. However, other boats were immediately launched and sent on the battlefield. Meanwhile, the whaler succeeded in killing with one stroke of the spear the biggest of the two enemies; and soon after the third one, which had been wounded by the same shot which killed the first animal, went away and was hunted for an hour by the steam launch, but was lost after all. This incident turned into a real battle, which lasted an hour, and in which four boats and 17 men had to be engaged. The spot was about 20 miles off Monaco. During the same cruise I got also one of three other pretty large cetaceans that I met and which were of the *Grampus griseus* species. I returned to the harbour towing these three animals, and I succeeded in pulling them on the beach, where crowds of people soon gathered and where thousands came on the next days. The *Orca* was about six mètres long and weighed between three-and-a-half and four tons; the others were four and three mètres long. They were properly studied by Richard; the skeletons of two of them are now being prepared, and the whaler made half a ton of oil out of the blubber. Monaco is now a whaling station, as I am preparing a place to receive a whale and other big cetaceans next winter."

(1918.) The *Orca gladiator*, a school of which was engaged and beaten by the Prince in the boats of his yacht, is the lion or tiger of the ocean, carrying jaws filled with formidable teeth for attack and animated with dauntless courage. The biggest of the three, which was killed by one stroke of Wedderburn's harpoon, now forms the centre-piece of the collection of cetacean skeletons in the Musée Océanographique of Monaco. A fight with these animals was worthy of the truly Homeric description of it given by His Highness.

[1] See Contents, p. xl.

No. 17. [*From The Times, September* 25, 1902.]

THE CRUISE OF THE "PRINCESSE ALICE" (1902)

THE steam yacht "Princesse Alice," with her owner, H.S.H. the Prince of Monaco, on board, has just returned from her fourth annual scientific cruise. Although nothing sensational occurred, much useful work was done, and a short account of it may interest the readers of *The Times*.

It is to be observed that, though only the fourth cruise of the present yacht, it must be something like the thirtieth cruise of the Prince himself. For many years all his sounding and dredging was done by hand in the "Hirondelle," a schooner of about 200 tons. Later, the first "Princesse Alice," of 550 tons, fully fitted with steam appliances, was used, and since 1898 the present magnificent ship, built by Messrs Laird, of Birkenhead, has been in use. Every year several fascicules are issued by the Press at Monaco; and already the reports of the scientific work of the Prince in his various yachts occupy a large number of volumes, which contain original material of the greatest value; while for beauty of production, type, paper, and plates they have no rival.

The yacht left Monaco on the 18th of July of this year, and made a straight course to Gibraltar. No work was done on the way, and the passage was not interrupted except by the pursuit and capture of an *Orca gladiator*, an interesting species of the whale tribe which is pretty abundant in the Mediterranean as well as in the North Atlantic. Its distinctive feature is that it preys on other cetaceans, and being bolder it is more easily approached, and when struck it is apt to give more trouble than the others. Fortunately, it is not as large as the sperm or other well-known whales.

Having filled up with coal, the ship left Gibraltar on the evening of the 23rd of July, and shaped a course westwards towards the Azores. As the waters in the straits have often been explored by the Prince on previous occasions, nothing was done until well outside of the line from Cape Spartel to Cape Trafalgar, which delineates pretty accurately the "lip" of the basin of the Mediterranean.

On the 24th of July, being in the position where a remarkable haul of large crustaceans was made on the homeward journey last year, the trawl was put over in 800 fathoms. As so often happens in deep-sea work, the successful experience of one year is not repeated in the next. That we are usually unable to give any good reason for this shows how much we have yet to learn. Although the trawl failed to bring up the rich harvest of animals expected, it worked quite well, and brought up what it found, which was mainly mud. In the sounding made previous to putting over the trawl the temperature of the bottom water was found to be remarkably high—namely, 9·4° C. In the absence of confirmatory evidence, this temperature would certainly have been rejected, but the large quantity of mud brought up in the bag of the trawl was found to have a temperature of 8·75° C., so that it was legitimate to take 9° C. as a close approximation to the actual temperature at the bottom. The position of the station, lat. 36° 6′ N., long. 7° 56′ W. (Greenwich), is a long way outside of the basin of the Mediterranean, where all the water at a greater depth than 250 fathoms has a nearly uniform temperature of about 13° C. On the other hand, in the North Atlantic Ocean the water at a depth of 800 fathoms could not have a higher temperature than 4·5° C. It was evident, therefore, that we had here struck one of the main drains of overflow from the abysmal regions of the Mediterranean; and, taking the above temperatures as bases, we calculate that the water at the bottom in this locality consisted, roughly, of 50 per cent. of Mediterranean and 50 per cent. of Atlantic water. From a purely oceanographical point of view this was one of the most interesting occurrences of the cruise. In order that it may be thoroughly appreciated I may remind your readers that the Mediterranean is situated in a region which is relatively dry, and that during the year more water leaves its surface by evaporation than the combined amount which falls upon it in the form of rain along with that which is delivered into it by rivers. If shut off by a dam at the Straits of Gibraltar it would shrink in size and increase in saltness until it attained a condition analogous to that of the Dead Sea. The deficiency due to over-evaporation is made up by the surface current of Atlantic water which enters through the Straits and is so well known to navigators. But every gallon of Atlantic water brings with it about 6 oz. of salt, which remains in the sea when the water evaporates. As the influx of water at the Straits goes on day after day, year after year, and century after century, some means must be found for removing the salt. To effect this nature does exactly what the marine engineer does when he feeds his boiler with sea-water—she *brines* her Mediterranean *down*. The outflow of brine is naturally at the bottom on account of its high density, and after passing the lip of the basin between Capes Spartel and Trafalgar, it must follow the deepest channels outwards until it is lost in the ocean by mixture and diffusion. These local rivers of relatively warm and salt water are probably very narrow; and it is quite reasonable to suppose that in 1901 the trawl was dropped by the side of one of them, while this year it was dropped into the middle of it.

Proceeding westwards, two remarkable banks or oceanic shoals lie on the route to the Azores, the Gorringe or Gettysburg bank and the Josephine bank. The Gorringe bank was examined by the former "Princesse Alice" on July 25, 1894, and on July 25, 1902, her successor continued the examination, but found nothing very new. It was, however, remarkable that the presence of the bank was shown quite unmistakably by the swirls and ripples on the surface of the sea to an extent that I have never witnessed on any other oceanic shoal having the same depth of water over it. July 27 was spent on the Josephine bank. Its extent must be very considerable, but, in the time at disposal, it was not possible to delimit it. An area of some three miles square was sounded over, showing an unusual uniformity of depth, always within a fathom or so of 120 fathoms. Between the Josephine bank and the Azores the greatest depth obtained was 2340 fathoms, the temperature of the water at the bottom being 3° C.; and the ship arrived at Ponta Delgada, in the island of St Michael's, on the afternoon of July 31. In the 29 years which have elapsed since the "Challenger" cast anchor in the same place many changes have occurred. The breakwater, which was then only begun, has been finished, and a large portion of it has again been demolished by a storm. The orange trade has disappeared, and its place has been taken by that in pineapples. The countless glasshouses in which the pines are grown give the country quite a new appearance. Otherwise things are much the same. Life in the islands has the merit of tranquillity. The population is small in relation to the area and fertility of the soil, so that, while perhaps no one is very rich, none are very poor, and all have enough to eat.

On leaving Ponta Delgada the yacht proceeded to the channel which separates the islands of Terceira and St Michael's. The only existing deep sounding in this district was one of 1900 fathoms made many years ago by the Prince in the "Hirondelle." A few miles from this spot a sounding was now made; and a depth of 1645 fathoms was found, which confirmed the existence of deep water in this locality. A remarkable feature of this sounding, besides the depth, was the comparatively high temperature of the bottom water. The thermometer which was sent down on the sounding-line showed a temperature of 5° C. In the open water of the North Atlantic the temperature at this depth would not be higher than 3° C. It was evident that we had here sounded in an enclosed basin, shut off from the general waters of the surrounding ocean by a "lip," situated at such a depth below the surface that the *minimum* temperature of the water which can gain access to it from the outside is 5° C. It need hardly be pointed out that in the latitude of the Azores such a temperature cannot be obtained from the surface. This result was confirmed by a number of subsequent soundings and temperature determinations. A sounding about 40 miles north of this position in 1650 fathoms gave a bottom temperature of 3·2° C., and one about 30 miles south of it in 1285 fathoms gave a bottom temperature of 3·5° C., so that the existence of the enclosed basin is placed beyond all doubt, and it was appropriately named the Hirondelle Deep. The temperature of the secluded water, 5° C., enables us, by comparison with temperature observations at different depths outside, to fix within pretty narrow limits the depth below the surface of the ocean where the lowest "col" occurs which leads from the interior of the basin across the lip to the outer oceanic world, and it must be between 850 and 900 fathoms.

There is no doubt that other and similar enclosed basins occur amongst the Azores; but they remain for future investigation. Chapters might be written about this interesting feature which is common to all archipelagoes, as witness the West Indies and the seas of the further East, such as the Celebes and the Sulu seas. Nearer home we have the Mediterranean with the original archipelago and the Red Sea. Indeed the further we proceed with the detailed sounding of the ocean the more such features do we discover. But, as the normal change of temperature with increase of depth decreases, it becomes more and more difficult to detect such features by temperature observations alone. It is, however, quite evident that the bottom of the ocean is not a smooth spherical surface; and, if it is not, then it must consist of heights and hollows.

As a complimentary feature to the Hirondelle Deep the Prince discovered a few years ago the Princesse Alice bank, a very extensive shoal lying to the south-west of the island of Fayal. A few days were now spent in making further soundings on it. For some years it has proved a most productive fishing ground; and this year all the trapping resources of the "Princesse Alice" were put forth, but in vain. Hardly anything of any kind was caught by any means. This was certainly remarkable, because the extent of the bank and its distance from the islands are both so great that there can be no question of over-fishing. Moreover, oceanic shoals are always good fishing grounds, and fishing on them reduces itself to putting out and hauling in lines as fast as possible. Yet here the fish would neither go into nets nor take hooks, and it almost seemed reasonable to suppose that they had gone elsewhere.

This year's cruise having specialized itself with the investigation of exceptional deeps and shoals, the Prince determined to finish it in the same sense. On finally quitting the islands the ship was steered northwards towards a place where the chart bears two soundings of 70 and 48 fathoms respectively and a note attributing them to the ship "Chaucer" in the year 1850. The entry on the chart is accompanied by the letter *D* to indicate doubt either as regards the depth or the position, or both.

In the year 1850 deep sounding was not understood, and extravagant depths were from time to time reported. But in 1850, as in 1750, a seaman knew perfectly well when he struck bottom with his deep-sea lead in water of 50 or 70 fathoms, and he will know no better in 1950. Indeed, the ordinary merchant ship at the time of the "Chaucer" depended perhaps more on his lead than on his chronometer, if he had one, in approaching such places as the mouth of the channel. When, therefore, he reports two soundings of such moderate depth as 70 and 48 fathoms, these depths are, in the first instance, to be accepted. The matter is a little different when we come to the other element—namely, the geographical position of his soundings. The "Chaucer" found 70 fathoms in lat. 43° 0′ N., long. 29° 0′ W. (Greenwich), and 48 fathoms in lat. 42° 45′ N., long. 28° 35′ W. It may be taken that his latitudes are fairly correct; but, in the absence of any information as to how his longitudes are arrived at, they must be considered to be somewhat uncertain. The soundings made by the "Princesse Alice" on September 5 and 6 are—1450 fathoms in lat. 42° 24′ N., long. 28° 15′ W. (Greenwich); 1360 fathoms in lat. 42° 50′ N., long. 28° 38′ W.; 1250 fathoms in lat. 42° 55′ N., long. 28° 47′ W.; 1345 fathoms in lat. 42° 53′ N., long. 28° 31′ W.; 1078 fathoms in lat. 42° 57′ N., long. 28° 27′ W.; and 1192 fathoms in lat. 42° 57′ N., long. 28° 22′ W. The first of these soundings is four miles and the last is 14 miles east of the mean position of the two "Chaucer" soundings. These soundings may be taken to prove that the bank does not lie immediately to the eastward of its reported position. On the other hand, the shoaling from 1345 fathoms to 1078 fathoms in a distance of less than five miles, taken in connection with the nature of the ground disclosed by the trawl, furnishes a clue which, if followed, might lead up to the Chaucer or an independent bank. It must be remembered that a set of 14 miles to the eastward in a day's run would in these latitudes be nothing extravagant. Admitting that the bank does not lie to the eastward of its reported position, there remains the possibility that it may lie on its reported position or to the westward of it. In view of the lateness of the season no more time could be spared, and the further elucidation of this question is postponed to a future occasion. Meantime, it is perhaps not impossible that the captain of the "Chaucer" may be still alive, and that these lines may meet his eyes. If so, it would be interesting if he would furnish some information about the soundings which stand in his name.

On the second day of the search, September 6, the trawl was put over in water which according to the sounding should have been over 1300 fathoms, but it dragged into shallow water and hooked on to the bottom. For long it appeared to be certain that it would have to be left there; but, thanks to the Prince's great experience in these matters and to his unwearied patience, it was at last disengaged, and furnished a remarkable haul of black manganese-covered coral, which reminded me of the first haul in the "Challenger" on similar ground to the south-west of the Canary Islands. It is such ground that forms the foundation of oceanic shoals. This I was able to observe and to study in the s.s. "Dacia" belonging to the India Rubber Company of Silvertown, in the year 1883, and an account of the investigations appeared in the columns of *The Times* at the end of that year. I had always been unable to follow the sedimentary theory of the foundation of coral islands beyond the point where the accumulation of sediment reaches such proportions as to oppose resistance to the tidal wave; because resistance to a wave produces a current, and a current prevents the deposition of sediment, and this takes place even in very deep water. But the banks then discovered and investigated showed that the current which prevents the accumulation of dead sediment brings food to living settlers, such as deep sea polyps and corals, and favours their existence. When once started, they grow upwards, often in perpendicular pillars, the living building upon the dead, until in time they reach the surface or near to it. If the conditions, principally those of temperature, are favourable, the reef-building species settle on it and in time produce a coral island; if they are unfavourable, we have banks or shoals such as the Dacia bank, the Seine bank, the Gorringe, the Josephine, and, when found, the Chaucer bank.

The route was now for home. Only one more station was made in the deep water of over 3000 fathoms recently discovered by the Telegraph Construction Company's steamer "Britannia," and on the 17th of September the yacht returned to Havre after a very successful cruise of exactly two months.

No. 18. [*From Nature, January* 9, 1896, *Vol.* LIII. *pp.* 223–5.]

THE SPERM WHALE AND ITS FOOD

THE services which H.S.H. the Prince of Monaco has rendered to the science of oceanography, during the last ten or twelve years, are familiar to every one interested in that department of research. First in the small schooner "Hirondelle," with no power but the strong arms of his Breton crew, and later, in the large and perfectly equipped auxiliary steam yacht "Princesse Alice," there is no branch of the science which has not been enriched by his enlightened enterprise and his unwearied perseverance. It may be interesting to the readers of *Nature* to know something of what was achieved in the summer cruise of 1895 in the waters of the North Atlantic, chiefly in the vicinity of the Azores. The dredging and other deep-sea operations conducted on board the yacht herself were very successful, and produced an abundant harvest. The most interesting result of the cruise, however, was due to the lucky chance of a cachalot or sperm whale being pursued by the whale-fishers of Terceira, and killed almost under the bows of the "Princesse Alice," and to the prompt measures taken by the Prince to utilise this rare opportunity, the importance of which for science he immediately and intuitively perceived. The preliminary reports of the investigation of the material thus collected by the Prince, in collaboration with the Portuguese whalers, go to show that an almost new and unsuspected animal kingdom has been opened to the zoologist.

A general account of the nature of the results has just been given, under the title "Prise d'un Cachalot," to the Société des Naturalistes, at its meeting in the amphitheatre of the Museum of the Jardin des Plantes, on December 24; and two communications were made, on December 30, to the Academy of Sciences by the Prince, of which one was from Prof. Joubain, of Rennes, dealing especially with the specimens of gigantic cephalopods obtained. It will be convenient to give the proceedings of the yacht, during and after the capture of the cachalot, in the form of an abstract of the Prince's own communications, and to deal with the specific details of the animals, collected, in the form of an abstract of Prof. Joubain's paper.

Proceedings of the Yacht.

On July 18 of last year, about nine o'clock in the morning, I[1] observed, while engaged in operations in the deep waters of the south of the island of Terceira, two boats leave the coast under sail, and about half an hour later two other boats proceeded in the same direction from another point. It was evident that they were not bound on an ordinary fishing expedition, and I quickly perceived that they were in pursuit of a school of sperm whales or cachalots; and I finished with the greatest speed the work in hand, in order to be able to take full advantage of the rare occasion of being able to assist at the capture of one of these interesting animals, should such be the result of the exertions of the whalers. About eleven o'clock I observed a whale spouting at a distance of about two miles, and I perceived that one of the whalers was approaching it cautiously. I was careful to remain at the same distance, in order not to run the risk of interfering with the whaler, and I closely followed all the manœuvres. The officer or coxswain of the boat stood erect in the stern, steering the boat with an oar. The harponeer stood in the bows, and I distinctly saw him strike the whale. I then approached the group at full speed, while the other whalers dispersed in pursuit of other members of the school. When I had arrived within one or two hundred metres, the cachalot had already towed the whaler attached to it by the harpoon, and the whole length of this line, to a considerable distance, and the harponeer had just succeeded in giving the animal the thrust of the lance, which terminates the struggle if skilfully delivered. The spray thrown out by the animal had become pink, and soon became quite red, while a pool of blood extended itself more and more

[1] The Prince of Monaco.

on the surface of the water around. The "Princesse Alice" was lying at about one hundred metres from the animal when it turned slowly round, lashed out with its tail, and then came straight for the yacht at a speed of ten or twelve knots. As there are many records of whaling ships having been sunk by the cachalot under similar circumstances, it will not be wondered at if I confess to having experienced some anxiety during the approach of the whale, and when powerless to avoid it. Just, however, at the moment when I expected the shock, the whale sounded, passed under the keel without touching it, and re-appeared on the other side in the agony of death. The rescue of the yacht from certain damage, if not from destruction, would have been impossible had the life of the whale been spared for a little longer. The cachalot was now floating alongside, with its head at a distance of about fifteen metres from the rudder, when its jaws opened and allowed several objects to escape, which I quickly recognised as cephalopods; but, notwithstanding the speed with which a boat was got away, in order to secure these animals, of the inestimable value of which I had already a presentiment, I perceived that they had begun to sink. On the spur of the moment, I started the engines very slow astern, and the coveted remains circulated slowly in the vortices produced by the propeller until they were secured by the boat.

The vessel now floated in a sea of blood of some acres in extent, and the whalers fixed one of our hawsers to the head of the dead animal; for they had gladly accepted my offer to tow it to El Negrito, where they have their installation for harvesting the oil from the whales that they are lucky enough to catch. The towing operation was not an easy one. The tail, acting as a rudder, caused the animal to swerve so violently from one side to another, that it was necessary to desist from the attempt to tow it head foremost, and to shift the tow-rope to the tail, after which the operation was completed without difficulty. The creek, which was the final destination of the whale, was not a suitable place for the yacht to remain; so, after landing the zoologists, MM. Richard and Lallier, and the artist, M. Borel, she left for the anchorage of Angra, while these gentlemen remained to assist at the breaking-up of the whale, with all the materials necessary for preserving the interesting matter which it promised to furnish.

For four days, under a burning sun, the whalers worked at removing the blubber and transferring it to the neighbouring house, where it was boiled down. At the same time they endeavoured to assist me in every way in securing the portions of the animal which interested me, more especially the brain. But the work was so difficult that it was only at the end of the fourth day that the skull was penetrated, and then the brain was found in a too advanced state of decomposition to be of use for preservation. It was impossible to approach the brain sooner, except by sacrificing the spermaceti, of which the volume was more than a cubic metre, and the commercial value very great. For half a day several men stood up to their middles in the cavity of the head which contains the spermaceti, and ladled it out. It must be remembered that the whale, which was stranded at high water, could only be worked at after the tide had ebbed considerably.

A large number of parasites were collected from the stomach, the digestive organs, the blubber, and the skin of the animal. M. Richard discovered on the lips of the whale certain round impressions, which he identified as the marks of the suckers of the great cephalopods. One can imagine the struggles of the giants which take place deep under the surface of the ocean. Notwithstanding his activity the cephalopod is seized by the cachalot, who, by means of the formidable teeth of the lower jaw, and the corresponding recesses in the upper jaw, holds the body of the animal without hope of escape. The cephalopod, in its defence, envelopes the face and head of the whale with the crown of its tentacles, the suckers of which leave deep impressions on its lips, and other parts where they have fastened. Meantime the cachalot makes efforts to swallow the portion of the cephalopod of which it has really taken possession, with the effect that the part of comparatively small calibre connecting the body with the head gives way; the body is swallowed, and the head dies and either drops off or is eaten by the whale.

Zoological Details from Professor Joubain's Paper.

The sperm whale or cachalot (*Physeter macrocephalus* Lacepede), caught on July 18, 1895, measured 13·7 metres in length. While in the act of death it ejected several large cephalopods which it had only just swallowed, as was evident from their perfect state of preservation. Amongst them were three large specimens, each over one metre in length, of a species, probably new, of the little-known but interesting genus *Histioteuthis*. The bodies of two other immense cephalopods were collected at the same time. When the stomach of the cachalot was opened, it was found filled with a quantity, estimated at over one hundred kilogrammes, of the partially digested *débris* of these cephalopods, all of them of enormous size. Amongst this *débris* may be noticed the crown and tentacles of a cephalopod, the body of which could not be found, belonging probably to the genus *Cucioteuthis*, hitherto known only by a few fragments. The muscular arms, which, though much shrunk and contracted by the preserving liquid, are as thick as those of a man, were covered with great suckers, each armed with a sharp claw, as powerful as those of the larger carnivora. More than one hundred of these suckers remain adhering to the arms.

The bodies of the two great cephalopods constitute one of the most interesting novelties of the scientific cruise of H.S.H. the Prince of Monaco. Their structure and their appearance are so different from all that is known amongst these animals, that it is impossible to place them in any species, genus, or family of this order. I propose for them the name of *Lepidoteuthis Grimaldii*, hoping that the discovery of complete specimens may permit of their affinities being more perfectly defined. One of these animals, half digested, is useless for study; the other, though headless, is much better preserved. It is a female, of which the body or visceral sac, after prolonged immersion in formol and alcohol, still measures 90 centimetres in length, from which it may be concluded that the length of the complete animal would exceed two metres. The surface of the sac is covered with large, solid, rhomboidal scales, arranged spirally like those of a pine cone. The fin (*nageori*) is very powerful, and forms one-half of the length of the body. It is not furnished with scales.

The stomach of the cachalot contained, besides, another cephalopod of large size, provided with a large fin, the skin of which enclosed certain photogenic organs. The head is wanting, so that it is impossible to affirm with certainty that it belongs to a new species, which is made very probable by the form of the body. Finally, the stomach of the cachalot contained a large number of beaks and rays or plumes, the difficult digestible residue of former repasts.

The cachalot which was killed by the whalers of Terceira, almost under the keel of the "Princesse Alice," seems as if it had been guided, in the pursuit of its food, by a desire to devour nothing but animals which, up to the present, are completely unknown, and in addition are of the highest importance for the morphology of the cephalopods. These cephalopods are all powerful swimmers, and very muscular. They appear to belong to the fauna of the deep intermediate waters, which is almost completely unknown, at least as regards the larger animals. They never come to the surface, nor do they lie on the bottom of the sea. Their great agility enables them to avoid every attempt to take them by nets; and it would appear that, for the present, the only means of capturing these interesting and gigantic animals is to commission a bigger giant to undertake the task, and to kill him in his turn when he has performed the service.

(1918.) Twenty-two years have passed since the natural history of the intermediate waters of the ocean was illuminated by the prompt and skilful use made by His Highness the Prince of Monaco of an opportunity which, as a mere matter of probability, cannot be expected to present itself twice in the lifetime of one man. During all these years the Prince has never lost sight of the great problem, but opportunity has failed; and it seems that the grand fauna of these depths cannot be caught except vicariously through, and by the death of, the cachalot.

The sperm whale fishing is a very old occupation and it is

certain that the circumstances and accidents, attending the capture of one of these animals, must repeat themselves.

The matter was investigated in the latter part of the eighteenth century and forms the subject of a most interesting memoir entitled "An Account of Ambergrise," by Dr Schwediawer, which was published in the *Philosophical Transactions of the Royal Society* in the year 1783 (Vol. 73, pp. 226–241).

This account was prepared at the desire and with the assistance of Sir Joseph Banks then President of the Royal Society; who accompanied Captain Cook on his first voyage to the South Seas in H.M.S. "Endeavour" in the years 1768–71.

In his[1] *Journal*, p. 65, under date 3rd March, 1769, he writes: "I found also this day a large Sepia or cuttlefish lying in the water, just dead, but so pulled to pieces by the birds that its species could not be determined. Only this I know, that of it was made one of the best soups I ever ate. It was very large; and its arms, instead of being like European species, furnished with suckers, were armed with a double row of very sharp talons, resembling in shape those of a cat, and like them, retractable into a sheath of skin, from whence they might be thrust at pleasure."

It is probable that this observation, and perhaps others made during the voyage of the "Endeavour," induced Sir Joseph Banks to assist Dr Schwediawer in the research reported in his paper.

As it deals largely with the food of the sperm whale, as well as with ambergrise, which is the by-product of a disease to which the animal is subject, the following extracts cannot fail to interest the reader.

Extract from Dr Schwediawer's Paper.

Ambergrise is found swimming upon the surface of the sea or in the sand near the sea coast.

It is also sometimes found in the abdomen of whales by the whale-fishermen, always in lumps of various shapes and sizes, weighing from half an ounce to an hundred and more pounds. The piece, which the Dutch East India Company bought from the King of Tydor, weighed 182 pounds.

[1] *Journal of the Right Hon. Sir Joseph Banks*, edited by Sir Joseph D. Hooker. London. Macmillan & Co. 1896.

An American fisherman from Antigua found some years ago, about 52 leagues south-east from the Windward Islands, a piece of ambergrise in a whale, which weighed about 130 pounds, and sold for five hundred pounds sterling.

We are told by all writers on ambergrise, that sometimes claws and beaks of birds, feathers of birds, parts of vegetables, shells, fish, and bones of fish, are found in the middle of it, or variously mixed with it; but of a very large quantity of pieces which I have seen, and which I have carefully examined, I have found none that contained any such thing, though I do not deny, that such substances may sometimes be found in it; but the circumstances which to me seems to be the most remarkable, is, that in all the pieces of ambergrise of any considerable size, whether found on the sea, or in the whale, which I have seen, I have constantly found a considerable quantity of black spots, which, after the most careful examination, appear to be the beaks of the *Sepia octopodia.* These beaks seem to be the substances which have hitherto been always mistaken for claws or beaks of birds, or for shells.

The persons who are employed in the spermaceti-whale fishery, confine their views to the *Physeter macrocephalus.* They look for ambergrise in all the spermaceti-whales they catch, but it seldom happens that they find any. But whenever they discover a spermaceti-whale, male or female, which seems torpid and sickly, they are always pretty sure to find ambergrise. They likewise generally meet with it in the dead spermaceti-whales which they sometimes find floating on the sea.

It is observed also, that the whale, in which they find ambergrise, often has a morbid protuberance; or, as they express it, a kind of gathering in the lower part of its belly, in which, if cut open, ambergrise is found. It is observed, that all these whales, in whose bowels ambergrise is found, seem not only torpid and sick, but are also constantly leaner than others; so that, if we may judge from the constant union of these two circumstances, it would seem that a larger collection of ambergrise in the belly of the whale is a source of disease, and probably sometimes the cause of its death.

As soon as they hook a whale of this description, torpid, sickly, emaciated, they immediately either cut up the above-mentioned protuberance, if there be any, or they rip open its bowels from the orifice, and find the ambergrise, sometimes in one, sometimes in different lumps of generally from three to twelve and more inches in diameter, and from one pound to twenty or thirty pounds in weight, at the distance of two, but most frequently of about six or seven feet from the orifice, and never higher up in the intestinal canal; which, according to their description, is, in all probability, the *intestinum coecum,* hitherto mistaken for a peculiar bag made by nature for the secretion and collection of this singular substance[1].

Having discovered, as I just now mentioned, beaks of the cuttle-fish in all the pieces of ambergrise I had an opportunity of examining, it now

[1] It may be taken from this that the disease from which the ambergrise-bearing whale suffers is allied to, if not identic with *Appendicitis.*

remained to be ascertained, how those beaks became so constantly mixed with ambergrise? In prosecuting this enquiry, I had the satisfaction to learn from the same persons who gave me the information above-mentioned, that the *Sepia octopodia*, or cuttle fish, is the constant and natural food of the spermaceti-whale, or *Physeter macrocephalus*. Of this they are so well persuaded, that whenever they discover any recent relics of it swimming on the sea, they conclude that a whale of this kind is, or has been, in that part. Another circumstance which corroborates this fact is, that the spermaceti-whale on being hooked generally vomits up some remains of the *Sepia*.

It will not be improper here to remark, to what an enormous size this species of *Sepia* grows in the ocean. One of the gentlemen who was so kind as to communicate to me his observations on this subject, about ten years ago hooked a spermaceti-whale that had in its mouth a large substance with which he was unacquainted, but which proved to be a dentaculum of the *Sepia octopodia*, nearly 27 feet long: this dentaculum however did not seem to be entire, one end of it appearing in some measure corroded by digestion, so that in its natural state it may have been a great deal longer. With regard to its being a dentaculum of the cuttle fish, the fishermen could not have been mistaken, as they themselves often feed upon the smaller sort of the same *Sepia*. When we consider the enormous bulk of the dentaculum of the *Sepia* here spoken of, we shall cease to wonder at the common saying of the fishermen, that the cuttle-fish is the largest fish of the ocean.

No. 19. [*From Nature, November* 3, 1910, *Vol.* LXXXV. *pp.* 7—11.]

THE OCEANOGRAPHICAL MUSEUM AT MONACO[1]

In the history of the development of the study of the sea all the sciences find an application, and all were worthily represented at the inauguration of the Oceanographical Museum of Monaco on March 29 of this year. The ceremonies and festivities incident to the occasion have already been chronicled in the columns of *Nature* (April 14, vol. LXXXIII. p. 191). It is proposed here to give an impression of the life-work of the Prince of Monaco, which found expression in the solemnities of that occasion. The accompanying illustrations afford an idea of the magnificence of the building and of the richness of the collections. Fig. 1 gives a view of the museum from the sea. The scale on which it is built can be judged from the fact that the height of the roof above the lowest masonry is 75 metres. Fig. 2 is the statue of the Prince standing on the bridge of his yacht. It is an artistic work, and a good portrait. It gives fine expression to the modesty as well as to the power of the creator of the great monument in the centre of which it stands.

The museum and the vessels attached to it, with their staffs and general organisation, are only one-half of the great enterprise which is entitled, "Institut Océanographique Fondation Albert Ier Prince de Monaco." Its seat is in Paris, where is possesses its own buildings and a rich endowment, both of them the gift of the Prince. It has professors of physical and biological oceanography and of the physiology of marine animals, and the lectures delivered during last year had the most numerous attendance of any in Paris. During the life of the Prince he exercises supreme authority. Both in Paris

[1] Note by the Editor of *Nature*:—For the illustrations of this article we are indebted to the courtesy of the proprietor of the *Naturwissenschaftliche Wochenschrift*. They are reproduced from photographs by Prof. Döflein of Munich, and illustrate an article by him in that periodical.

and at Monaco there is complete organisation for giving effect to his wishes, and, in the event of his death, for carrying on the work without interruption, and on the lines inaugurated by himself. Thus continuity and permanence have been assured.

It will be readily realised that the establishment of these two great institutions has not been accomplished without the

Fig. 1. General view of the Oceanographical Museum at Monaco as seen from the sea.

expenditure of large sums of money and the devotion of much time and labour to it. It is almost impossible for anyone to realise the greatness of the work which is being accomplished without having been intimately connected with it, and even with this advantage the development of the conception is slow. As with all great achievements, it will take at least a generation before it is thoroughly understood and adequately appreciated.

The museum at Monaco bears testimony at every turn to the great lines on which the Prince has himself worked, and in which his work is fundamental. Thus, in the purely hydrographical department, we see his bathymetrical chart of the world, on which all the trustworthy deep soundings are entered. This great document may be said to be the foundation-stone

Fig. 2. Statue of Prince Albert I[er] of Monaco in the Museum.

of oceanographical work. Another and much earlier piece of hydrographical work is the current chart of the North Atlantic, which gives the result of his laborious work on board the schooner "Hirondelle." By the methodical dispersion of floats, especially constructed to expose the least possible surface above water, along different lines radiating generally from the group of the Azores, by patiently awaiting their recovery, and by then combining their records, he furnished the demonstration that this portion of the ocean is practically a lake, bounded, not by land, but by the motion of its own peripheral waters, thus enclosing a roughly circular portion of the sea, part of which is generally associated with the Sargassum weed and called the

Fig. 3. Skeleton of the great *Orca* killed by the Prince of Monaco near Monaco.

Sargasso Sea[1]. The water, thus self-confined in the warm, dry subtropical region, is exposed to powerful evaporation, and to a considerable annual variation of temperature at the surface. The combination of these two thermal factors furnishes the mechanical power by which the deeper layers of the water obtain more heat and attain a greater density in this sea than they do in any other part of the open ocean, as was pointed out

[1] See Chart, p. 106, where the Sargasso Sea is shown as a "Downthrow."

by the writer in a paper "On the Vertical Distribution of Temperature in the Ocean," read before the Royal Society on December 17, 1874, and published in its *Proceedings*, vol. XXIII. p. 123.

In the great hall to the left of the entrance the visitor is at once struck by the magnificent collection of skeletons of Cetaceans, which includes those of many species. These are skeletons of individual whales, nearly all of which have been killed by the Prince himself, and each is complete, every bone in the animal being accounted for. From all points of view this collection is at once the most attractive and the most interesting in the museum, and in it we see the Prince reflected as a hunter and as a naturalist.

In Fig. 3 we have the *Orca*, with its formidable double row of teeth. It preys on other Cetaceans, and always shows plenty of sport. The specimen[1] figured belonged to the leader of a school of three, which was met with a few miles outside of Monaco. They fought to the death, and when killed they were towed in and beached on what is now the new harbour of Monaco.

Not far from the *Orca* is a skeleton, Fig. 4, of the best known of the toothed Cetaceans, the cachalot or sperm whale. It was not taken by the Prince himself, but he was present at its capture, and his scientific instinct enabled him to seize an opportunity which would probably have been missed by another. The cachalot had been struck by a crew of whalers from Terceira, one of the islands of the Azores. The Prince followed the chase in his yacht, and was close to the animal when it became evident that its end had come. At this moment these animals always charge whatever they see, and in their death agony they usually render whatever they have last eaten. This animal charged the "Princesse Alice," but the charge did not get home. The animal stopped, and a large mass of something came out of its mouth close to the yacht and began slowly to sink. The Prince at once jumped into the dingy, and, with a long landing net, retrieved the object before it sank out of sight. The object is represented

[1] This is the whale which Wedderburn succeeded in killing by one stroke of the spear, as related p. 260.

in Fig. 5, and is a unique piece. It is a fragment of the gigantic scaled cephalopod which Prof. Joubain, who described it, named *Lepidoteuthis Grimaldii.*

A healthy cachalot is valued for the spermaceti, or wax, which is contained in its head, and a sick one is still more valued for the ambergrise which it may contain. This curious substance, which has at all times been so highly esteemed in

Fig. 4. Skeleton of the Cachalot which furnished the fragments of gigantic Cephalopods.

pharmacy and perfumery, forms the subject of a very interesting "Account of Ambergrise" by Dr Schwediawer, which was read before the Royal Society on February 13, 1783, and published in the *Philosophical Transactions*, vol. LXXIII. p. 226. From his investigations it appears that ambergrise is a by-product of an inflammation of the intestinal canal, most probably of the *caecum*, which has been started by the "beaks" of the cephalopods which it has swallowed, for these are the invariable and characteristic ingredient of all genuine ambergrise. He further states that the whalers are convinced that the cachalot feeds only on squids, which, when unmutilated, must be of great size. One whaler reported a case where the whale in its death-throe rendered a single tentacle, which, though incomplete from having been partially digested, still measured 27 feet in length, and he held that this justifies the common saying of the whalers that the squids are the biggest fish in the sea.

The work of the Prince amongst the toothed Cetaceans has

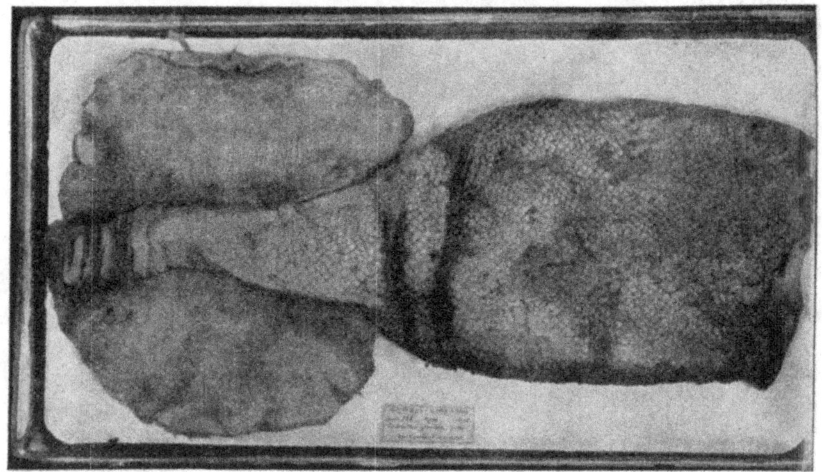

Fig. 5. The principal fragments of *Lepidoteuthis Grimaldii* Joubain.

had an interesting sentimental result. The combat of the "thrasher" and the whale, so dear to the nautical mind, seems to be nothing but the violent and desperate resistance of the giant squid to being swallowed when brought to the surface by the cachalot.

The whalebone whale, shown in Fig. 6, was struck by the Prince in May, 1896, not many miles from Monaco, but it escaped. Its carcase was washed ashore in September of the same year, near Pietra Ligure, on the Italian Riviera. A remarkable feature of this skeleton is the evidence of fracture and repair of a number of ribs on its left side. This has been ascribed to collision with

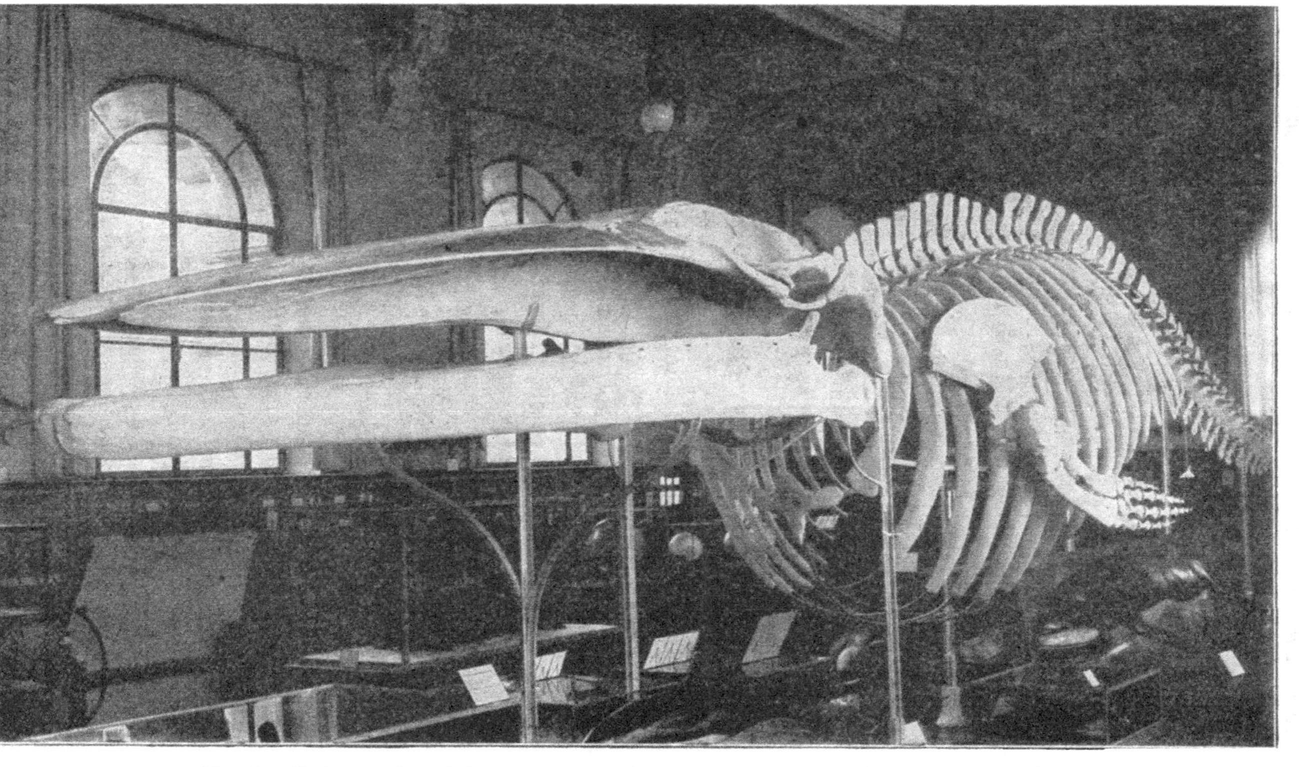

Fig. 6. Skeleton of whalebone whale the ribs of which have been broken and mended.

a steamer, but it is very unlikely that such an experience would leave its mark in nothing but a number of perfectly repaired ribs. It would seem to point to a type of accident to which whales are certainly exposed, and from which they perhaps not infrequently suffer.

The habitat of the whale is the air and the water, and its functional economy has to be adapted to life in both elements, or rather, to life sometimes in the one, at other times in the other element.

In one of the Prince's recent cruises in the Mediterranean the yacht was found to be steaming in the wake of a whale, which was evidently making a passage, and in a leisurely way. The Prince seized the opportunity to follow the animal without pursuing it, and this was done with such skill that it remained unconscious of being followed. It kept a steady course, and, to "keep station" with it, the "Princesse Alice" had to steam at a speed of about ten knots. In these conditions the whale came up to breathe at regular intervals of between ten and eleven minutes, the intervals between the spouts being the same almost to a second. This experiment supplies an important constant in the natural history of the whale. It looks very simple, but it will not be readily repeated, except perhaps by the Prince himself. As the whale was on passage, it is unlikely that it went far below the surface, but there is abundant evidence that, in the search for food or to escape enemies, it penetrates to very considerable depths. In these excursions its body is exposed to rapid and great variations of pressure. These have to be borne by the structural frame of the animal, of which the ribs are an important part.

It is generally assumed that, before sounding, the whale fills its lungs with air, but this, being at atmospheric pressure, is of no use in assisting the body to resist the external pressure of a column of water equivalent, it may be, to many atmospheres. How the power of resistance is, in fact, provided, I am not anatomist enough to know, but it must be finite, and it is easy to imagine conditions in which the animal, whether in the pursuit of prey or in the endeavour to escape being made itself a prey, may strain it beyond its limits, and the ribs of one side, whichever is the weaker, may give way. In such an accident, beyond being broken, the ribs need not be seriously disturbed, and with the return to the surface or more moderate depths, they would fall into their places again, and that all the more easily because there is little or no pressure of one part on another, every part of the body of a totally immersed animal being water-borne. In such conditions recovery would be rapid and the joints perfect, as can be seen to be the case in the skeleton in the museum.

The accident to this whale is very suggestive. In a well-known experiment, Paul Bert rapidly reduced the pressure of the air in the lungs of a dog by a not very large fraction of an atmosphere, when the thorax immediately collapsed, every rib being broken. When a whale is struck and sounds, if only to a depth of one hundred metres, the pressure on its body is increased tenfold in a few seconds. How does its body stand it?

It is certain that the cachalot finds its prey in water of considerable depth. When it has seized it, can it swallow it *in situ*, in a medium of water under very high pressure? The dentition of this animal, a formidable row of teeth in the lower jaw fitting into corresponding sockets in the upper jaw, makes it certain that, when it has seized its prey it can hold it indefinitely. It has been observed that the cachalot sometimes takes its prey to the surface and swallows it there. Is this accidental or habitual? If habitual is it not another link with the far-back time when its habitat was the air and the land? These are some questions suggested by an attentive visit to the Museum of Monaco.

In the museum, room is provided for a department of meteorology, a science which, especially as regards its application to the higher regions of the atmosphere, owes much to the participation of the Prince in its development. Until he directed his attention to it, the *ballons-sonde*, carrying their freight of valuable instruments, were very frequently lost. Now, thanks to the method of keeping the "dead reckoning" of the balloon, developed and brought to perfection on the "Princesse Alice," if it is followed for a few minutes during its ascent, it may disappear in the clouds, and its recovery, when it descends at sea, is almost a certainty. This department of investigation has been prosecuted outside the Mediterranean, and in the Prince's cruises of the last two or three years it has been carried from the Cape Verde Islands in the heart of the tropics to the north of Spitsbergen, within five hundred miles of the Pole.

Besides the collections of animals and the instruments for their capture and study, there is in the lower part of the museum an aquarium, remarkable for its size and the completeness of its installation. This already commands a constant flux of visitors, chiefly the curious, but it is also frequented by men of science for serious study. It is already proposed to enlarge it considerably. The storey above the aquarium is divided into separate laboratories, fitted with a service of both fresh and sea water, and everything else required for chemical, physical, and biological study. In these laboratories the occupant has all that a laboratory can supply, and at any time fresh material from the sea, collected by one of the small steam tenders of the museum.

Any notice of the museum of Monaco would be incomplete without an acknowledgment of what it owes to its director, Dr Richard. None of the many men of science who have enjoyed the hospitality, either of the museum or the yacht, will require to be reminded of this, nor will they forget what they individually owe to Dr Richard's never-failing courtesy and helpful aid. Personally, I have more thanks to offer than I can express for the countless services that he has rendered me during our friendship of twenty years. The Prince was fortunate in being able to attach him to his service in the early days of the "Hirondelle." Since that time Dr Richard has been his never-failing aid and assistant. It is not too much to say that without Dr Richard's strenuous and unselfish work during these many years the museum with its rich collections and complete equipment would not be, as it is now, the greatest institution of the kind in the world.

No. 20. [*From Encyc. Brit. 9th ed.* 1882, *Vol.* XIV. *p.* 216.]

LAKES

When a stream in its course meets with a depression in the land it flows into it and tends to fill it up to the lip of its lowest exit. Whether it succeeds in doing this or not depends on the climate. In the British Islands, and in most temperate and equatorial regions, the stream would fill the depression and run over, and the surplus water would flow on towards the sea. Such a depression, with its contents of practically stagnant water, constitutes a lake, and its water would be fresh. In warm dry regions, however, such as are frequently met with in tropical latitudes, it might easily happen that the evaporation from the surface of the depression, supposed filled with water, might be greater than the supply from the feeding stream and from rain falling on its surface. The level of the waters in the depression would then stand at such a height that the evaporation from its surface would exactly balance the supply from streams and rain. We should have as the result a lake whose waters would be salt. Lakes of the first kind may be considered as enlargements of rivers, those of the second kind as isolated portions of the ocean; indeed, salt lakes are very frequently called seas, as the Caspian Sea and the Dead Sea. The occurrence of freshwater lakes and salt lakes in the same drainage system is not uncommon. In this case the salt lake forms the termination. Well-known examples of this are Lake Titicaca and the Desguadero in South America, and Lake Tiberias and the Dead Sea on the Jordan.

Distribution of Lakes. Although there are few countries where lakes are entirely absent, still it requires little study to see that they are much more thickly grouped in some places than in others. Of the larger lakes, for instance, we have the remarkable group in North America, which together form the greatest extent of fresh water in the world. A similar group of immense lakes is found in Central Africa: Lakes Victoria Nyanza and Albert Nyanza, whose overflow waters go to form the Nile; Lake Tanganyika, at the source of the Congo; and Lake Nyassa, on a tributary to the Zambesi. In Asia the largest freshwater lake is Lake Baikal, on the upper waters of the Lena. All these freshwater lakes of great size are at the sources of large and important rivers; the salt lakes in which Asia also abounds are at the mouths of large rivers, as the Caspian at the mouth of the Volga, and Aral Sea at the mouth of the Oxus.

Passing from the consideration of these larger lakes, which from their size may be considered inland oceans, and which therefore necessarily occur in small number, we find large numbers of lakes of comparatively small dimensions, and when we consider them attentively we find that they are reducible to a small number of species, and, as in the case of plants and animals, the distribution of these species is regulated chiefly by climate, but also by geological conditions. Perhaps the most important and remarkable species of lakes is that to which the Scottish lakes belong. They are generally characterized by occupying long narrow depressions in the valleys of a mountainous country in the neighbourhood of the sea, and in a temperate climate. On the sea-coast, lakes of this character are found in Norway, Scotland, Newfoundland, Canada, the southern extremity of South America, and the south end of the middle island of New Zealand; somewhat removed from the sea we have the Alpine lakes of Switzerland and Tyrol, and the great Italian lakes, all of which display the same features as those of Scotland or of Norway. In many flat countries lakes are extraordinarily abundant, as for instance in the north part of Russia and Finland, in the southern part of Sweden, in the northern parts of Canada, and on a small scale in the Hebrides.

Lagoons, found on all low sandy coasts, owe their origin to the shifting of the sand under the influence of the wind and tide. They are found at the mouths of large rivers, as on the Baltic and at the mouth of the Garonne.

In volcanic regions lakes are not uncommon, generally of a more or less circular form, and either occupying the site of extinct craters or due to subsidences consequent on volcanic eruptions; such are the Maare of the Eifel in Germany, and many lakes in Italy and in the Azores.

Lakes are not only widely distributed in latitude and longitude, they also occur in all elevations. Indeed, as a certain elevation above the sea produces an effect as regards climate equivalent to a certain increase of latitude, we find lakes existing in the centre of continents, and on high plateaus and mountain ranges, in latitudes where they would be speedily dried up if at the level of the sea. Many of the lakes in Scotland (as Lochs Lomond, Morar, Coruisk), of Norway, of British Columbia, and of southern Chili are raised only by a few feet above the level of the sea, and are separated from it often by only a few hundred yards of land, while in the Cordilleras of South America we have Lake Titicaca 12,500 feet, and in Asia Lake Kokonor 10,500 feet above the sea. Many lakes whose surface is raised high above the level of the sea are so deep that their bottom reaches considerably below that level.

Dimensions of Lakes. The principal measurements connected with a number of lakes in different parts of the world, presented in the table, p. 292, will give a more precise idea of the size of the lakes than could be given by description alone.

From this table it will be seen that by far the largest continuous sheet of fresh water is the group of North American lakes, and of these Lake Superior is more than double the size of any of the others; this is principally due to its great breadth, as it is very little longer than Lake Michigan. Lake Superior communicates with Lakes Michigan and Huron, which are really branches of one and the same lake, by the St Mary's river the fall being 49 feet from Superior to Huron. Huron empties itself into Erie by the St Clair river, Lake St Clair, and finally the Detroit river. Lake Erie overflows by the Niagara river and falls into Lake Ontario, whence the water finally is conveyed to the sea by the St Lawrence. The area of the lakes together is in round numbers 100,000 square miles, and, if that of the St Lawrence and its estuary be added, the water

Name of lake	Mean latitude	Length	Breadth (max.)	Depth (max.)	Height in feet above the sea of		Temperature of water at bottom
					surface	bottom	
		miles	miles	feet			° F.
Superior	47° 45′ N.	350	100	978	627	−351	38·8
Michigan	44° N.	320	80	840	594	−246	—
St Clair	42° 30′ N.	18	22	20	570	+550	—
Erie	42° N.	220	48	204	564	+360	—
Titicaca	16° 30′ S.	90	30	924	12,500	+11,576	54·6
Kokonor	37° N.	91	42	—	10,500	—	—
Baikal	53° N.	330	40	4,080	1,360	−2,720	—
Balkash	46° N.	280	25	238	72	−166	—
Caspian	42° N.	600	50	3,600	−85	−3,685	44·6
Dead Sea	31° 30′ N.	45	10	1,308	−1,272	−2,580	—
Tanganyika	6° S.	330	40	1,000	2,700	—	—
Como	46° N.	48	2·5	1,356	670	−686	—
Geneva	46° 25′ N.	45	8·7	1,092	1,218	+126	41·7 to 43·5
Constance	47° 40′ N.	35	8	394	1,300	+906	39·6
Lomond		20	4	630	25	−605	41·4 to 42
Morar		11	1·5	1,020	30	−990	40·8 to 41·4
Ness		23	1·3	774	50	−724	41·2 to 42·4
Lochy		10	1	480	93	−387	42 to 44
Katrine	56° to	7	0·8	480	364	−116	41·4
Tay	57° 30′ N.	14·5	1	450	390	−60	43·9
Rannoch		9·4	1	378	668	+290	43·9
Ericht		14·5	0·8	330	1,153	+823	44·7
Tummel		2·5	0·5	120	450	+330	45·5
Garry		2·5	0·3	102	1,330	+1,228	53·9

area will be about 150,000 square miles, while the whole drainage area is only 537,000 square miles. Hence the water conveyed by the St Lawrence to the sea, rather more than one-fourth falls on the surface of the water itself. Looking to their great extent, we should have suspected them to be much deeper than is found to be the case. The deepest, Lake Superior, is no deeper than Loch Morar in Inverness-shire. Comparatively shallow, however, as they are, the bottoms of them all, with the exception of Erie, are several hundred feet below the level of the sea. It has been supposed that in former times this chain of lakes formed an arm of the sea similar to the Baltic in Europe, and in support of this view we have the fact of the discovery of marine forms in Lake Michigan.

In Asia, Lake Baikal is in every way comparable to the great Canadian lakes as regards size. Its area of over 9000 square miles makes it about equal to Erie in superficial extent, while its enormous depth of over 4000 feet makes the volume of its waters almost equal to that of Lake Superior. Although its surface is 1360 feet above the sea-level, its bottom is 2720 feet below it. A former connection with the ocean has been claimed for this lake, owing to the fact that seals inhabit its waters. Other large lakes in Asia are mostly salt, and some lie wholly below the level of the sea. Thus the surface of the Caspian lies 85 feet below that of the Black Sea, and the bottom at its greatest depth is 3600 feet deeper. The Dead Sea is over 1300 feet deep, and its surface is 1272 feet below the Mediterranean, so that its bottom is 2580 feet below the level of the sea. In the Caspian seals are found. A former connection with the Red Sea has been claimed for the Dead Sea, but this is disallowed by Peschel and others. The Jordan valley, and the Sea of Tiberias and the Dead Sea, lie on the line of an extensive fault, and it is claimed that this depression in the surface occurred with the production of the fault. Further evidence in support of the statement that the Dead Sea was never connected with the sea is of a negative character, and consists chiefly in the fact that marine forms have not been found in the waters of the Jordan or of Lake Tiberias, and that silver is absent from the waters of the Dead Sea.

A former connection with the ocean is claimed for a number of the Swiss and Italian lakes by Dr Forel and Professor Pavesi, and the Norwegian lakes by Loven and Sars, on the ground of the occurrence of marine forms of the crustaceans and other classes. For a summarized account of these researches see Pavesi, *Arch. de Genève*, 1880, III. I.

Temperature of Lakes. The earliest reliable temperature observations in lakes or seas are those of Saussure, and they are to be found in his charming *Voyage dans les Alpes.* He was the first to obtain thoroughly trustworthy observations in the deeper waters of the lakes. He used for this purpose an ordinary thermometer whose bulb was covered over with several thicknesses of cloth and wax, so as to render it very slowly conducting. He was in the habit of leaving it down fourteen hours, and then bringing it up as quickly as possible and immediately reading the temperature. He did not, however, trust to his thermometer not changing its reading while being brought up, but by an elaborate series of experiments he obtained corrections, to be applied when the thermometer had to be drawn through more or less water of higher temperature. His observations are collected in the following table along with those of Jardine in some of the Scottish lakes, at the beginning of the century:

Name of lake	Date	Temperature of		Depth	Height above sea
		surface	bottom		
		° F.	° F.	feet	feet
Geneva ..	20th Feb. 1780	42·1	41·6	1013	1230
Neuchâtel	17th July 1780	73·7	41·4	346	1304
Bourget ..	October 1784	64·0	42·1	256	—
Annecy ..	14th May 1780	57·9	42·1	174	1426
Joux.....	—	55·6	51·3	85	350
Bienne ...	—	69·3	44·4	231	1419
Constance	25th July 1784	64·6	39·6	394	1250
Lucerne ..	28th July 1784	68·4	40·8	640	1380
Thun	7th July 1783	66·2	41·0	373	1896
Brienz ...	8th July 1783	68·0	40·5	533	—
Maggiore .	19th July 1783	78·1	44·1	357	—
Lomond..	8th Sept. 1812	59·5	41·5	600	25
Katrine ..	7th Sept. 1812	57·3	41·0	480	364
	3rd Sept. 1814	56·4	41·3	—	—

An exceedingly important and valuable series of observations was made by Fischer and Brunner[1] in the Lake of Thun throughout the course of a whole year (March 1848 to February 1849). They used, after Saussure's method, thermometers protected by non-conducting envelopes, which were pulled up as quickly as possible. The depth of the water where they observed was 540 feet, and they made a series of observations of the temperature at that depth, at the surface, and at eleven intermediate depths, and repeated the series of observations at eight different dates over the year. From these series, which afford the first information of the yearly march of temperature at different depths, we learn that the lake as a whole gains heat till the end of September, then loses it until the month of February, when it begins to warm again, though slowly. The maximum temperature occurs in October at depths from the surface to 70 feet, in November at depths from 70 to 120 feet, in December from 120 to 200 feet, and in February at 500 feet. As the whole yearly variation of the temperature at 200 feet is less than a degree, the epoch at which the greater depths attain their maximum and minimum temperatures cannot be certainly deduced from one year's observations. The minimum temperature of depths from the surface to 80 feet is attained in the month of February, at greater depths in the month of March. During the course of the whole year the temperature at the bottom varied between 40·7° and 40·9° Fahr., and in the month of February the whole of the water from the surface to the bottom was between 40·7° and 41° Fahr.

These and other observations showed that, from depths of 400 feet, the variation of temperature with increasing depth is quite insignificant, so that even though the lake might be 1000 feet deep the temperature at 400 feet is only one or two tenths of a degree different from that of the bottom; further, on many of the thermometers recently used, it is impossible to distinguish with certainty temperature differing by less than half a degree, consequently it was not difficult to believe that in all deep lakes there is a considerable stratum of water which

[1] *Mém. Soc. Phys. Genève*, XII. p. 255.

remains constantly at the same temperature, all the year and every year, and that in winter this stratum thickens so as often to fill the lake, and gets thinner again in summer. By the improvement of the instruments both of these suppositions have been shown to be erroneous. In summer and in temperate latitudes, however deep the lake may be, its temperature falls as the depth increases, first rapidly and then very slowly, and the bottom temperature observed in any summer depends on the nature of the winter which preceded it, and may vary from year to year by one or two degrees. It was also believed that the deep water of a lake preserved constantly the mean winter temperature or the mean temperature of the six coldest months of the year in the locality. This was deduced from some observations by Sir Robert Christison in Loch Lomond, who found the bottom temperature at Tarbet to be 41·4° Fahr., agreeing with the mean of the six winter months as observed at Balloch Castle, which, however, is about 15 miles distant. Although the theory may be accidentally true for Loch Lomond, it has been proved not to hold for other lakes. Thus Simony (*Wien. Sitz. Ber.* 1875, LXXI. p. 435) gives the following table, comparing the temperature of the bottom water in the G'münder See with the winter (October to March) air temperature[1]:

	Winter period, Mean temperature			Summer period, Mean temp.	Bottom temp. Gmünder See	Date of observation of bottom temperature
	Oct.–Mar.	Dec.–Feb.				
	° F.	° F.		° F.	° F.	
1867–68	37·5	32·9	1868	64·4	40·5	6th Oct. 1868
1868–69	40·1	36·8	1869	63·1	40·5	1st Oct. 1869
1869–70	35·0	29·3	1870	60·8	40·2	26th Sept. 1870
1871–72	35·2	27·8	1872	62·2	40·0	3rd Oct. 1870
1872–73	41·0	35·0	1873	60·2	40·5	5th Oct. 1873
1873–74	39·0	32·7	1874	61·9	40·4	25th Sept. 1874
1874–75	33·8	28·2	1875	—	39·1	10th April 1875

It will be seen that, with the exception of the end of 1872,

[1] These air temperatures are those of the observatory at Vienna corrected for difference of level.

the mean winter temperature is below that of the bottom water, and generally very markedly so.

During 1877–81 observations have been made by the present writer on the distribution of temperature in lakes forming part of the Caledonian Canal. The monthly mean temperatures at Culloden and at Corran Ferry lighthouse, which cannot differ much in climate from Loch Ness and Loch Lochy respectively, have been supplied by Mr Buchan of the Scottish Meteorological Society. The bottom temperatures are those observed in the deepest part of the lakes, namely, 120 fathoms in Loch Ness, and 80 fathoms in Loch Lochy. The connection between bottom temperature (as observed in the second week of August) and winter temperature can be judged of from the following table, where the mean temperatures of October to March, and also of November to April are given:

	Loch Ness		Culloden		Loch Lochy		Corran	
	Surface	Bottom	Oct. to March	Nov. to April	Surface	Bottom	Oct. to March	Nov. to April
	°	°	°	°	°	°	°	°
1877	53·0	42·4	40·2	40·0	55·0	44·0	42·3	40·8
1878	59·0	42·3	41·6	40·9	61·0	43·7	42·7	42·5
1879	51·4	41·2	37·2	35·8	54·0	42·0	38·9	37·5
1880	57·0	42·4	41·0	40·8	57·6	43·8	42·0	41·9
1881	53·1	41·45	36·1	36·2	54·0	42·25	38·6	38·7

From this table it is apparent that the bottom temperature, even of lakes as deep as Loch Ness, is subject to considerable variation from year to year, that it depends on the temperature of the previous winter, and that it is usually higher than that temperature. The difference between the bottom temperature and the mean winter temperature is greater the lower the winter temperature is. It is further interesting to notice that the mean winter temperature of 1878–79 was about one degree higher than that of 1880–81, yet the bottom temperatures were 0·25° lower in 1879 than in 1881, and this is no doubt due to the fact that the cold of 1878–79 was more continuous than that of 1880–81, when the actual temperatures observed were much lower. The temperature of the bottom water depends not only on the temperature of the previous winter, and on the depth of the lake; it also depends on the nature of the country where it lies, and especially on its exposure to winds. Winds drive the surface water before them, and if there were no return current it would be heaped up at the further end. The effect is to accumulate surface water at one end, and to draw on deeper water to make up the deficiency at the other end. Hence the prevailing direction of the wind impresses itself on the distribution of temperature in the water; and this is well shown in the distribution of temperature as determined from observations at five stations on the same day in Loch Ness in a summer after a warm winter, and in one after a cold winter. In Scotland, warm weather is associated with southerly and westerly winds, and cold weather with northerly and easterly winds. In the warm years we have accumulation of surface water at the north-eastern end, and of bottom water at the south-western end, producing in summer a higher mean temperature of water at the north-east, and a lower mean temperature of water at the south-west end. In cold years the reverse is observed. Thus in 1879, after a cold winter, the mean temperature of the first 300 feet of water at the south-west end of Loch Ness was 48·8°, and at the north-east end 44·96°, a difference of nearly four degrees. In 1880, after a comparatively mild winter, it was 48·13° at the south-west end, and 47·95° at the north-east end, or nearly identical temperatures. Even at stations a few hundred yards from each other, great differences are often observed in the temperatures observed at the same depth, and it is evident that the difference of density so produced must cause a certain amount of circulation. There can be but little doubt that, under the influence of the varying temperature of the seasons, and of the winds, the water of a lake is thoroughly mixed once a year. In lakes which do not consist of a single long trough like Loch Ness, but of several basins as Loch Lomond, the bottom temperature is different in the different basins, even when the depth is the same. Loch Lomond consists of three principal basins of very unequal depth: the large expanse of water studded with islands at the lower end, the Balloch basin; the middle or Luss basin; and the upper and deepest or Tarbet basin. In the last we have 600 feet of water, in the Luss basin 200 feet, and in the Balloch basin a maximum of 72 feet of water. On 23rd September 1876 the bottom temperature in the Tarbet basin was 41·4°, and in the Luss basin 46·4°. Loch Tummel, a much smaller lake, consists of three basins, each of them being from 100 to 120 feet deep, and in them we have bottom temperatures of 46·3°, 46·9° and 45·2°, the lowest temperature being nearest the outlet.

It might have been expected that the bottom temperature in lakes similar as regards size and depth would be lower at greater elevations and higher nearer the sea-level. This does not, however, hold universally; thus Lochs Tummel and Garry are very similar in size and depth; they are only 12 miles from each other, but Loch Tummel is 450 feet and Loch Garry 1330 feet above the sea; yet at 102 feet in Loch Garry the temperature on the 18th August 1876 was 53·9°, and in Loch Tummel at the same depth on the 16th August 1876 it was 45·4°. The difference of elevation is nearly 900 feet, and, instead of the higher lake holding the colder water, its water is 8·5° warmer than that of the lower one. Similarly in Loch Ericht, 1153 feet above the sea, the bottom temperature at 324 feet was 44·7°, and in Loch Rannoch, 668 feet above sea, at the same depth it was 44·0°. These examples will suffice to show that many circumstances concur in determining the temperatures of the waters of lakes. There is one factor which is often neglected, namely, the amount of change of water. This depends on the drainage area of its tributary streams, and necessarily varies greatly.

In comparing the bottom temperature in lakes with the mean temperatures of the coldest half of the year, we find that the two approach each other more nearly the higher these temperatures are. When the temperature of the air falls for a lengthened period below the temperature of maximum density of water (39·2° Fahr.), then the mechanical effect produced is much the same as if the temperature had been raised. For, in virtue of the cooling above, the water will have no tendency to sink; it will rather tend to float as a cold layer on the surface of the warmer and denser water below. Were a lake comparable with a glass of water, that is, were its depth equal to or greater than its length or breadth, it would be possible to realize this ideal condition of things, which, until recently, was supposed to represent what really takes place when a lake is covered with ice, namely, that after the water has all been cooled to a uniform temperature of 39·2° Fahr. further cooling affects only a small surface layer, which consequently rapidly freezes. If this were the case, we should expect to find the temperature of the water below the ice of a frozen lake increasing rapidly from 32° where it is in contact with the ice to 39·2° at a short distance from it, and we should expect to find the remainder of the water down to the bottom at the same temperature. In fact, however, the depth of even the deepest lakes bears an insignificant proportion to their superficial dimensions, and temperature observations in summer show that the effective climate, that is, the climate in so far as it is effective for the purpose under consideration, varies much over the surface of even very small lakes. The variations in distribution of temperature produce variations in density which of themselves are sufficient to produce convection currents. Then, as a factor of climate, there are the winds, which are the main mixing agents, and also the movement in the waters caused by the inflow of water at different points and the removal of the excess at one point. The effect of these mechanical agents, winds and currents, is to propagate the air temperature at the surface to a greater depth than would otherwise be the case. At the same time it must be remembered that in seasons of great cold there is rarely much wind. If we reflect, however, on what must take place when there is a large expanse of open water in the middle of a country covered with snow, and exposed to the rigours of a winter night, we see that the air in contact with the surface of the water must get warmed and form an ascending current, its place being taken by fresh air drafted from the cold land surface, which not only cools the water but forces it out towards the middle; thus establishing a circulation consisting in broad lines of a surface movement from the sides to the middle of the lake, and a movement in the opposite direction below the surface. Even if the current of air were not sufficient of itself to produce a surface current in the water, it would do it indirectly. For, as it first strikes the water at the edges, the water would get cooled most rapidly, and under suitable circumstances would form a fringe of ice; the water so cooled would be lighter than the warmer water farther out, and would have a tendency to flow off towards the middle, or with the current of air. Now, although, when compared with other seasons, there is in a hard frosty winter not much wind, still, even in the calmest weather there is almost always sufficient motion in the atmosphere to enable the meteorologist to state that the wind is from a particular quarter; this will assist the circulation which has just been described as taking place in a calm lake, though it will somewhat distort its effects. It will produce excessive cooling at the side nearest the wind, and, when the lake freezes, it will have a tendency to begin at the windward side.

The extent to which this circulation affects the deeper waters of a lake depends on local circumstances, and generally we may say that the more confined a lake is the more easily will it freeze, and the higher will be the mean temperature of its waters. In the very cold winter 1878–79 the writer was able to make observations on the temperature of the water under the ice in Linlithgow Loch and in Loch Lomond. In the following winter, which, though mild in Scotland, was excessively severe in Switzerland, Dr Forel made observations in the Lakes of Morat and Zürich, confirming the writer's observations of the unexpectedly low temperature of the water. The freezing of so deep a lake as that of Zürich was a fortunate circumstance, because in it the bottom is actually at the temperature of maximum density. The majority of the lakes which freeze are so shallow as to admit of the whole of their water being cooled considerably below the temperature of maximum density.

The distribution of temperature in frozen lakes will be

apparent from the table given below. Of the Lakes of Zürich and Morat and Loch Lomond the mean temperatures are in the order of their depth. Linlithgow is altogether peculiar. Its high temperature, which increased steadily all the time it was covered with ice, was due to chemical action amongst the filth which has been allowed to accumulate at its bottom. When the ice broke up the dead fish were taken away in carts.

Table of Temperature in Frozen Lakes.

Depth (in feet)	Temperature in degrees Fahr.				
	Zürich 25th Jan. 1880	Morat 23rd Dec. 1879	Lomond 29th Jan. 1879	Linlithgow 11th Jan. 1879	Linlithgow 25th Jan. 1879
3	—	—	33·00	35·90	36·00
6	—	—	33·50	36·30	36·80
18	—	35·06	33·95	36·90	37·80
(Bottom) 48	36·95	36·14	35·20	39·85	42·05
(Bottom) 65	37·25	36·30	36·30	—	—
100	37·76	36·68	—	—	—
(Bottom) 150	38·39	37·04	—	—	—
200	38·66	—	—	—	—
300	38·84	—	—	—	—
(Bottom) 435	39·20	—	—	—	—
Mean......	38·40	36·00	34·46	37·22	38·28

Dr Forel gives the following particulars about the frozen Swiss lakes. "The Lake of Morat has a surface of 27·4 square kilometres and a maximum depth of 45 metres (147 feet); it is 1425 feet above the sea; and its mean latitude is 49° 56′ N. The ice overspread its whole surface suddenly in the night of the 17th to the 18th December, and it remained frozen till the 8th March. The Lake of Zürich has a superficies of 87·8 square kilometres, a maximum depth of 468 feet and altitude of 1338 feet, and a mean latitude of 47° 16′ N. Its congelation was gradual, and not sudden like that of the Lake of Morat. First the upper part of the lake was covered with ice between Mannedorf and Wadensweil. At the end of December the 28th, the ice covered it entirely, but only for a single day. On the 29th it thawed, and the lake remained partially free of ice until the middle of January. It froze over completely on the 22nd January, and on the 25th the ice was four inches thick in the centre of the lake." Of the larger Swiss lakes, Morat, Zürich, Zug, Neuchâtel, Constance, and Annecy were frozen in 1880; Thun is known to have been frozen four times, namely in 1363, 1435, 1685, and 1695; Brienz has only once been frozen, in 1363; Lucerne freezes partially in very severe winters, and Geneva in its western and shallower part, whilst Wallenstadt and Bourget are not known to have ever been frozen.

For further information on the temperature of frozen lakes, see Buchanan, *Nature*, March 6, 1879; Forel, *Arch. de Genève*, 1880, IV. I; Nichols, *Proc. Boston Soc. of Nat. Hist.* 1881, XXI. p. 53.

Changes of Level. As the water supply of lakes depends on the rainfall, and as this varies much with the season, and from year to year, we should expect, and indeed we find, fluctuation of level in all lakes. There are, however, other changes of level which are independent of the water supply, and which resemble tides in their rhythmic periods. They have long been known and observed in Switzerland, and especially on the lake of Geneva, where they are known by the name of "seiches." The level of the lake is observed to rise slowly during twenty or thirty minutes to a height which varies from a few centimetres to as many decimetres; it then falls again slowly to a corresponding depth, and rises again slowly, and so on. These movements were observed and much studied at the end of last century by Jallabert, Bertrand, and Saussure, and at the beginning of this century they formed the subject of an instructive memoir by Vaucher, who enunciated the following law connecting the seiches with the movements of the barometer. "The amplitude of seiches is small when the atmosphere is at rest; the seiches are greater the more variable is the atmosphere's pressure; they are the greatest when the barometer is falling." Vaucher recognised the existence of seiches in the Lakes of Geneva, Neuchâtel, Zürich, Constance, Annecy, and Lugano, and Dr Forel of Morges, from whose papers, published principally in the *Bibliothèque Universelle et Revue Suisse* during the last five years, the facts regarding the seiches have been taken, has observed them in every lake where he had looked for them. It is in every way likely that they are to be found in all lakes of notable extent and depth. They have been studied principally on the Lake of Geneva, where Dr Forel, at Morges, about the middle of the lake on the north shore, and M. Plantamour, at Sécheron, about a mile from Geneva on the north shore, have had self-registering tide gauges in operation for a number of years. In the writings of the Swiss observers the seiche is the complete movement of rise above and fall below the mean level, the amplitude is the extreme difference of level so produced, and the duration of the seiche is the time in seconds measured from the moment when the water is at the mean level until it is again at the mean level, after having risen to the crest and sunk to the trough of the wave. The amplitude of the seiches is very variable. At the same station and on the same day successive seiches are similar. When the seiches are small they are all small, when they are large they are all large. At the same station and on different days the amplitudes of the seiches may vary enormously. For instance, at Geneva, where the higher seiches have been observed, they are usually of such a size as to be imperceptible without special instruments; yet on the 3rd August 1763 Saussure measured seiches of 1·48 metres, and on the 2nd and 3rd October 1841 the seiches observed by Vénié were as much as 2·15 metres. They are greater at the extremities than at the middle of lakes, at the head of long gulfs whose sides converge gently than at stations in the middle of a long straight coast, and in shallow as compared with deep lakes or parts of a lake. They also appear to increase with the size of the lake. The duration of the seiches is found to vary considerably, but the mean deduced from a sufficient number of observations is fairly constant at the same locality. Thus, for Morges, Dr Forel has found it to be for the half seiche 315 ± 9 seconds. At different stations, however, on the same lake and on different lakes it varies considerably.

Thus on the Lake of Geneva it is, for the complete seiche, 630 seconds at Morges, and 1783 seconds at Veytaux; on Lake Neuchâtel it is 2840 seconds at Yverdon, and 264 seconds at Saint Aubin.

The curves traced by the gauge at Geneva have been subjected to a preliminary harmonic analysis by Professor Soret, and he has decomposed them into two undulations, the one with a period, from crest to crest, of 72 minutes, and the other with a period of 35 minutes, or a little less than half the larger period. As the amplitudes of the composing curves vary much, there is great variety in the resultant curves. Besides these two principal components, there are others which have not yet been investigated.

With regard to the cause of the phenomenon, Dr Forel attributes the ordinary seiches to local variations of atmospheric pressure, giving an impulse the effect of which would be apparent for a long time as a series of oscillations. The greater seiches, such as those of 1·5 metres, he attributed to earthquake shocks; but, as a very sensible earthquake passed over Switzerland quite recently without leaving the slightest trace on the gauge, he has abandoned this explanation, and is inclined to attribute them to pulsation set agoing by violent downward gusts of wind, especially at the upper end of the lake. M. Plantamour, who has devoted much attention to the same subject, assured the writer, in the summer of 1881, that he was completely at a loss for a satisfactory explanation of them.

Seiches have not been observed on the Scottish lakes, though there is little doubt that they would be found if sought for. There are, however, records of disturbances of some of the lakes, especially in Perthshire, of which the following may be cited as an instance.

A violent disturbance of the level of Loch Tay is reported in the *Statistical Account of Scotland* (1796), XVII. p. 458, to have occurred at Kenmore on 12th September 1784, continuing in a modified degree for four days, and again on 13th July 1794. Kenmore lies at the north-eastern end of the lake, where the river Tay issues from it. It lies at the end of a shallow bay. "At the extremity of this bay the water was observed to retire about five yards within its ordinary boundary, and in four or five minutes to flow out again. In this manner it ebbed and flowed successively three or four times during the space of a quarter of an hour, when all at once the water rushed from the east and west in opposite currents,...rose in the form of a great wave, to the height of five feet above the ordinary level, leaving the bottom of the bay dry to the distance of between 90 and 100 yards from its natural boundary. When the opposite currents met they made a clashing noise and foamed; and, the stronger impulse being from the east, the wave after rising to its greatest height, rolled westward, but slowly diminishing as it went, for the space of five minutes, when it wholly disappeared. As the wave subsided it flowed back with some force, and exceeded its original boundary four or five yards; then it ebbed again about 10 yards, and again returned, and continued to ebb and flow in this manner for the space of two hours, the ebbings succeeding each other, at the distance about seven minutes, and gradually lessening, till the water settled into its ordinary level. During the whole time that this phenomenon was observed, the weather was calm. On the next and four succeeding days an ebbing and flowing was observed nearly about the same time and for the same length of time, but not at all in the same degree as on the first day."

The above is the account given by the Rev. Thomas Fleming, at the time minister of Kenmore, who was an eye-witness. It resembles in all essential particulars the descriptions of waves which accompany actual earthquakes, yet in his account he goes on to say: "I have not heard (although I have made particular inquiry) that any motion of the earth was felt in this neighbourhood, or that the agitation of the wave was observed anywhere but about the village of Kenmore." It is well known that there were great seismic movements observed in Perthshire at the time of the Lisbon earthquake, and there is a tradition in the neighbourhood that Loch Lubnaig near Callander was largely increased in extent by the dislocations which took place.

In all lakes there are changes of level corresponding with periods of rain and of drought. They are the more considerable the greater the extent of country draining into them, and the more constrained the outflow. In the great American lakes, which occupy nearly one-third of their drainage area, the fluctuations of level are quite insignificant; for Lake Michigan the United States surveyors give as the maximum and minimum yearly range 1·64 and 0·65 feet. In the Lake of Geneva the mean annual oscillation is five feet, and the difference between the highest and the lowest waters of this century is 9·3 feet. The most rapid rise has been 3·23 inches (82 mm.) in 24 hours. A very remarkable exception to the rule that large freshwater lakes are subject to small variations of level is furnished by Lake Tanganyika in Central Africa. Since its discovery travellers have been much perplexed by the evidence and reports of considerable oscillations of level of uncertain period, and also by the apparent absence of visible outlet, while the freshness of its waters was of itself convincing evidence of the existence of an outlet. By the careful observations of successive explorers the nature of this phenomenon has been fully explained, and is very instructive. It has recently been visited by Captain Hore of the London Missionary Society, and it appears from his reports that the peculiar phenomena observed depend on the fact that the area of country draining into the lake is very limited, so that in the dry seasons the streams running into it dry up altogether, and its outlet gets choked by the rapid growth of vegetation in an equatorial climate. A dam or dyke is thus formed which is not broken down until the waters of the lake have risen to a considerable height. A catastrophe of this kind happened whilst Captain Hore was in the neighbourhood, and he noted the height of the water at different times near his station at Ujiji, and observed it fall two feet in two months. It continued to fall until in seventeen months it had fallen over ten feet. Taking the length of the lake at 330 miles, and the mean breadth at 30 miles, its surface is 9900 square nautical miles. If this surface be reduced two feet in 60 days, the water will have to escape at the rate of 137,500 cubic feet per second. The mean rate of discharge of the Danube is 207,000 cubic feet per second. Hence, without taking into account water which would be brought into the lake by tributaries during the two months, we require for outlet a river at least two-thirds of the size of the Danube, and in the Lukuga such a river is found. When Stanley visited it the Lukuga was quite stopped up with dense growth, and no water was issuing; the lake was then rising; when Captain Hore visited it the lake was falling rapidly, and the Lukuga was a rapid river of great volume. One of the chief affluents to the lake was found to be discharging at the rate of 18,750 cubic feet of water per second; a few months later it was dry and the mouth closed with vegetation. During the dry season too the lake, with its 10,000 square miles of surface, is exposed to the evaporating action of the south-east trade wind, and when the supply is so insignificant this must be sufficient of itself to sensibly lower the level. Ordinarily then we might expect the lake to be subject to a yearly ebb and flow corresponding to the periods of drought and rains; and, from what we learn of the great fluctuations of rainfall one year with another, we should expect that during a series of dry years the obstructions to the outflow would gain such a head that the rains of several wet seasons would have to accumulate before forcing a passage. The result would be a tide of a period corresponding to the recurrence of series of wet or dry years. Were the lake situated at or near the level of the ocean, its equatorial position would give it such a preponderance of rain over the whole year as to keep its outlet constantly open; but its actual position, 2700 feet above the sea, produces an alteration in climate, equivalent to an increase of latitude, which would place it in the trade wind region rather than in that of equatorial calms and rains. That such is actually the effect is shown by the range of temperature, which is moderate (59° to 83° Fahr.), and the rainfall (27 to 30 inches), which is almost exactly that of London. The Central African lakes, from their immense size and from their equatorial position, possess a peculiar interest for the physical geographer, and it is to be hoped that before long we shall have sufficient soundings to give a general idea of the size of their basins, and also temperature observations to show the effect of a vertical sun on large bodies of water at a moderate elevation, and removed from the disturbing influence of oceanic circulation.

As might be expected, in salt lakes which have no overflow, the yearly rise and fall is often considerable. In the Great Salt Lake in Utah, the greatest depth of which is 56 feet, changes of level are accompanied by great changes in water surface, and also in saltness of water. In the rainy season the Dead Sea stands ten or twelve feet higher than in the dry season. The table, p. 310, shows the chemical composition of the waters of various salt lakes, that of the sea-water in the Suez Canal being added for comparison.

This table embraces examples of several types of salt lake. In the Kokonor, Aral, and open Caspian seas we have examples of the moderately salt, non-saturated waters. In the Kara-bugas, a branch gulf of the Caspian, the Urumieh, and the Dead Sea we have examples of saturated waters containing principally chlorides. The Van Sea is an example of the alkaline seas which also occur in Egypt, Hungary, and other countries. Their peculiarity consists in the quantity of carbonate of soda dissolved in their waters, which is collected by the inhabitants for domestic and for commercial purposes. The chemical reader will be struck by the quantity of magnesia salt dissolved in water which contains so much carbonate of soda. The analysis in the table is by Abich, quoted by Schmidt in his interesting "Etudes Hydrologiques," published in the *Bulletin de l'Académie de St Pétersbourg*. Another analysis by De Chancourt, quoted by Bischof, omits all mention of sulphate of magnesia, but inserts the carbonate.

The limits of this article do not admit of the discussion of the many interesting phenomena connected with salt lakes. With regard, however, to a former connection of the Caspian with the Black Sea, which has been so often suggested, it seems improbable, both on chemical and on physical grounds, that they were ever connected as seas, that is, in the same way as the Black Sea is connected with the Mediterranean; but, if we consider the topography of the Caucasus district, we see that the lowest summit level of the land between the two seas is

	Kokonor Sea	Aral Sea	Caspian Sea		Urumieh Sea	Dead Sea	Van Sea	Suez Canal Ismailia
			Open	Karabugas				
Specific gravity	1·00907	—	1·01106	1·26217	1·17500	—	1·01800	1·03898
Percentage of salt	1·11	1·09	1·30	28·5	22·28	22·13	1·73	5·1
Name of salt	Grammes salt in 1000 grammes water							
Bicarbonate of lime	0·6804	0·2185	0·1123	—	—	—	—	0·0072
,, iron	0·0053	—	0·0014	—	—	—	—	0·0069
,, magnesia	0·6598	—	—	—	—	—	0·4031	—
Carbonate of soda	—	—	—	—	—	—	5·3976	—
Phosphate of lime	0·0028	—	0·0021	—	—	—	—	0·0029
Sulphate of lime	—	1·3499	0·9004	—	0·7570	0·8600	—	1·8593
,, magnesia	0·9324	2·9799	3·0855	61·9350	13·5460	—	0·2595	3·2231
,, soda	1·7241	—	—	—	—	—	2·5673	—
,, potash	—	—	—	—	—	—	0·5363	—
Chloride of sodium	6·9008	6·2356	8·1163	83·2840	192·4100	76·5000	8·0500	40·4336
,, potassium	0·2209	0·1145	0·1339	9·9560	—	23·3000	—	0·6231
,, rubidium	0·0055	—	0·0034	0·2510	—	—	—	0·0265
,, magnesium	—	0·0003	0·6115	129·3770	15·4610	95·6000	—	4·7632
,, calcium	—	—	—	—	0·5990	22·4500	—	—
Bromide of magnesium	0·0045	—	0·0081	0·1930	—	2·3100	—	0·0779
Silica	0·0098	—	0·0024	—	—	0·2400	0·0761	0·0027
Total solid matter	11·1463	10·8987	12·9773	284·9960	222·7730	221·2600	17·2899	51·0264

in the Manytsch valley, 86 feet above the Black Sea. Were the climate of the Caspian to change only very slightly for the moister, its waters might easily rise the 196 feet which would enable it to overflow towards the Mediterranean, while a relapse towards dryness would be followed by the retreat of the waters, which would be then confined as they are now to the basin of the sea. It is important, therefore, to bear in mind that no terrestrial dislocations are required to produce enormous changes in the level of salt lakes; we require only changes of climate, and these very slight. There can be little doubt that, if the climate of the Black Sea extended across the isthmus to the Caspian, the latter would now stand 200 feet higher, would be fresh, and would overflow into the Sea of Azov.

No. 21. [*From Proc. Roy. Soc. Edin.* 1885, *Vol.* XIII. *pp.* 403—408.]

ON THE DISTRIBUTION OF TEMPERATURE IN LOCH LOMOND DURING THE AUTUMN OF 1885

IN the course of the autumn of this year (1885) I have taken several occasions to determine the distribution of temperature in the water of Loch Lomond. The results of these observations are interesting, as indicating the march of temperature in the different layers at different localities in the lake, and also the gain and loss of heat with the changing seasons.

Loch Lomond is divided naturally into three basins[1]. If the level of the water were reduced by about eight fathoms, it would form three lakes—the upper and largest extending from the head of the loch to Rowardennan, the middle one from Rowardennan to the chain of islands stretching from Luss to Balmaha, and the third, and shallowest, from these islands to Balloch. The ridges which separate these basins are covered in the present state of the lake by from five to eight fathoms of water. The lowest, or Balloch basin, is of great extent and comparatively shallow, having a maximum depth of 13 fathoms. The middle, or Luss basin, is also of considerable extent, and has a maximum depth of 35 fathoms. The upper, or Tarbet basin, is long and narrow, and very deep, the maximum depth being 105 fathoms. At the upper end of this basin is a subsidiary one, which I call the Ardlui basin, with a maximum depth of 34 fathoms, and separated from the main basin by a ridge with a probable maximum depth of 17 fathoms.

The general direction of the lake is north and south, so that the prevailing westerly and south-westerly winds blow across it, and, as is always the case in mountainous districts, they are diverted into squalls, which blow sometimes up and sometimes

[1] See Chart of Loch Lomond, p. 339.

down the lake. At Tarbet there is a deep rift in the mountains separating Loch Lomond from Loch Long, which gives access to the westerly winds to this part of the lake. On the whole, the geographical position of the lake tends to neutralise the effect of the prevailing winds.

Extended temperature observations were made on the following days: 18th August, 5th and 22nd September, 15th October, and 14th November. On the 18th August observations were made only in the Tarbet basin, and only down to a depth of 30 fathoms. On the 5th September observations were made in the Luss basin, at four stations in the Tarbet basin, and at one station in the Ardlui basin. On the 22nd September observations were made in the Luss basin, and at two stations in the Tarbet basin. On the 15th October observations were made at the same stations as on 5th September, omitting Culness; and on 14th November observations were made in the Luss basin and at Inversnaid.

The observations were made with an improved form of protected six's thermometer, having a millimetre scale on the stem and a Fahrenheit scale on slips at the side. The average length of a degree Fahrenheit was three millimetres, and all the thermometers had been carefully and repeatedly compared with each other, and with a Kew corrected standard. The temperatures given are all in terms of the Kew standard. As a rule, the same thermometer has been sent to the same depth. Further, the same sounding-line was used on all occasions.

The results are collected in tables, and in some cases they are represented graphically by curves.

If we represent the distribution of temperature graphically[1] by a curve, having depths measured along the horizontal line of abscissae and temperatures along the ordinates, the winter distribution will be represented by a straight line parallel to the line of abscissae, such as *A*. As the spring advances and the meridian altitude of the sun daily increases, the temperature of the surface rises rapidly. The heat received at the surface is, during this season, propagated downwards, chiefly by conduction, which, in water, is a comparatively slow process,

[1] See Fig. 1, p. 314.

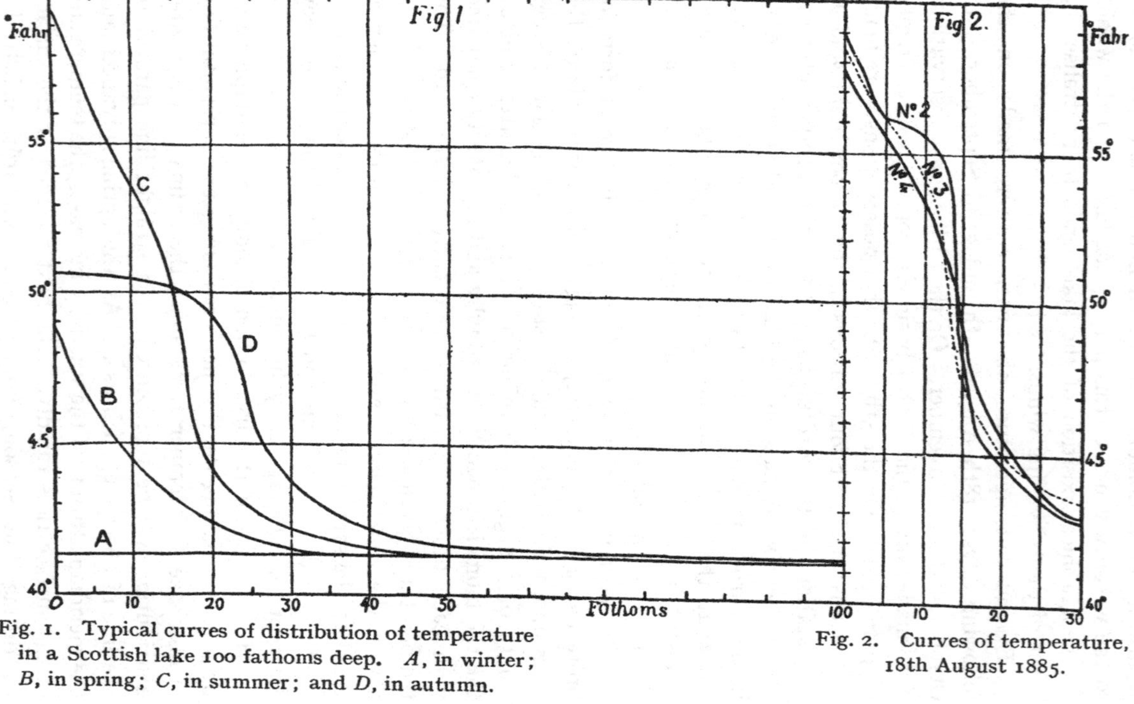

Fig. 1. Typical curves of distribution of temperature in a Scottish lake 100 fathoms deep. *A*, in winter; *B*, in spring; *C*, in summer; and *D*, in autumn.

Fig. 2. Curves of temperature, 18th August 1885.

hence the temperature of the surface, and of the layers near it, rises much more rapidly than that of those below it, and consequently the curve representing the vertical distribution takes the form *B*, which, from the bottom to within about 15 fathoms of the surface, preserves its parallelism to the line of abscissae, but then bends sharply upwards, presenting a well-marked convexity to the origin. This convexity of the curve is the distinctive feature of a *vernal* distribution of temperature. As the summer advances the temperature of the surface no longer increases at the same rate as before, indeed it tends always more and more to become constant. The heat of the surface layers is, however, always being propagated downwards by conduction, and when the temperature of the surface layer has become nearly constant, it follows that, at some depth a little below the surface, the temperature will be rising more quickly than in the layers above, and this produces a slight bulge in the curve *C*, representing the distribution. This part of the curve presents a concavity to the origin which, combined with the pronounced convexity below and the less marked convexity above, produces the typical *summer* distribution. When the autumn has set in, and the surface temperature falls from day to day, heat is still being propagated downwards by conduction and convection, the curve takes the typical autumnal form *D*, consisting of a horizontal piece near the surface united to another horizontal piece near the bottom by the summer concavity and the vernal convexity. Hence in the autumn the waters of a deep lake are exposed to all the different conditions of the four seasons of the year. In the deeper layers heat is propagated downwards most rapidly in the first half of the month of October. As the winter progresses heat leaves the water so rapidly by the surface that conduction downwards is checked and the deeper waters derive but very slight benefit from the summer heat at the surface.

Observations on 18th August 1885. These observations were confined to the Tarbet basin, and were only carried to a depth of 30 fathoms. The weather was perfect for sounding operations, being quite calm, so that the steam launch, which was used for the work, remained in position without trouble.

The sun was very powerful all day, and its heating effect may be judged from the fact that at 10 A.M. the surface water in the channel off Camstradden was 58·9°, and at 5.30 P.M. in the same position it was 62·6° F., indicating a rise of 3·7° in the course of the day. As an assistance to finding the positions of the stations on the Chart, p. 339, I give the distance in nautical miles in a straight line from Balloch Pier; as a rule, they were made in the deepest part of the loch in the locality.

TABLE I. *Observations in Loch Lomond, 18th August 1885.*

Locality		Rowardennan	Rowcreeshie	Culness	Rob Roy's Cave	Range of temperature at the same depth
Miles from Balloch Pier		9	10	12¾	15	
Depth at station		25	70	100	60	
Hour of day		noon	3.30 P.M.	1 P.M.	2 P.M.	
No. of station		1	2	3	4	
Depth fathoms	No. of thermometer	Temperature (Fahr.)				
		°	°	°	°	°
0	—	57·5	58·7	58·4	57·7	1·2
5	21	56·75	56·2	56·3	55·5	1·25
10	9	56·0	55·75	54·3	53·45	2·55
15	23	49·45	48·85	47·95	50·0	2·05
20	47	44·5	45·0	45·2	45·8	1·3
25	79	43·35	43·85	44·0	43·8	0·65
30.	80	—	42·95	43·35	43·0	0·4
0 to 30 mean ...		—	50·90	49·74	49·8	0·35
Steepest gradient	Degs. per fm.	1·31	1·38	1·26	0·84	—
Steepest gradient	Depth	13	13	12	17	—

The distribution in series Nos. 2, 3, and 4 is graphically represented in Fig. 2, and it will be seen that the curves have a marked summer character. The steepest gradients are between 10 and 15 fathoms. At Culness (No. 3) the fall of temperature is 6·9° in this interval, or 1·38° per fathom. The least steep gradient is on the most northerly station, No. 4, indicating greater mixture of the layers of water towards the head of the loch. The curves show also, in a remarkable manner, the

TABLE II. *Observations in Loch Lomond, 5th September* 1885.

Name of basin ...		Luss	Tarbet				Ardlui
Locality........		Ross Mill	Rowardennan	Stuckgowan	Culness	Inversnaid	Doune Farm
Miles from Balloch Pier		7½	9	11	12¾	14	17
Depth at station ..		33	37	87	100	100	34
Hour of day		11 A.M.	Noon	1 P.M.	2 P.M.	3 P.M.	4.20 P.M.
No. of station		5	6	7	8	9	10
Depth fathoms	No. of thermometer	Temperature (Fahr.)					
		°	°	°	°	°	°
0	—	56·2	56·5	56·4	—	56·0	56·5
5	9	55·9	55·1	55·3	—	55·8	56·3
10	80	55·75	54·8	54·8	—	55·75	55·75
15	79	49·1	51·25	49·0	—	49·8	48·1
20	47	48·3	45·6	45·8	—	45·6	46·5
30	21	47·0	43·2	43·5	—	43·2	45·25
35	47	—	—	—	42·4	42·7	—
40	79	—	—	42·5	—	—	—
45	79	—	—	—	42·15	42·2	—
50	9	—	—	42·05	—	—	—
65	9	—	—	—	41·85	41·8	—
70	80	—	—	42·0	—	—	—
80	47	—	—	41·75	—	—	—
85	80	—	—	—	41·8	41·7	—
Btm. 87	21	—	—	41·8	—	—	—
Btm. 100	21	—	—	—	41·8	41·8	—
0 to 30 mean......		53·93	51·23	50·87	—	51·32	51·57
Steepest gradient: Degs. per fathom		1·33	1·13	1·16	—	1·19	1·53
Steepest gradient: Depth		13	16	13	—	15	13

difference in temperature of the water at the same depth in different localities. The greatest difference is found at ten fathoms, at which depth the temperature at Rowardennan exceeds that at Rob Roy's Cave by 2·55°. Although the distribution varies at the different stations, there is little difference in the mean temperature of the 30 fathoms, it is higher at Rowcreeshie than farther north. From the surface to ten fathoms the highest temperatures are at the lower end of the basin, from 15 to 30 they are nearer the upper end. At 15 fathoms the temperature at Culness is lower than either north or south of it.

5th September 1885. All day the weather was most favourable for experimenting, except perhaps at Culness, when it threatened for a few minutes to blow and rain. Otherwise it was almost quite calm, with overcast sky, so that there was no overheating by the sun or cooling by the wind.

Positions—No. 5. Outer Ross Island bears S. 27° E. (true), distant 0·74′.

No. 6. Rowardennan Lodge bears N. 102° E. (true), distant 0·43′.

No. 7. Stuckgowan Lodge bears S. 67° W. (true), distant 0·32′ to 0·37′.

No. 8. Tarbet Pier bears S. 43° W. (true), distant 1·3′.

No. 9. Inversnaid Inn bears N. 34° E. (true), distant 0·7′.

No. 10. Stuckindroir House bears N. 60° W. (true), distant 0·38′.

All the places mentioned in this paper are to be found on the Chart of the lake, page 339.

Owing to the overcast state of the sky, there is little variation in the temperature of the surface. Below the surface there is again considerable variation, but the maximum range, 2·25° at 15 fathoms, is less than was observed on 18th August. The character of the distribution is distinctly autumnal. On 18th August the observations were confined to the Tarbet basin and were limited to 30 fathoms; to-day they extend to the three deep basins of the loch. The Ardlui and the Luss basins resemble each other in that their maximum depth is about the same 34 fathoms; but the Ardlui basin is separated from the Tarbet one by a ridge of probably 17 fathoms, while the ridge shutting off the Luss basin has a maximum depth of only eight fathoms, cold deep water is thus enabled to penetrate from the Tarbet basin into the Ardlui basin, but not into the Luss basin. Further, the Ardlui basin receives, for its size, a much greater supply of the affluent waters from the land, so that its waters in winter are probably colder than those of the lake lower down.

In accordance with the autumnal character of the distribution, the temperature of the first ten fathoms approaches uniformity at all the stations. It is highest at Ardlui, being 56·2°, and lowest at Rowardennan, being 55·1°. The steepest gradients are all between 10 and 20 fathoms. They are steeper in the shallow basins than in the deep ones; in the Ardlui basin the average gradient is 1·53° per fathom between 10 and 15 fathoms.

On 7th September the Inversnaid station was revisited, and the temperatures on the gradients accurately ascertained by sending thermometers to every fathom, from 13 to 17 inclusive, with the following result:

Observations at Inversnaid, 7th September 1885.

Depth fathoms	No. of thermometer	Temperature degs. Fahr.	Gradient, degs. per fathom
		°	°
13	21	52·0	—
14	47	51·25	0·75
15	79	49·8	1·45
16	9	48·8	1·0
17	80	47·25	1·55

The mean gradient in these four fathoms of water is 1·19° per fathom, the maximum is 1·55° between 16 and 17 fathoms. It is therefore probable that the actual maximum gradient in the Ardlui basin may be as much as 2° per fathom.

Owing to a mistake, 105 fathoms of line were paid out at Culness and Inversnaid instead of 100, which accounts for the irregular intervals between the thermometers at the deeper depths. From 30 to 70 fathoms the temperature of the water is slightly higher at Stuckgowan than at Inversnaid. From

70 fathoms to the bottom the temperature of the water at the three deep stations is sensibly uniform, namely, 41·8°. On 7th September the three thermometers, Nos. 47, 79, and 80, were sent down together to 60 fathoms at Inversnaid, and their corrected temperatures were 41·9°, 41·9°, and 41·85°. There is, therefore, a fall of 0·1° between 60 and 100 fathoms at this season of the year. The same thermometers were also sent to 30 fathoms in the Luss basin, and their corrected readings were 47·0°, 47·0°, and 47·0°.

TABLE III. *Observations in Loch Lomond, 22nd September* 1885.

Name of basin		Luss	Tarbet	
Locality		Ross Mill	Rowar-dennan	Inver-snaid
Miles from Balloch Pier		7½	9	14
Depth at station		33	37	100
Hour of the day		1.30 P.M.	2.30 P.M.	4 P.M.
No. of station		11	12	13
Depth fathoms	No. of thermometer	Temperature (Fahr.)		
		°	°	°
0	—	53·6	53·6	53·7
5	9	53·5	53·35	53·7
10	80	53·45	53·4	53·65
15	79	53·4	52·3	52·25
20	47	48·8	45·1	47·2
30	21	47·55	42·9	43·5
35	47	—	—	42·6
45	79	—	—	42·25
65	9	—	—	42·0
85	80	—	—	41·8
Btm. 100	21	—	—	41·8
0 to 30 mean		51·94	50·49	51·08
Steepest gradient	Degs. per fath.	0·9	1·44	1·01
	Depth	18	17	18

22nd September 1885. The weather was very stormy, and had been so for a fortnight, with much rain, so that the level of the lake was very high. Wind fresh from the south-west. The weather was so squally that it was difficult to keep station. At the Ross Mill station I kept the launch head to wind with a couple of oars out, but it was not very successful. At the other two stations I kept her stern to wind, with a steer-oar out over the bow and occasionally driving the engine astern. At Inversnaid the wind was so strong that I was able to keep the engine going continually dead slow astern, and kept station well.

The character of the distribution is pronouncedly autumnal, cooling at the surface is going on rapidly while heat is being propagated into the lower layers. The surface layer of approximately constant temperature is now about 15 fathoms thick at all the stations, and the steepest gradients are found between 15 and 20 fathoms, the maximum being 1·44° per fathom at Rowardennan. The bottom temperature has not sensibly altered at Inversnaid since 5th September. In the Tarbet basin the greatest difference of temperature at the same depth is 2·1° at 20 fathoms.

15*th October* 1885. A very fine day, with moderate north-easterly wind. For three days a strong northerly wind had been blowing with dry weather, and for a fortnight before there had been a succession of cyclonic gales, with much rain, so that at the beginning of the week the level of the lake was higher than I have ever seen it, and quite three feet above its usual summer level. The operations were all successfully carried out, the launch being kept stern to wind.

Amongst the salient features of the distribution of temperature at this date may be mentioned the rapid cooling which has taken place in the interval since 22nd September, the mean temperature of the first 30 fathoms having fallen about 3° F. at Ross Mill and Inversnaid. It will be seen that in the Luss basin the distribution has become nearly quite uniform from surface to bottom, and at Ardlui the water is rapidly approaching the same condition. The temperature of the water from the surface to 20 fathoms is practically uniform. At Ross Mill, in the Luss basin, and at Rowardennan and Stuckgowan, in the Tarbet basin, it is much alike, namely,

49·12°, 49·04°, and 49°. At Inversnaid it has fallen to 48·24°, and at Ardlui it is as low as 46·58°. The depth of water at Ross Mill and at Ardlui is nearly identical (34 fathoms), but

TABLE IV. *Observations in Loch Lomond, 15th October 1885.*

Name of basin		Luss	Tarbet			Ardlui
Locality		Ross Mill	Rowardennan	Stuckgowan	Inversnaid	Doune Farm
Miles from Balloch Pier		7½	9	11	14	17
Depth at station......		33	37	91	100	34
Hour of day		11 A.M.	Noon	1 P.M. 4.30 P.M.	1.30 P.M. 3.30 P.M.	2.30 P.M.
No. of station		11	12	13	14	15
Depth fathoms	No. of thermometer	Temperature (Fahr.)				
		°	°	°	°	°
0	—	49·4	49·1	49·1	48·7	47·8
5	9	49·2	49·0	49·05	48·4	46·7
10	80	49·05	49·05	49·0	48·2	46·3
15	79	49·0	49·0	49·0	48·1	46·3
20	47	49·1	48·95	48·8	47·8	46·2
30	21	48·9	42·44	42·9	45·3	45·9
35	47	—	—	—	44·2	—
40	79	—	—	42·4	—	—
45	79	—	—	—	42·8	—
50	9	—	—	42·2	—	—
65	9	—	—	—	42·2	—
70	80	—	—	42·1	—	—
80	47	—	—	42·0	—	—
85	80	—	—	—	42·0	—
90	21	—	—	42·0	—	—
Btm. 100	21	—	—	—	42·0	—
0 to 30 mean		49·0	48·37	48·37	47·9	46·47
0 to 20 ,, ...		49·12	49·04	49·00	48·24	46·58

the Luss basin is large, broad, and situated near the outflow of the lake; while the Ardlui basin is small and confined, and situated close to the head of the loch, so that it might almost be taken as forming part of the embouchure of the river Falloch. On the 5th September the temperature of the water down to

15 fathoms was nearly alike in the two basins, but below 15 fathoms it was much colder at Ardlui than at Ross Mill, the difference being 1·75° at 30 fathoms.

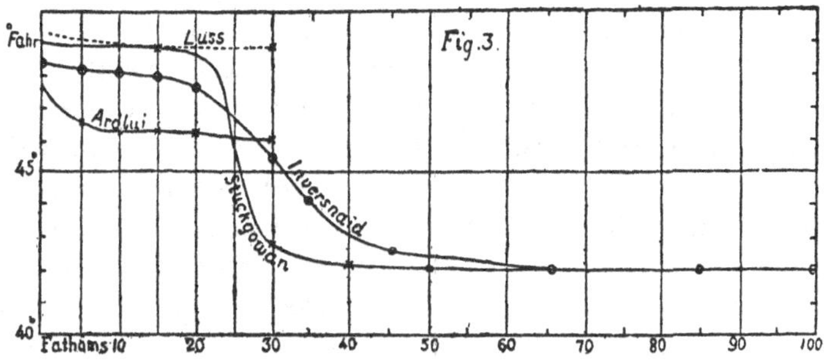

Fig. 3. Curves of temperature, 15th October 1885.

In the following table the mean temperature of the water found between the surface and 20 fathoms, the surface and 30 fathoms, and between 20 and 30 fathoms, are given for Luss and Ardlui as observed on 5th September and 15th October:

TABLE V. *Comparative Table of the Mean Temperature of certain intervals in the waters of the Ardlui and the Luss Basins on 5th September and 15th October* 1885.

Interval	Surf. to 20 fms.		Surf. to 30 fms.		20 to 30 fms.	
Date	Sept. 5	Oct. 15	Sept. 5	Oct. 15	Sept. 5	Oct. 15
	°	°	°	°	°	°
Mean temp. at Luss...	53·24	49·12	51·37	49·00	47·65	49·00
Mean temp. at Ardlui .	52·90	46·58	50·33	46·40	45·85	46·05
Difference	0·34	2·54	1·04	2·60	1·80	2·95

A comparison of these data brings out very clearly the difference in conditions obtaining in the two basins notwithstanding their likeness in depth. At Ardlui the predominating influence is that of the important tributary, the Falloch, which influences the temperature of the lake water in its neighbourhood most, while its temperature is lower than that of the lake. On 16th October the temperature of the water of the Douglas was

44·6° F., while that of the surface of the lake in its neighbourhood was 49·0°. On 13th November the temperature of the stream at Inversnaid was 40°, and that of the lake surface 46°. From its rise and course the Falloch is more likely to be colder than warmer than these streams, so that even in October it must have begun to spread its cooling influence over the lower waters of the Ardlui basin. When the water of the stream is warmer than that of the lake surface, it passes away with the drainage, and imparts as much of its heat to the atmosphere above it as to the water below. When its temperature is lower than that of the lake surface, and in all probability it is so for more than half the year, it sinks into the body of the lake, and imparts its cold entirely to its deeper waters. It is obvious then that during the time that it is colder than the lake, the water of the Falloch must produce a much greater effect on it than during the opposite season; hence the position of the Ardlui station, with respect to the principal tributary of the lake, renders it natural to expect that its waters would be colder than they are found to be in the Luss basin, which, from its size and position, is comparatively exempt from the direct influence of tributary waters.

Both in September and in October the temperature of all these bodies of water is lower at Ardlui than at Luss, but the contrast is much greater in the colder than in the hotter month.

In the Tarbet basin a salient feature is the greater mixture of waters at Inversnaid than at either Rowardennan or Stuckgowan, the curves at the latter localities being much steeper than at Inversnaid. The same phenomenon was observed on the 18th August. But, perhaps, the principal feature in the Tarbet basin is that both at Inversnaid and at Stuckgowan the water at and near the bottom has risen in temperature by 0·2° F., or from 41·8° to 42·0° since the 22nd September. As the temperature of the bottom water at these localities was 41·8° on the 5th September, it is probable that it had been so during the summer, and it is only by the end of September or beginning of October that the summer heat begins to have any effect on the water near the bottom. The steepest gradients are at Rowardennan and Stuckgowan between 20 and 30 fathoms. With a view of further investigating this body of water, I returned on 16th October to the Rowardennan and Stuckgowan stations, and took the temperature at 20, 22½, 25, 27½, and 30 fathoms, using, at 20 and 30 fathoms, the same thermometers as had been used the day before. As the temperatures at 20 and 30 fathoms were found very different from those observed the day before, the observations in the Rowardennan locality were repeated close to the east side of the loch, the usual station being nearer the west side. The following table gives the temperatures observed on 16th October and also the corresponding ones of 15th October.

TABLE VI. *Temperatures on Steepest Gradient at Rowardennan and Stuckgowan, 16th October* 1885.

Locality		Rowardennan			Stuckgowan	
		West side		East side	West side	
Date		Oct. 15	Oct. 16	Oct. 16	Oct. 15	Oct. 16
Depth fathoms	No. of thermometer	Temperature (Fahr.)				
		°	°	°	°	°
0	—	49·1	49·0	48·9	49·1	49·0
20	47	48·95	47·8	47·6	48·8	48·4
22½	9	—	47·4	46·4	—	47·3
25	79	—	46·55	46·0	—	47·0
27½	80	—	45·3	45·5	—	46·75
30	21	42·45	45·2	45·25	42·9	44·5

These results go to accentuate the fact borne out by all the observations quoted in this paper, and by all the observations which I have been able to make in other lakes, namely, that at any one date, especially during the warm half of the year, the isothermal surfaces, even at depths where there is no rapid change of temperature, are not planes, but have many curvatures and unevennesses. These unevennesses are particularly accentuated in the region of most rapid change of temperatures or on the steepest gradient of the temperature curve. Here the movements of the thermometer a few yards in a horizontal direction may place it in water of very different temperature. On both days a fresh breeze was blowing, and, though it was possible to keep station very satisfactorily from a nautical point of view, the station kept was an average one—that is, instead of being a point, it was an area, and an area perhaps

TABLE VII. *Observations in Loch Lomond, 14th November 1885.*

Name of basin		Luss	Tarbet
Locality		Ross Islands	Inversnaid
Miles from Balloch Pier .		7	14
Depth at station		34	100
Hour of day		3 P.M.	1 P.M.
No. of station		16	17
Depth fathoms	No. of thermometer	Temperature (Fahr.)	
		°	°
0		46·6	46·0
10	Negretti's overturning thermometers	46·3	45·8
20	Negretti's overturning thermometers	46·3	45·8
30	Negretti's overturning thermometers	46·3	44·3
30	Negretti's overturning thermometers	—	44·2
40	Negretti's overturning thermometers	—	42·3
50	Negretti's overturning thermometers	—	42·2
65	9	—	42·15
84	80	—	42·1
100	21	—	42·1

20 to 30 yards long by 10 to 20 yards wide. The investigation of the body of water having the steepest temperature gradient is very interesting, but it should be attempted only under the most favourable circumstances—either the weather should be perfectly calm, or the boat should be anchored. In future work the minute delineation of the steepest part of the temperature gradient should have an important place.

14th November. It had been raining all night, but cleared when I arrived at Balloch. I was accompanied by Mr Morrison, from the Scottish Marine Station at Granton, who brought with him three overturning thermometers. A very heavy snow squall occurred on the way up the loch, but it cleared off before we arrived at Inversnaid. While at Inversnaid the weather was very favourable, and we had no difficulty in getting this series of observations. The temperatures at 65, 85, and 100 fathoms were taken with my thermometers, the others were taken with the overturning thermometers of the Negretti type. In the afternoon a series of temperatures was taken with the overturning thermometers in the Luss basin west of the Ross Islands.

The salient feature at both stations is the great cooling which has taken place since 15th October. The whole body of water in the Luss basin has been cooled 2½°. The temperature of the bottom water at Inversnaid has risen 0·1°, and it has probably reached its maximum.

Having described and discussed the observations made at the different stations at the same date, it will be useful to consider some of the stations with respect to the variation in the distribution of temperature with changing season, and for this purpose it will be convenient to take two typical stations, namely, that at Ross Mill in the Luss basin, which represents a shallow lake, and that at Inversnaid, representing a deep one.

The Luss Basin. Observations were made in this basin on 5th September, 22nd September, 15th October, and 14th November. Already on the 5th September the curve has almost lost its summer feature, and is becoming pronouncedly autumnal. On 15th October it has assumed the winter form, and between that date and 14th November the nearly uniform temperature of the water has fallen about 2·6° F. The surface temperature falls from 55·2° on 5th September to 53·6° on 22nd September, to 49·4° on 15th October, and 46·6° on 14th November. The autumnal zone of nearly uniform temperature in the surface layer extends on 6th September to 15 fathoms, and on 15th October to the bottom. Therefore, until some date between the 22nd September and 15th October the upper

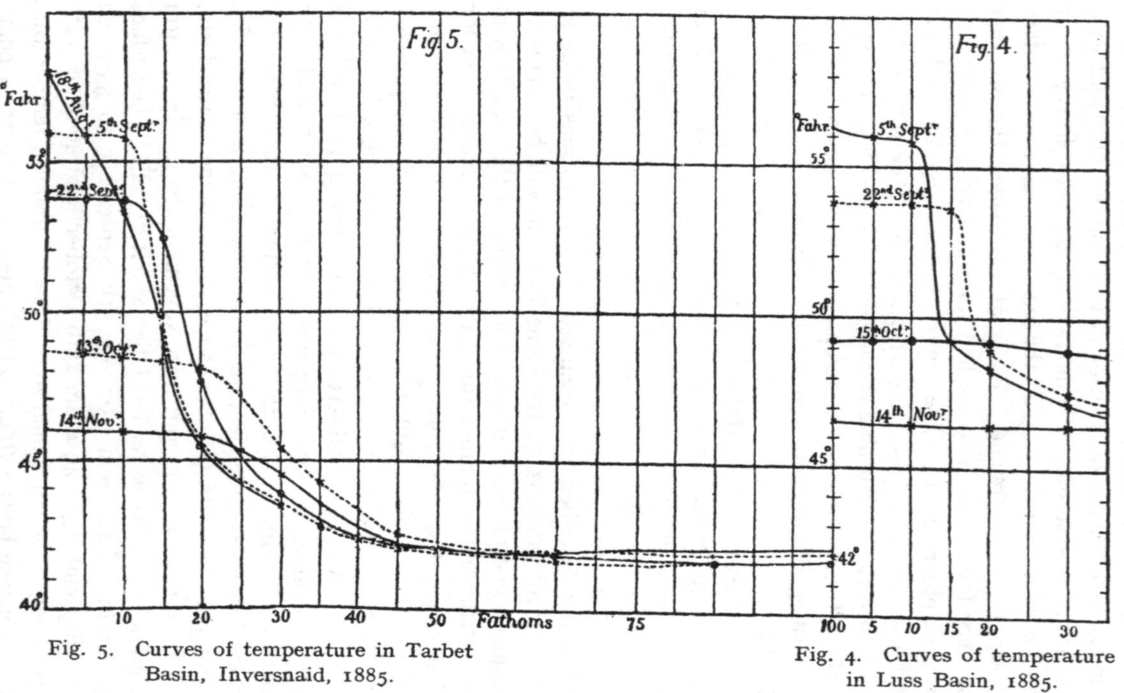

Fig. 5. Curves of temperature in Tarbet Basin, Inversnaid, 1885.

Fig. 4. Curves of temperature in Luss Basin, 1885.

stratum of water is being cooled on both sides—that is, it loses heat by radiation and convection upwards into the atmosphere, and by conduction and convection downward into the lower strata of the water. By the 15th October the loss downwards has ceased, and the heat lost between that date and 15th November has escaped by the surface.

TABLE VIII. *Collected Observations in Luss Basin*, 1885.

Date....	Sept. 5	Sept. 22	Oct. 15	Nov. 14
Depth	Temperature (Fahr.)			
	°	°	°	°
0	56·2	53·6	49·4	46·6
5	55·9	53·5	49·2	—
10	55·75	53·45	49·05	46·3
15	49·1	53·4	49·0	—
20	48·3	48·8	49·1	46·3
30	47·0	47·55	48·9	46·3

Taking the depth at the Ross Mill station as 35 fathoms, we have the following values for the mean temperature of the whole of the water:

Date	5th Sept.	22nd Sept.	15th Oct.	14th Nov.
Mean temperature	50·71°	50·71°	49·03°	46·3°

The mean temperatures on the 5th and 22nd September are identical, hence the epoch of maximum heat in the water must fall between these two dates, and as cooling goes on more rapidly than heating, it will fall nearer the 22nd than the 5th, and may with safety be said to occur in the third week of September. The curves of the 5th and 22nd September intersect at a depth of 13·5 fathoms, hence the temperature at 13·5 fathoms was the same on both these dates. But on the earlier one the temperature of the water at that depth was rising, whereas at the later one it is falling, therefore it must have attained a maximum some time between these dates, and no doubt in the third week of September. The curves of 22nd September and 15th October intersect at 19·5 fathoms; here the temperature of the water at both dates was 49·0°. The maximum temperature in this layer must have been attained towards the end of the first week of October. Similarly, from

TABLE IX. *Gain and Loss of Heat above and below the Depth corresponding to the intersections of each pair of Curves for the Station in Luss Basin* (1885).

Dates..........	Sept. 5	Sept. 22	Sept. 5	Oct. 15	Sept. 22	Oct. 15	Oct. 15	Nov. 14
Interval, days ...	17		40		23		30	
Depth of intersection, fathoms	13·5		16		19·5		none	
Mean temp. of water above intersection	55·7°	53·45°	55·0°	49·2°	52·89°	49·1°	49·03°	46·30°
Heat in fathom-degrees in water above intersection	780	748	880	787	1031	957	1716	1620
Loss of heat in interval	32		93		74		96	
Mean temp. of water below intersection	46·28°	47·85°	47·10°	48·90°	48·07°	48·95°		
Heat in fathom-degrees in water below intersection	995	1028	895	929	745	759		
Gain of heat in interval	33		22		14			
Percentage of heat passed downwards	100		24		19		0	
Heat passed into the air per day	0		1·78		2·61		3·2	
Heat passed down to water per day	1·94		0·55		0·61		0	
Mean temp. of the 35 fathoms	50·71°	59·71°	—	49·03°	—	—	—	46·30°

the intersection of the curves of 5th September and 15th October, we should infer that the maximum temperature at

16 fathoms occurs in the last days of September. It is probable that temperature at the bottom (35 fathoms) was about its maximum on 15th October.

The points of intersection of the curves indicate depths where the temperature of the water was found to be the same on both days. Although the temperature of the water at this depth has risen and fallen in the interval, it has returned to the same thermal state at the end of the interval as at the beginning. In Table IX the distribution of heat in the water above and below these points is indicated for the different pairs of dates. As a linear heat unit is required the *fathom-degree*[1] has been adopted.

From the table we see that between the 5th and 22nd September the water as a whole has lost as much heat as it has gained, so that all the heat that the layer above the intersection at 13·5 fathoms has lost appears in the deeper water below the intersection. On an average over the whole period, no heat has been dissipated to the atmosphere. This, of course, is only true on an average; for during the first portion of the period the water has been receiving heat, and during the second portion it has been dissipating it at the surface. During this interval the heat passed downwards to the deeper layers has been at the rate of 1·94 fathom-degrees per day. Between the 22nd September and 15th October the amount of heat transmitted downwards has been 14 per cent. of the total loss above the intersection. The heat dissipated per day to the atmosphere is 2·61 units, and that passed downwards 0·61 unit. The curve of 14th November does not cut that of 15th October, and the loss of heat is at the rate of 3·2 units per day, all of which has gone out into the atmosphere. On the 23rd November 1876 I found the water of the Luss basin to have a uniform temperature 47·8° from surface to bottom. In the year 1876, therefore, the water had a temperature 1·5° higher on the 23rd November than it has in 1885 on 14th November, or nine days earlier. For the same date the

[1] The fathom-degree is one fathom heated 1° F. If the fathom has a sectional area such that the volume of water weighs one pound, then the fathom-degree is the same as the ordinary heat unit. See Contents, page xlv.

water is this year at least 2·5° colder than it was in 1876, and with this rapid and severe autumnal cooling it will require no very hard or long-continued frost to freeze the great shallow Balloch basin of the loch. In fact, it will be found that in years when Loch Lomond has been frozen there has always been an exceptionally cold autumn. The prolonged action of the low temperature of a cold autumn prepares the water for any severe frost which may occur in the long nights about Christmas. Under its influence the water rapidly freezes. When the Balloch basin was frozen in the winter of 1878–79, I took the temperature of the water beneath the ice in several places, the deepest water being 11 fathoms. The temperature of the whole of the water was under 35° F., and varied from 32° in contact with the ice to 34·5° at the bottom. In the long clear nights of a severe winter the temperature of the atmosphere may often be 30° below that of the surface of the water, so that both by radiation and convection cooling goes on with great rapidity, notwithstanding the fact that water expands in being cooled below 39° F.; and it is remarked by people dwelling by the side of the lake that the moment ice begins to freeze it spreads with great rapidity, so that the whole basin is usually frozen over in a night, and as soon as a skin of ice has been formed it very rapidly becomes thick enough to bear.

Epochs of Maximum Temperature. From the curves and their intersections we see that the maximum temperature of the surface water occurred before the 5th September. At 13·5 fathoms it occurred some time between the 5th and 22nd September, and as cooling goes on faster than heating, the date of maximum temperature at this depth will be nearer the 22nd than the 5th September, probably about the 15th. Similarly at 16 fathoms it occurs between the 5th September and 15th October, and probably about 30th September. At 19·5 fathoms the maximum temperature occurs about 6th October, and in the second week of October (in 1885) all water from 20 fathoms to the bottom attained and passed its maximum temperature. The maximum temperature of the surface occurs about the middle of August, and that of the bottom in 35 fathoms about the middle of October.

Inversnaid Station. The observations at this station at the different dates have, in many respects, greater interest than those made in the Luss basin. The depth here is 100 fathoms, and is so great that for a thickness of 40 or 50 fathoms above the bottom the change of temperature during the course of the season amounts only to a fraction of a degree, and is so slight as to elude detection, except by using very delicate thermometers. The observations made at Inversnaid are collected in Table X, and the results are expressed graphically in the curves (Fig. 5). On the 18th August no observations were made exactly on the Inversnaid station, but observations were made at two neighbouring stations, Rob Roy's Cave, about a mile north, and Culness, about a mile south of it. The curve has been drawn from the means of the temperatures observed at these two stations.

TABLE X. *Inversnaid Station, Collected Observations.*

Date ..	Sept. 5	Sept. 22	Oct. 15	Nov. 14
Depth	Temperature (Fahr.)			
	°	°	°	°
0	56·0	53·7	48·7	46·0
5	55·8	53·7	48·4	—
10	55·75	53·65	48·2	45·8
15	49·8	52·25	48·1	—
20	45·6	47·2	47·8	45·8
30	43·2	43·5	45·3	44·3
35	42·7	42·6	44·2	—
40	—	—	—	42·3
45	42·2	42·25	42·8	—
50	—	—	—	42·2
65	41·8	42·0	42·2	42·15
85	41·7	41·8	42·0	42·1
100	41·8	41·8	42·0	42·1

Table XI gives the analysis of the results of the observations at Inversnaid, and is a form of *heat account* for the period—that is, it gives the receipt and expenditure of heat during the various intervals with reference to the points of intersection of the curves, where receipt and expenditure exactly balance each other over the period under consideration. Between the

TABLE XI. *Gain and Loss of Heat above and below the Depths corresponding to the intersections of each pair of Curves for the Station at Inversnaid* (1885).

Dates	Aug. 18	Sept. 5	Sept. 5	Sept. 22	Sept. 22	Oct. 15	Oct. 15	Nov. 14
Interval, days ..	18		17		23		30	
Depth of intersection, fathoms	5		14		19·5		65	
Mean temp. of water above intersection	57·0°	55·9°	55·62°	53·62°	52·65°	48·24°	45·21°	43·93°
Heat in ditto (fathom-degrees)	285·0	279·5	779	751	1027	941	2939	2856
Loss of heat in interval	5·5		28		86		83	
Mean temp. of water below intersection	43·67°	43·93°	42·74°	43·05°	42·51°	43·08°	42·0°	42·1°
Heat in ditto (fathom-degrees)	4148	4174	3676	3703	3422	3468	1470	1473·5
Gain of heat in interval	26		27		46		3·5	
Percentage of heat passed downwards	473		96		53		4·2	
Heat passed out to air per day	0·30		1·65		3·74		2·77	
Heat passed downwards per day	1·44		1·59		2·00		0·12	
Mean temp. of 100 fathoms	44·33°	44·53°	44·55°	44·54°	44·49°	44·09°	44·09°	43·30°

18th August and the 5th September, an interval of 18 days, during which the temperature of the layer immediately at the surface had begun to fall, five times as much heat was conveyed downwards as was dissipated from the surface. Between the dates 5th September and 22nd September the two quantities almost exactly balance one another. Between 22nd September and 15th October the amount transmitted downwards is only half what leaves the surface; and between the 15th October and 14th November the amount transmitted downwards is quite insignificant—not more than 4 per cent. of what escapes to the air. The activity in the heat exchange has been greatest between 22nd September and 15th October when it has been dissipated at the surface at the rate of 3·74 fathom-degrees, and conveyed downwards at the rate of 2 fathom-degrees per day. The crest of the heat wave passes from surface to bottom in about three months, the height of it decreasing very rapidly as the depth increases.

Dates of Maximum Temperature at Different Depths. The intersections of the curves give, as above shown, an indication of the date of maximum temperature at the particular depth. In Table XII will be found the depths corresponding to the

TABLE XII.

Depth	Dates of intersecting curves		Mean date
5	18th August	5th September	27th August
9	18th August	22nd September	5th September
14	5th September	22nd September	14th September
16	5th September	15th October	26th September
19½	22nd September	15th October	4th October
23	22nd September	14th November	19th October
65	14th October	14th November	31st October

principal intersections and the mean dates. The actual dates of maximum temperature will always be a day or two later than the mean dates. The number of observations is too small to enable us to say what the maximum temperature at these depths has been.

As the autumn of 1885 has been a cold one, the above mean

dates may be taken as the earliest dates of maximum temperature at the depths indicated.

If we compare the mean temperature of the whole column of 100 fathoms of water at Inversnaid on the 5th and the 22nd September, we have on the 5th the mean temperature 44·54° F., and on the 22nd September 44·52°, the difference being 0·02° F. Remembering that heat is lost more quickly in autumn than it is gained in spring, and considering that these temperatures are nearly identical, we shall not be far wrong if we put the epoch of maximum heat in the water of the deepest part of the lake as occurring some time in the third week of September. This brings it very near the date of the equinox, and it seems natural *to expect the heat to accumulate in the water so long as the day is longer than the night, and to decrease so soon as the conditions are reversed.* This conclusion is supported by the observations of Fischer, Forster, and Brunner, who made a most interesting series of observations on the distribution of temperature in the Lake of Thun, in Switzerland, during the years 1848 and 1849. They found hardly any increase of heat between 3rd February and 28th March, but after the latter date the influx of heat was very rapid. It is probable, therefore, that the temperature of the bottom water in our deepest lakes depends chiefly on the temperature of the air between the preceding autumnal and vernal equinoxes. Two causes combine, namely, the greater meridian altitude of the sun and the greater length of the day in the summer than in the winter half year. The latter cause has the effect that the water is exposed to heating for a greater portion of the 24 hours than it is to cooling, while the former cause ensures a greater supply of heat per minute during the day in summer than in winter.

Further, the rate of loss of heat, due to radiation alone, must be very much greater during a winter night than during a summer one.

The rise of temperature in the bottom water and deeper layers is made more apparent by considering the actual readings of the thermometer employed as given by the millimetre scale on the stem.

The position of the index in the thermometers when referred to this scale can be fixed almost to one-tenth of a millimetre, certainly to one-fifth. In thermometer No. 9, 1 millimetre = 0·285° F.; in No. 80, 1 millimetre = 0·345° F.; and in No. 21, 1 millimetre = 0·31° F. The difference of 0·1 millimetre between the readings of No. 21 on 5th and on 22nd September cannot be depended on as real. It is probable that between these dates and during the whole of the summer the temperature had been quite constant. Between 22nd September and 14th November the whole rise of temperature at the bottom is only 0·25°, and it may be confidently affirmed that at a depth

Depth		65 fathoms			85 fathoms			100 fathoms		
No. of thermometer		9			80			21		
Date	Interval	Reading	Difference		Reading	Difference		Reading	Difference	
	days	mm.	mm.	° F.	mm.	mm.	° F.	mm.	mm.	° F.
Sept. 5	—	51·0	—	—	70·1	—	—	60·5	—	—
	17	51·6	0·6	0·17	70·4	0·3	0·10	60·4	−0·1	−0·03
Oct. 15	23	52·4	0·8	0·23	71·0	0·6	0·21	60·9	+0·5	+0·16
Nov. 14	30	52·2	−0·2	−0·06	71·2	0·2	0·07	61·2	0·3	0·09

of 100 fathoms the whole range of temperature during a single season does not exceed 0·3° F. By season is meant the summer and winter half of the year, or the period between the date when heating begins in the spring, and that at which the summer heat has been almost wholly lost, and when the water begins again to assume a sensibly uniform temperature from top to bottom. At 85 fathoms the temperature had begun distinctly to rise on 5th September, and by 14th November it had evidently reached about its maximum. At this depth, therefore, the summer range is at least 0·4° F. At 65 fathoms the water had begun to cool between 16th October and 14th November. In order to determine the range at this depth, it will be necessary to have observations earlier than 5th September.

The bottom temperature has been determined in the deepest part of Loch Lomond by several observers in different years, and the results show that it varies from year to year. James Jardine found it to be 41·1° F. on the 8th September 1812. Sir Robert Christison found it 42·0° F. in 1871. I found it to be 40·2° in April 1872, and 41·4° F. on 23rd September 1876. About these figures there is always some uncertainty, from the want of comparison between the thermometers. The best evidence of the variation of the temperature of the deep water of our lakes, from year to year, is furnished by my observations in Lochs Lochy (80 fathoms) and Ness (120 fathoms) in five consecutive years in the second week of August. These were all made with the same thermometers, and are to be relied on to one-tenth of a degree. They are—

Years	1877	1878	1879	1880	1881
	°	°	°	°	°
Loch Lochy	44·0	43·7	42·0	43·8	42·25
Loch Ness	42·4	42·3	41·2	42·4	41·45
Mean winter temp. of air at Corran	42·3	42·7	38·9	42·0	38·6

The mean winter temperature of the air is the mean temperature of the months of October to March (inclusive) preceding the dates of observation in the lakes. The place of observation is Corran lighthouse, on Loch Linnhe. It is too remote from the lochs themselves to give more than an indication of the climate at the surface of the lakes. Still the bottom temperature of Loch Ness does follow very closely the mean winter temperature at Corran. In the two severe winters, 1878–79 and 1880–81, the mean winter temperature fell below 39°, and as might have been expected the temperature of the deep water of the lake was slow to follow it, owing to the change in the properties of water at this temperature. Hence the higher the mean temperature of the cold months of the year is, the more closely is it reproduced in the deep water of the lake. Forty years ago Aimé showed that the temperature which he observed in the abyssmal regions of the Western

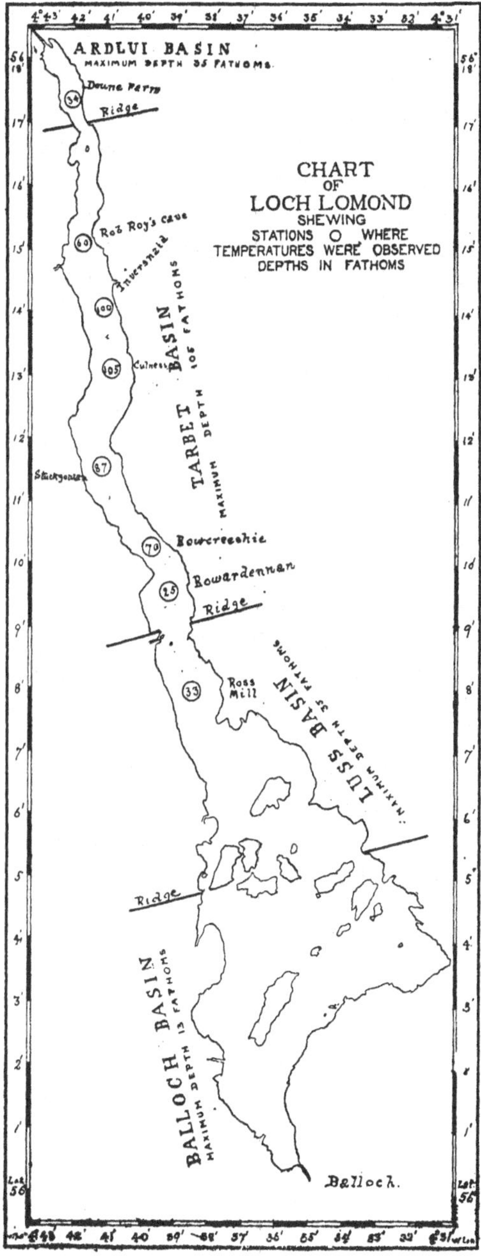
ARDLUI BASIN
MAXIMUM DEPTH 35 FATHOMS.
Doune Ferry
Ridge
CHART
OF
LOCH LOMOND
SHEWING
STATIONS O WHERE
TEMPERATURES WERE OBSERVED
DEPTHS IN FATHOMS
Rob Roy's cave
Inversnaid
TARBET BASIN
MAXIMUM DEPTH 105 FATHOMS
Culness
Rowardennan
Ridge
Ross Mill
LUSS BASIN
MAXIMUM DEPTH 35 FATHOMS
Ridge
BALLOCH BASIN
MAXIMUM DEPTH 13 FATHOMS
Balloch.
34
60
100
105
87
70
25
33

Mediterranean agreed sensibly with the mean winter temperature of the air at its surface. The later view, which ascribed the temperature of the deep water in the Mediterranean to the Atlantic water flowing over the ridge at Tarifa, which has the same temperature as the Mediterranean water within the ridge, left out of account the fact that there is on the whole an outflow of pure Mediterranean water at the bottom which affects the temperature and density of the Atlantic water outside. In sea water, owing to its saltness, convection currents are set up more actively by cooling than is found to be the case in fresh water. In lakes, however, and especially in those situated in mountainous districts, the production of convection currents is powerfully assisted by local differences of climate. These are due to differences of exposure, to radiation, and to prevailing winds. Such local differences of climate produce local differences of temperature, and consequently of density in the superficial layers. If we compare, for instance, the observations made at Ardlui and at Inversnaid on 15th October, we find that the mean temperature of the first 20 fathoms is 48·24° at Inversnaid, and only 46·58° F. at Ardlui. The stations are only three miles apart, and yet there is a difference of 1·66° F. in the mean temperature of the first 20 fathoms. Another cause affects the distribution of temperature in a lake, namely, the drainage, but this more particularly affects shallow lakes or basins. It is probable that the most powerful means of supply and removal of heat is direct radiation. The most powerful mixing agency is the wind.

No. 22. [*From Nature, December* 5, 1907, *Vol.* LXXVII. *pp.* 100—102.]

THE WINDINGS OF RIVERS[1]

At the meeting of the British Association at Edinburgh in 1892 I read a paper on the subject of the winding of rivers before the geographical section. It was illustrated by a large number of diagrams, but, as these could not be included in the report of the meeting, only the title of the paper appeared. It may not be out of place to give a short account of it, as the subject is now attracting some attention.

In Fig. 1 the courses of three streams are shown. These are distinguished by the letters *A*, *B*, *C*, without indication of

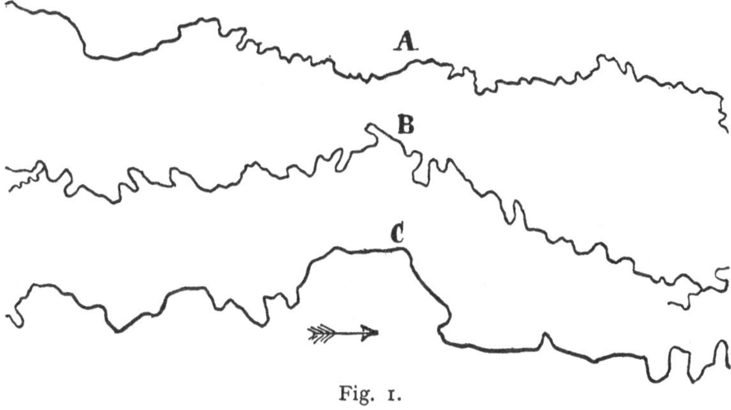

Fig. 1.

their identity or of the scale on which they are drawn. If anyone were to try to select the one which represents the largest or the smallest of these streams he might do so correctly, but he would not be surprised if he were told that he had guessed wrongly, for it could only be a guess. The length of each tracing is the same. In nature it represents in *A* nine English miles, in *B* 216 miles, and in *C* one-and-a-half miles.

[1] For Introduction, see Contents p. xlvi.

Tracing *B* represents the part of the Mississippi between the mouth of the Arkansas River and that of the Red River. Tracing *A* represents the Devon Water, a tributary of the Forth, and tracing *C* represents a quite insignificant brook called the Catter Burn, a tributary of the Endrick, one of the principal affluents of Loch Lomond. These tracings, and indeed the maps of all countries, show clearly the great family likeness exhibited by rivers in all parts of the world. This likeness rests on the fact that in all rivers the relation between the length of an arc or bow and that of its chord is nearly the same. It is a racial rather than a family likeness, and resembles that which exists between dogs of different breeds or builds.

The following table shows, for a selection of well-known rivers, the degree in which the above relation holds good:

River on the stretch		Length of stretch		Ratio	Number of bows	Average length of bows
From	To	Direct	Along windings			
Mississippi		Miles	Miles			Miles
Columbus	Memphis	124	204	1·65	23	8·87
Memphis	Natchez	270	490	1·83	62	8·10
Natchez	Baton Rouge	88	133	1·51	18	7·39
Baton Rouge	Carrolton	72	124	1·72	20	6·20
Columbus	Carrolton	554	951	1·72	123	7·64
Thames						
Marlow	Walton	18·7	30·0	1·61		
Teddington	Isle of Dogs	16·8	26·5	1·58	11	2·4
Danube	Near mouth	11·4	22·5	1·97	13	1·73
Rhine		kilom.	kilom.			kilom.
Germersheim	Mannheim	34·1	69·2	2·03	11	6·3
Main		114	144	1·27	44	3·27
Neckar						
Heilbronn	Mannheim	80	112	1·40	26	4·30
Lahn						
O. Lahnstein	Limburg	17·2	29·7	1·73		
Mosel	Near Coblentz	7·0	10·0	1·43		
Ahr						
Altenahr	Ahrweiler	4·6	9·6	2·09		

From the table it will be seen that over a length of nearly one thousand miles of the Mississippi the average length of the stream, following the windings, is 1·72 times greater than the direct distance. In the Lahn we find almost the same factor. The Lower Danube, the Rhine, and the Ahr show a factor approximating to 2. The Main, Mosel, Neckar, and Thames have lower factors. The mean of all the factors is 1·68. For a certain number of the rivers the number of "bows" is given with their average length. The size of the bows stands in some relation to the volume of the river. What that relation exactly is I am not able to state. To arrive at it will require a careful study of the flood waters of the river in connection with the form of its bed. It is the flood waters which form the bed. When the river falls to low-water level we often see it cutting out a secondary bed on a much smaller scale, which is obliterated by the next following flood.

It may be taken that the mean track of a stream traces the line of lowest level in the valley. Consequently, the ground must rise on both sides of it. The cross-section of the valley through the river resembles that through the middle of a watch glass, rising at first very slightly on both sides of the stream, then more rapidly as the confines of the valley are approached. It is evident that water displaced to one side of the river will, in returning to it, tend to pass to the other side, and to oscillate about the lowest point.

If the bed of a stream flowing through alluvial ground were rectified so as to direct the water along a straight trough cut in the material, it might preserve a straight course for a time, but a stream following such a course is in a state of unstable equilibrium. The smallest accident or obstruction disturbs the uniform rectilinear motion of the water, and tends to induce oscillations, both longitudinal and transverse. These begin immediately to cut into the banks, if they are yielding, and take larger and larger dimensions until *they reach a limit when they have produced a course of the sinuosity which corresponds to the laws of the harmonic motion of its waters.*

No attempt was made to arrive at these laws *a priori*. The method of investigation used was purely empirical. Curves were traced according to all kinds of harmonic specification until some were obtained which resembled the courses of actual rivers. Fig. 2 shows one specimen out of many which were exhibited at the meeting of the British Association.

It is assumed that the rhythmic motion set up in a mass of water which is disturbed in its uniform rectilineal motion will be reducible to two reciprocating motions, one in the direction of the fall of the stream and the other at right angles to it. When the gradient of the stream is very steep and the nature of the bed homogeneous, as it is in the case of water flowing down the front of a glacier, the longitudinal oscillation is swamped by the powerful and continuous action of gravity, which does not affect the transverse component. In these circumstances we often meet with small streams which describe an almost perfect simple harmonic curve.

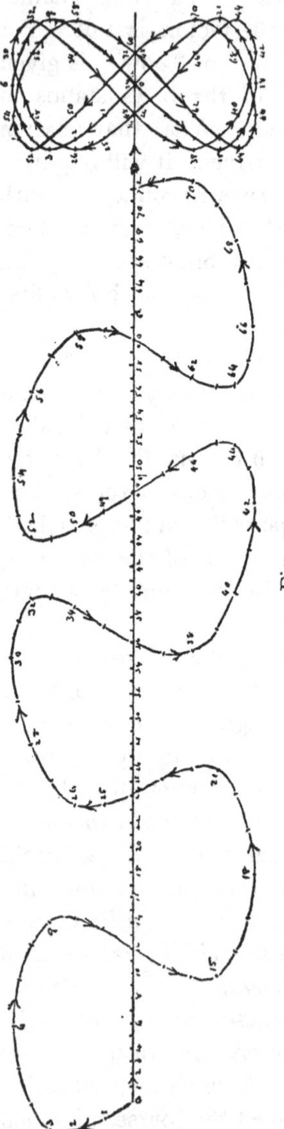

Fig. 2.

In the ordinary stream of the meandering type the gradient is very small, in the case of the Mississippi from two inches to four inches per mile, so that the longitudinal pulse can produce its full effect. When the two oscillations are simple pendulum motions and have the same period they produce an ellipse, which, when combined with the steady onward flow due to gravity, produces cycloidal sinuosities which are unlike those of actual streams. When the period of the transverse oscillation is twice that of the longitudinal one, their combination produces a Lemniscate or figure of eight (8). When a figure of eight is combined with steady forward motion so that both are travelled over in the direction of the arrows in the figure, then it delineates a Lemniscoid which may resemble the course of an actual stream. This is illustrated in Fig. 2. In it the sinuous curve falls into three parts, each consisting of a double bow, corresponding to a complete excursion of the tracing point round one of the figures of eight. The horizontal line indicates the path of undisturbed flow of the stream running from left to right in the direction of the arrows. It is divided into 72 equal spaces, each of which represents the distance which would be covered by the undisturbed stream in the interval of time in which the circle which generates the transverse reciprocating motion describes one twenty-fourth of a revolution, so that the undisturbed stream passes over 24 spaces in the time that the tracing point passes once round the figure of eight. The resultant path of the tracing point is the sinuous curve, which cuts the horizontal line at 12 and 24 when the symmetrical 8 is used, and in 36, 48, or 60, 72 when one of the other two figures is used. It is an essential condition that the tracing point shall go round the 8 in the direction of the arrows, so that it shall be moving in the same sense as the undisturbed water when it traverses the outside parts of the figure which are approximately parallel to the path of undisturbed flow. In describing the sinuous line it is convenient to draw the figure of eight on tracing paper. Then, when the centre of the 8 is placed over any mark on the horizontal line numbered, say, 9, the point on the periphery of the 8 numbered 9 must be superposed on the point on the sinuous curve also numbered 9.

The description of the sinuous line is a simple case of mechanical drawing, and presents no difficulty. By varying the harmonic composition of the figure of eight and the rate of undisturbed flow of the water, an infinite number of different individual curves can be produced which are all covered by the same generic specification. It is an interesting occupation, in leisure moments, to compose curves of this kind and to compare them with those traced by actual rivers on the face of the earth.

No. 23. [*From Nature, August* 1, 1872, *Vol.* VI. *pp.* 260, 261.]

VOLCANOES AND EARTHQUAKES

It is generally admitted that an earthquake is due to the passing of a sensible wave through the earth's crust. It has also been observed that the occurrence of earthquakes is frequently associated with eruptions from volcanic vents, usually in the neighbourhood, but not unfrequently at considerable distances. It is evident—and it has struck nearly all observers—that there must be some connection between the two classes of phenomena. The nature of this connection has been differently explained by different writers. But the purpose of this note is not to criticise existing theories, but to propose one, which I believe to be new, and to be capable of explaining why a sudden volcanic eruption must ordinarily be accompanied by earthquake shocks of greater or less violence (not necessarily always sensible), and why earthquakes may occur without any contemporaneous outburst.

In the preface of his *Physikalische Geologie* Bischof suggests that the phenomena observed in the laboratory should be taken as our guide to explain what happens in nature. Let us see, then, if in the laboratory we meet with any phenomena analogous to volcanoes and earthquakes.

When a reaction has to be performed in a sealed tube, and it is expected that much gas will be evolved, and consequently the pressure in the tube be much increased, it is one of the commonest precautions to draw out the tube to a capillary orifice before closing it. When this precaution has been neglected, and even although the point be allowed to blow itself out in the flame of a lamp, an explosion not unfrequently attends the attempt to open it. Let us consider the circumstances.

We have a tube whose walls are being pushed out by a very high pressure from within, which, however, it resists; but at the moment that this pressure is suddenly relieved at one point, the tube bursts. What is the cause of the explosion?

It clearly cannot be the mere reduction of pressure. As long as the pressure was equally distributed over the walls of the vessel, we have seen that it was successfully resisted; so soon, however, as it was suddenly relieved at one point, a great inequality in the tension of the gas in the immediate vicinity of the point would be the result, the gas immediately at the opening assuming at once the atmospheric pressure, while that at, say, the eighth of an inch from it is at the tension of the gas in the tube. The practical effect of this sudden inequality of pressure would be to produce a tug on the mass of elastic fluid, which would cause the walls momentarily to tend to collapse, and this tendency to collapse would be transmitted through the glass as a wave. This wave would to a certain extent distort, and therefore weaken, the walls; and consequently, if the pressure inside were great enough, it would burst them; if not, the only effect would be that a shock would traverse the walls of the vessel, and the pressure would relieve itself by the orifice.

Let us suppose the vessel to be a subterraneous cavity containing an atmosphere of very great tension, and that suddenly the envelope gives way at one point, what will be the result? Just as in the case of the glass tube, the sudden relief of pressure will, in the way indicated above, cause the walls to experience a momentary collapsing impulse, which will be propagated as a wave until extinguished by the imperfect elasticity of the crust. The sudden outburst will be a volcanic eruption, and the consequent collapsing shock will be an earthquake, which either will or will not be accompanied by rending of the crust, according to the strength of the walls and the greatness of the pressure.

It is, however, not necessary that there should be a visible volcanic eruption. For, suppose two such subterraneous cavities at different pressures, separated from one another by a wall weaker than that which separates either of them from the outside of the earth; then, if the pressure in the one becomes so great as to burst the barrier between the two, the result will be an earthquake. And similarly, the pressure in the two united cavities may go on increasing until they burst into a third, and so on until they come to a vent, which is either open or weak enough to yield to the pressure. In this way an earthquake and an eruption may be in intimate connection with one another, although a considerable interval of time intervene between the occurrences, and the localities affected be at great distances from each other. And it is possible that some connection of this kind may have existed between the earthquake of Antioch and the eruption of Vesuvius, both having been extreme in their violence. Indeed, the whole series of disturbances, commencing with the earthquake in California and terminating with the eruption of Vesuvius, all of which occurred in April 1872, may possibly find an explanation under this theory.

The effect of sudden relief of pressure in weakening the walls of vessels explains many cases of explosion which otherwise appear anomalous. Thus, high-pressure boilers have been frequently observed to burst at the moment when the engineer turns on the steam.

Postscript. The subject of this paper was suggested by the extraordinary sequence of seismic and volcanic phenomena which took place in the month of April, 1872. It included disastrous earthquakes in California, on the Gold Coast of Africa, and at Antioch in Asia Minor, as well as the great eruption of Vesuvius, rendered more celebrated by the courage and tenacity of Professor Palmieri who remained in the observatory and continued his work, though completely isolated and exposed to the risk of being overwhelmed at any moment.

I was at that time preparing for my work on the "Challenger" expedition which sailed at the end of the year 1872. During it volcanic and perhaps seismic phenomena were likely to present themselves; but I little expected my claim for the seismic effect of sudden relief of pressure to be made good, and still less to be able to demonstrate it by direct experiment.

Yet this is what happened on 27th March, 1873. In the experiment of that date the pressure suddenly relieved amounted to over five hundred atmospheres and the body on which its seismic effect was exhibited was a stout copper tube.

The incident is described in the first page of the next paper.

No. 24. [*From Proc. Roy. Soc.* 1903, *Vol.* LXXII. *pp.* 88—92.]

ON A REMARKABLE EFFECT PRODUCED BY THE MOMENTARY RELIEF OF GREAT PRESSURE

THE effect is shown in the brass tube and the copper sphere which I have the honour to exhibit to the Society. It is also illustrated in Figs. 1–4 which accompany this paper.

The experiment was made for the first time on board the "Challenger" in the early part of the cruise. At that time the deep sea thermometers, with protected bulb, had only been recently introduced, and the effect of pressure on thermometer bulbs, whether protected or not, occupied public attention. In the deepest sounding made by the "Challenger" in the Atlantic, namely, that of March 26, 1873, when a depth of 3875 fathoms was reached, both the thermometers, which were sent to the bottom, collapsed. It, therefore, became a question what recommendation should be made to the thermometer makers to assist them in producing thermometers which shall be able to withstand the greatest pressure to which they are likely to be exposed in the work of ocean sounding.

For this purpose I immediately prepared the following experiment: I took glass tubes of three different calibres. The widest had about the calibre of the outer bulb of a *Millar-Casella* thermometer, the narrowest had an internal diameter of 6 mm., and the third had a diameter of about 10 mm. A length of 75 mm. of each was sealed up at both ends, and the three tubes were wrapped in a cloth and enclosed in the cylindrical copper case of a deep-sea thermometer.

The upper and lower ends of these cases are pierced with many holes in order to permit the passage of water through them. On the next day, March 27, 1873, the case was attached to the sounding-line and a depth of about 2800 fathoms was reached. When the case came up again it looked as if someone had struck it in the middle with a hammer. When it was opened and the cloth unfolded, it seemed for a moment as if it were full of snow, but a second look showed that what appeared to be snow was nothing but finely comminuted glass. The two wider tubes had collapsed, but the narrow one was intact At first sight the effect produced on the copper case was puzzling, but after a little study and reflection its nature became apparent.

No account of the experiment has been published, and when I was able, through the kindness of H.S.H. the Prince of Monaco, to join his yacht the "Princesse Alice" for her cruise of last summer, I determined to repeat it, and, if possible, to vary it. Owing to press of other work, nothing was done until near the end of the cruise. The brass tube (Figs. 1 and 2), above referred to, was the case for holding a piezometer which was accidentally broken. With it I repeated the experiment which I had made in the "Challenger," with this difference, that I used only one sealed glass tube. It was an ordinary pipette of 50 c.c., sealed up at both ends close to the body. It was wrapped in a piece of muslin and loosely packed with cotton waste so as to occupy the middle of the brass tube.

The length of the brass tube was 33 cm., and its diameter 4·13 cm. Its weight without the cover was 350 grammes. Both the top and the bottom are pierced with many holes so as to allow passage to the water.

Thus charged, it descended on the sounding-line to a depth of 3000 metres, and when it came up it was evident from its appearance that the experiment had succeeded. As in the experiment on board the "Challenger," the glass tube had been converted into a snow-white powder. The external effect also was confined entirely to that part of the brass tube which had been occupied by the sealed glass tube. Above and below it there was no disfiguration.

The copper ball (Figs. 3 and 4), is an ordinary 5-inch ball for the supply tap of a cistern. A spherical glass fractionating flask, having a diameter rather less than 1½ inch, was hermetically sealed close to the spherical body. It was

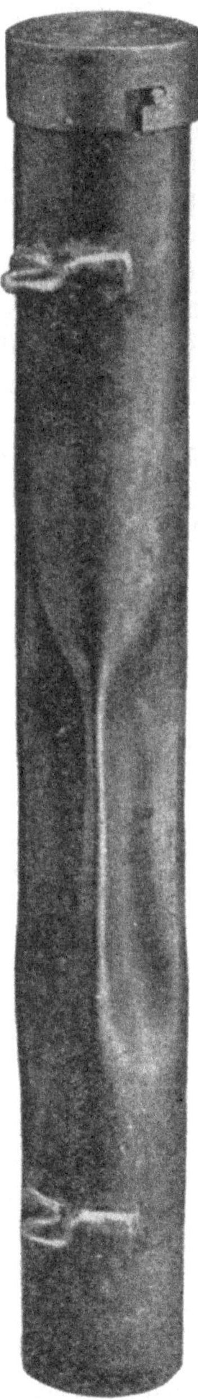

Fig. 1.

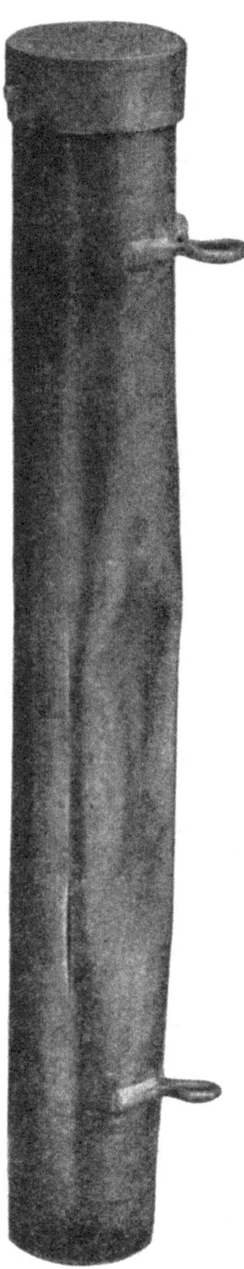

Fig. 2.

then wrapped in a piece of muslin, and with loose packing of cotton waste it was enclosed between the two copper hemispheres, which were then soldered together. The holes at the poles of the copper sphere gave free communication with the sea-water. The copper ball was then attached to the dredging cable, which took it to a depth of 3000 metres. When it came up no external effect was visible. I could not believe that even a small flask of the kind could support a pressure of 300 atmospheres, and I concluded that it had collapsed shortly after leaving the surface. Still, as the line was going to make a second excursion, and this time to 6000 metres, I re-attached the ball along with a larger one to it.

On returning to the surface the ball had the appearance which you see. If the soldered welt represent the equator, it will be seen that both polar areas are as they were. Perpendicularly to the equator a system of folds or creases runs northwards and southwards and extends very little beyond the tropics. The creasing is most accentuated at a part of the equator where there is a slight flattening. It is evident that the glass flask when it collapsed was relatively near this part of the ball. I did not open the ball, as I thought it would be more instructive to keep it as it is. The *débris* of the glass flask with the cotton waste is still inside it.

The effect of the sudden relief of pressure on the copper ball is distributed much more uniformly over it than is the case in the brass tube. In the latter the effect is very powerful and very local. In both cases the effects which we see have been produced in a moment of time, and are properly speaking, the effects of violent shock. It is remarkable that in the ball the equatorial zone which has the welt to stiffen it should be the field of all the disfigurement, while the polar areas which have no strengthening have not been exposed, or at least have not yielded to strain.

If we examine the brass tube, Figs. 1, 2, we see that, with the exception of the portion nearly in the middle which held the sealed glass tube, the case has perfectly preserved its cylindrical form. The distortion or crumpling affects only the part where the tube collapsed, and it is evident that it did

Fig. 3.

Fig. 4.

not occupy a truly axial position, but lay nearer that part of the brass envelope where the ears for attachment to the sounding-line are situated. Here a most formidable corrugation (Fig. 1) has been produced, the metal being pinched into a fold so as almost to meet inside. Besides this, there are two minor corrugations. A greater thickness of water intervened between this part of the brass envelope and the enclosed glass tube, and the small effect produced shows that the difference of pressure within and without the brass tube was here comparatively small. It will be observed that the butt-joint of the tube has been opened at Fig. 2; but this is a secondary effect due to the distortion.

The brass tube, as it stands, is a manometer or pressure gauge which records the distribution of pressure in it while filled with and immersed in water, during the instant of time when, while the pressure on all sides is very great, the pressure at a locality in the interior suddenly becomes nothing or very small. The effect of this sudden difference of pressure has been concentrated on the part of the brass tube nearest to which the glass tube was situated. Here the diminution of internal volume of the brass tube produced by the principal corrugation must, from rough measurements, be very nearly equal to that of the glass tube which collapsed. At first sight it appears remarkable that on the collapse of the glass tube, when it was free to the compressed sea-water to fill up the void with water through the two open ends, instead of doing so, it filled it by pinching up the stout brass of which the tube was made, to such an extent as to obliterate the void.

The experiment shows us that it was easier *in the time* to pinch the envelope of brass than to shove in the plugs of water at both ends. The complete absence of distortion or disfigurement of the upper and lower portions of the brass tube shows that the tension of the water in these two portions of the tube was not materially diminished in the time between the collapse of the glass tube and the occupation of its place by the corrugation of the envelope. In considering this experiment, we must distinguish between the tension and the pressure of the water. When the water is at rest they are equal. During a catastrophe of this kind the balance is destroyed, as in the case of air which is transmitting a sound wave. If water were incompressible, it could have no tension, however great the pressure to which it might be exposed. What pinched the brass tube was not the column of 2000 or 3000 metres subsiding on it, but the resilience of the unlimited supply of water in its neighbourhood at the high tension due to its compression by a pressure of 200 or 300 atmospheres. Relatively, this acts instantaneously; while, to put in motion a mass of water takes a definite time. The quantity of water contained in the brass tube is not sufficient for its resilience to produce any counter-vailing effect to the resilience of the mass of compressed sea-water outside.

In the case of the copper sphere it is otherwise. Its diameter is 5 inches, that of the sealed glass bulb inside of it was between 1 and 1½ inch, and certainly not greater than 1½ inch. If we assume it to have been 1½ inch, then its volume is to that of the copper sphere in the proportion $3^3 : 10^3 = 27 : 1000$. If we assume that the glass bulb succumbed at a depth of 5000 metres, or at a pressure of 500 atmospheres, then the resilience of the water inside of the copper sphere would have a very considerable effect in neutralising the crushing action of the water outside. At the low temperature found at great depths in the ocean the volume of a mass of distilled water is compressed by 2·5 per cent. by a pressure of 500 atmospheres. The compressibility of sea-water is nine-tenths of that of distilled water, therefore, it would be compressed by 2·25 per cent. The compression produced by a pressure of 500 atmospheres is equal to the expansion when the pressure is diminished by the same amount. But the volume of the glass bulb was not greater than 2·7 per cent. of that of the copper sphere, and it was probably less; therefore the water in the copper sphere would, at the moment of the collapse of the glass sphere, expand by very nearly the volume of the collapsed bulb, and the copper ball would then be filled, for the moment, with water having a tension equal to about atmospheric pressure. Its tension would then be brought up to 500 atmospheres by the entry of water through the holes at the two poles. The expansion of the mass of compressed water in the copper sphere takes off from the suddenness of the action, while it at the same time reduces, by at least one-half, the difference of pressures outside and inside the sphere at the instant of collapse, and this is the agent which deforms the metal sphere.

By altering the relation between the volume of the copper sphere and that of the glass sphere enclosed in it and the pressure to which the system is exposed, the effect produced may be varied at will. When experimenting in the sea, the volume of the compressed water outside of the copper sphere is practically infinite. If it is sought to reproduce these effects in the laboratory, then a very large pressure vessel must be used. If a pressure vessel of limited size be used and, altering the experiment, if the hermetically sealed glass sphere and the copper sphere with its polar perforations be placed in it separately, then when the pressure is raised to such a point that the glass sphere collapses, the copper sphere will burst outwards.

I was profoundly impressed at the time by the experiments which I made on board the "Challenger," and I connected them with another experiment which is familiar to chemists. When substances are set to re-act upon one another in a sealed tube, there is frequently disengagement of gas which produces a very high tension in the interior of the tube even when cold. If it is sought to open the tube by breaking off the sealed point, an explosion is almost sure to take place. This may have very serious consequences, and yet it has been produced by a *relief* of pressure. These examples of the destructive effect which can be produced by the sudden relief of pressure led me to believe that many shocks of earthquake may be due to similar relief of subterraneous pressure.

These experiments, whether made with the copper ball or with the brass tube, furnish striking demonstrations of the importance of the element of *time* in all physical considerations.

The collapse of the brass tube, under the peculiar circumstances of the experiment, is the exact counterpart of the experiment which is frequently, but unintentionally, made by people out shooting, especially in winter. If, from inattention or other cause, the muzzle of the gun gets stopped with a plug of even the lightest snow, the gun, if fired with this plug in its muzzle, invariably bursts. Light as the plug of snow is, it requires a definite time for a finite pressure, however great, to get it under way. During this short time the tension of the powder gases becomes so great that the barrel of the ordinary fowling-piece is unable to withstand it and it bursts.

No. 25. [*From Trans. Roy. Soc. Edin.* 1880, *Vol.* XXIX. *p.* 589.]

PRELIMINARY NOTE ON THE COMPRESSIBILITY OF GLASS

THE following experiments were undertaken with a view to determine by actual observation the effect produced on solids by hydraulic pressure.

The instrument was constructed according to my directions by Mr Milne of Milton House, about two years ago, but it is only now that I have been able to devote myself to its application to the purposes for which it was designed. It consists of a hydraulic pump, which communicates with a steel receiver capable of holding instruments of considerable size, and also with a second receiver of peculiar form. This receiver consists essentially of a steel tube, terminated at each end by thick glass tubes fitted tightly. It is tapped at the centre with two holes, the one to establish connection with the pump, and the other to admit a pressure gauge or manometer. The steel tube may be of any length, being limited only by the extent of laboratory accommodation at disposal. The tube which I am using at present has a length of a little over six feet, and an internal diameter of about three-tenths of an inch. The solid to be experimented on must be in the form of a rod or wire, and must, at the ends at least, be sufficiently small to be able to enter the terminal glass tubes, which have a bore of 0·08″ and an external diameter of 0·42″. The length of the solid is such that when it rests in the steel tube its ends are visible in the glass terminations.

When the joints have all been made tight, the experiment is conducted as follows: A microscope, with micrometer eyepiece, is brought to bear on each end of the rod or wire. These microscopes stand on substantial platforms altogether independent of the hydraulic apparatus. The pressure is now raised to the desired height, as indicated by the manometer, and the ends of the rod are observed, and their position with reference to the micrometer noted. The pressure is then carefully relieved, and a displacement of both ends is seen to take place, and its amplitude noted. The sum of the displacements of the ends, regard being had to their signs, gives the absolute expansion, in the direction of its length, of the glass rod when the pressure at its surface is reduced by the observed amount, and consequently also of the compression when the process is reversed. As in the case of non-crystalline bodies like glass there is no reason why a given pressure should produce a greater effect in one direction than in another, we may put the cubical compression at three times the linear contraction for the same pressure.

As yet I have only experimented on glass, and only on one sort, namely, that made by Messrs Ford & Co., of Edinburgh. It contains 56·29 per cent. of silica, 29·5 per cent. of oxide of lead, 6·52 per cent. potash, 3·36 per cent. soda, 3 per cent. alumina, and 1·05 per cent. lime. I have observed its compressibility up to a pressure of 240 atmospheres, and before proceeding to higher pressures I intend to determine the compressibilities of other solids, especially metals, at pressures up to 240 atmospheres. The reason for taking this course is, that having got two glass tubes to stand this pressure, I am anxious to utilise them as far as possible before risking them at higher pressures.

The pressure in these experiments was measured by a manometer, which consists simply of a mercurial thermometer with a stout bulb, which is immersed in the water under pressure, whilst its stem projects outside. The values of the readings of this instrument were determined by comparing it with a piezometer containing distilled water. This piezometer had been compared with others which had been subjected to the pressure of very considerable and measured columns of water on the sounding-line. The mean apparent compressibility of water in glass was thus found to be 0·0000486[1]; or,

[1] *Proc. Roy. Soc. Lond.*, 1876, p. 162.

multiplying by one thousand, to reduce the number of figures 0·04868 per atmosphere at temperatures from 1° to 4° C.

The manometer (No. 2) was compared with this piezometer. The temperature of the manometer was 12·5° C., while the piezometer was enveloped in ice in the receiver. The ice was thus melting under the same pressure as the instrument was undergoing, consequently the piezometer was not exposed really to precisely the same temperature at each succeeding experiment. For our present purpose, the effect of the possible variation in volume due to this thermic cause is negligible, and we assume that the indications of our piezometer are comparable with those obtained in deep ocean waters. In a future communication I hope to return to this point.

TABLE I. *Comparison of Manometer No. 2 at 12·5° C. with Piezometer K, No. 4, in Ice melting under Pressure.*

Piezometer *K*, No. 4, contains at atmospheric pressure 7·74 cub. cents. of water	Number of observations meaned	Pressure in divisions of manometer No. 2	Apparent contraction of water per 1000 in piezometer *K*, No. 4
		A	*H*
Temperature of manometer, 12·55° C.	4	26·08	4·0228
Piezometer immersed in ice melting under pressure represented by *A*.	4	30·28	4·6534
Probable temperature varying from	1	36·20	5·5972
−1° to 0° C.	5	40·08	6·1045
	3	50·08	7·6043
	3	60·20	9·1057
	3	70·08	10·5163
Total number of observations	23		
Mean reading of manometer		43·61	
Mean apparent contraction of water in piezometer			6·6495

In Table I we have in the first column the number of observations at each approximately identical pressure from which the average values of the manometer reading under *A*, and of the piezometer indication under H are computed. Manometer No. 2, when treated simply as a thermometer, showed at atmospheric pressure a rise of one division for a rise of 0·233° C. in temperature. Piezometer K, No. 4, was filled with distilled water, and contained 7·74 cub. centimetres at 0° and atmospheric pressure. It is made of Ford's glass, though not drawn at the same date as the experimental rod. Dividing the mean apparent contraction of the water in the piezometer by the apparent compressibility of water in glass 0·04868, we have for the pressure corresponding to a rise of 43·61 divisions on manometer No. 2 at 12·5° C.

$$P = \frac{H}{0{\cdot}04868} = \frac{6{\cdot}6495}{0{\cdot}04868},$$
$$= 136{\cdot}6 \text{ atmospheres.}$$

But this pressure produces a rise of 43·61 divisions on manometer No. 2. We have thus for the value of one division on the manometer

$$a = \frac{136{\cdot}6}{43{\cdot}61},$$
$$= 3{\cdot}132.$$

Hence, to convert readings of manometer No. 2 into atmospheres, we have to multiply by 3·132, the difference of the manometer reading under pressure and that at atmospheric pressure.

In another series of experiments piezometer K, No. 4, was compared with manometer No. 2, both being at a temperature of 12·5° C., and the following results were obtained as the mean of nineteen observations:

Mean rise of manometer No. 2 (A) . .	41·35 divisions
Mean apparent contraction per thousand of water in piezometer K, No. 4 (H) . .	5·8782 ,,

But from the results in Table I we have for the pressure in atmospheres

$$P = 3{\cdot}132 \times A = 3{\cdot}132 \times 41{\cdot}35,$$
$$= 129{\cdot}5 \text{ atmospheres,}$$

and the apparent compressibility of water in glass at this temperature (12·5° C.) in volumes per thousand per atmosphere, is

$$M = \frac{H}{P} = \frac{5 \cdot 8782}{129 \cdot 5},$$

$$= 0 \cdot 04539.$$

We see, then, that at pressures up to 240 atmospheres the property peculiar to water of diminishing in compressibility with rise of temperature is preserved unimpaired, and the amount of change corresponds closely with that found at low pressures in the experiments of Regnault and Grassi.

In Table II the details of the experiments on the effects of pressure on the glass rod are given. The length of the rod from point to point was 75·05″, at the temperature of the laboratory, 13° C. Its diameter was 0·28″, and was very uniform. The weight of the rod was 209·5 grammes. The substance of the rod was remarkably homogeneous, there being a complete absence of air-threads.

The micrometers used were, at the east end a photographic copy of Hartnack's eye-piece micrometer, and at the west end one of Merz's. They were both compared, and the value of their divisions, as used, determined by comparison with a stage micrometer of Smith & Beck, obligingly lent to me by my friend Dr William Robertson, who had very carefully verified its graduation. It was remarkable as a coincidence that the values of the divisions turned out to be identical in both, namely, 0·000417″.

Under *A* we have the manometric pressures, under *B* and *C* the micrometric determinations of the expansion at the east and the west end respectively, and under *D* the sum *B* and *C*, or the total expansion of the rod. It will be seen that while the values of *D*, or the total expansion, are very concordant in each series, those of *B* and *C* individually are not always so, the excess being sometimes at the one end and sometimes at the other. The effect of the rise of pressure is to extend the containing tube, and to compress the contained rod. On the relief of pressure the tube shortens again, and the rod recovers its length, and there is necessarily a sliding of the one on the other, and it depends entirely on minute local circumstances whether the rod finds it easier to return to its original relative position or to another. In some experiments made previously to the date of those quoted in Table II, the rod had greater freedom of motion longitudinally, and it happened several times that it crept bodily to the one end, necessitating the opening of the apparatus to replace it in a position suited to observation. Afterwards stops were placed in the tube, which, while setting limits to the crawling motion, did not in any way interfere with the expansion and contraction. The results of these previous experiments are not included in the table, because they were merely tentative in order to learn the details of the kind of experimentation; and further, because in the microscope at the east end the power used was very low, and the micrometer insufficiently delicate.

In the left hand columns the individual experimental data are given. The arithmetical means of the manometric pressures and of the total micrometric expansions are taken for each series. These mean results are then further developed on the right hand side of the table. First the temperature is given, T. This remains always very constant, as it was the temperature of the room, which varied very little. It was further controlled between each experiment by the reading of the manometer when the pressure was reduced to that of the atmosphere. The pressure in atmospheres (P) is obtained, as explained above, by multiplying the manometric pressure (A) by 3·13,

$$P = 3{\cdot}13 \times A.$$

The linear compression (F) for pressure (P) is given by multiplying the micrometric expansion by the value of a division, or 0·000417″,

$$F = 0{\cdot}000417'' \times D,$$

$$H = \frac{10^6}{75{\cdot}05} F.$$

H is the linear compression in inches of a rod one million inches long for pressure P.

TABLE II. *Details of Experiments on the Compression produced on a Glass Rod by Pressures up to 240 Atmospheres.*

Series	Pressure. Reading of manometer No. 2		Pressure. Divisions of manometer No. 2	Expansion in micrometer divisions			Length of glass rod .... 75·05″ Diameter of do. .... 0·28″ Weight of do. .... 209·5 grammes 1 division of micrometer .... =0·000417″			
				East end	West end	Total				
	A_1	A_2	$(A_1 - A_2)$ A	B	C	$B+C=$ D				
	33·0	13·0	20	5	6	11	Date, 3rd June 1880			
	33·0	13·1	19·9	5	6	11	Temperature (Centigrade)	T	—	13·5°
	35·0	13·8	21·2	6	6	12	Number of observations ..	—	—	8
	34·0	13·9	20·1	7	5	12	Pressure in atmospheres ..	P	$A \times 3{\cdot}13$	64
	34·0	13·9	20·1	5	6	11	Linear compression	F	$0{\cdot}000417D$	0·0047
I	33·7	14·0	19·7	4	6	10	Do. per million	H	$\frac{10^6}{75{\cdot}05}F$	62·6
	35·5	14·0	21·5	6	5	11				
	34·8	14·1	20·7	5	6	11	Do. do. per atmosphere ...	K	$\frac{H}{P}$	0·983
			20·40			11·13	Cubical compression, per 10^6 per atmosphere	N	$3K$	2·949
	51·5	14·1	37·4	9	10	19	Date, 3rd June 1880			
	51·7	14·2	37·5	9	11	20	Temperature (Centigrade)	T	—	13·7°
	52·0	14·2	37·8	9	11·5	20·5	Number of observations ..	—	—	8
	53·0	14·8	38·2	9	11	20	Pressure in atmospheres ..	P	$A \times 3{\cdot}13$	118
	52·3	14·9	37·4	9·5	10	19·5	Linear compression	F	$0{\cdot}000417D$	0·00826
II	52·2	15·0	37·2	16	4	20	Do. per million	H	$\frac{10^6}{75{\cdot}05}F$	110·0
	52·1	15·0	37·1	15	5	20				
	52·5	15·1	37·4	15	4·5	19·5	Do. do. per atmosphere ...	K	$\frac{H}{P}$	0·942
			37·50			19·81	Cubical compression, per 10^6 per atmosphere	N	$3K$	2·826

III	61·0	11·0	50·0	11	15	26	Date, 4th June 1880			
	61·0	11·0	50·0	13	14	27	Temperature (Centigrade)	T	—	12·8°
	61·0	11·1	49·9	12	16	28	Number of observations ..	—	—	13
	61·0	11·1	49·9	22	3·5	25·5	Pressure in atmospheres ..	P	$A \times 3{\cdot}13$	158
	61·5	11·2	50·3	15·5	11	26·5	Linear compression	F	$0{\cdot}000417D$	0·01119
	63·0	11·2	51·8	15	13	28	Do. per million	H	$\frac{10^6}{75{\cdot}05}F$	149·1
	62·2	11·5	50·7	19	8	27				
	63·0	11·8	51·2	15	12	27	Do. do. per atmosphere ...	K	$\frac{H}{P}$	0·946
	62·8	11·9	50·9	22	6	28				
	62·0	11·9	50·1	13	13	26	Cubical compression, per 10^6 per atmosphere	N	$3K$	2·838
	62·3	12·0	50·3	18	8	26				
	62·3	12·0	50·3	23	5	28				
	63·0	12·0	51·0	14	12	26				
			50·49			26·84				
IV	69·8	12·4	57·4	15	17	32	Date, 4th June 1880			
	69·8	12·6	57·2	32	0	32	Temperature (Centigrade)	T	—	12·5°
	69·7	12·6	57·1	27	4	31	Number of observations ..	—	—	8
	69·5	12·8	56·7	19	13	32	Pressure in atmospheres ..	P	$A \times 3{\cdot}13$	177
	69·8	13·1	56·7	19	13	32	Linear compression	F	$0{\cdot}000417D$	0·01334
	69·1	13·0	56·1	20	12	32	Do. per million	H	$\frac{10^6}{75{\cdot}05}F$	177·8
	69·5	13·0	56·5	14	18	32				
	70·3	13·0	57·3	16	17	33	Do. do. per atmosphere ...	K	$\frac{H}{P}$	1·002
			56·88			32·00	Cubical compression, per 10^6 per atmosphere	N	$3K$	3·006
V	76·3	13·3	63·0	27	9	36	Date, 7th June 1880			
	75·5	13·8	61·7	17	18·5	35·5	Temperature (Centigrade)	T	—	12·5°
	76·8	13·8	63·0	18	18·0	36	Number of observations ..	—	—	10
	77·0	13·9	63·1	18	18·0	36	Pressure in atmospheres ..	P	$A \times 3{\cdot}13$	197
	77·0	14·0	63·0	15	20·0	35	Linear compression.......	F	$0{\cdot}000417D$	0·1475
	77·6	14·0	63·6	17	19·0	36	Do. per million	H	$\frac{10^6}{75{\cdot}05}F$	0·1966
	77·0	14·0	63·0	16	18·0	34				
	77·3	14·0	63·3	16	18·5	34·5	Do. do. per atmosphere ...	K	$\frac{H}{P}$	0·998
	77·1	14·1	63·0	16	20·5	36·5				
	77·1	14·1	63·0	14	20·5	34·5	Cubical compression, per 10^6 per atmosphere	N	$3K$	2·994
			62·97			35·39				

TABLE II. *Continued.*

Series	Pressure. Reading of manometer No. 2		Pressure. Divisions of manometer No. 2	Expansion in micrometer divisions			Length of glass rod .... 75·05″ Diameter of do. .... 0·28″ Weight of do. .... 209·5 grammes 1 division of micrometer .... =0·000417″			
	A_1	A_2	$(A_1 - A_2)$ A	East end B	West end C	Total $B+C=D$				
	43·0	8·2	34·8	10	8	18	Date, 7th June 1880			
	42·9	8·3	34·6	12	6	18	Temperature (Centigrade)	T	—	12·2°
	43·0	8·6	34·4	14·5	3	17·5	Number of observations ..	—	—	11
	43·6	8·7	34·9	14·0	4	18	Pressure in atmospheres ..	P	$A \times 3{\cdot}13$	109
	43·8	8·9	34·9	14·5	3	17·5	Linear compression	F	$0{\cdot}000417D$	0·00753
	43·8	8·9	34·9	11·5	5·5(−)	17·0	Do. per million	H	$\frac{10^6}{75{\cdot}05}F$	100·3
VI	43·8	9·0	34·8	11·0	7·0	18				
	44·2	9·0	35·2	10·5 (+)	8·0	18·5				
	44·0	9·1	34·9	10·0	9·0(+)	19	Do. do. per atmosphere ...	K	$\frac{H}{P}$	0·924
	43·9	9·2	34·7	9·0	10·0	19	Cubical compression, per 10^6 per atmosphere	N	$3K$	2·772
	44·0	9·2	34·8	9·5	8·5	18				
			34·81			18·05				
	59·4	9·2	50·2	11·5	14	25·5	Date, 7th June 1880			
	59·8	9·3	50·5	18	8·5	26·5	Temperature (Centigrade)	T	—	12·3°
	60·0	9·6	50·4	16	11	27	Number of observations ..	—	—	10
	59·2	9·8	49·4	12	15·5	27·5	Pressure in atmospheres ..	P	$A \times 3{\cdot}13$	157
	60·0	9·8	50·2	12	15·5	27·5	Linear compression	F	$0{\cdot}000417D$	0·01126
VII	59·9	9·9	50·0	12	14	26·0				
	60·4	9·9	50·5	16	11·5	27·5	Do. per million	H	$\frac{10^6}{75{\cdot}05}F$	150·0
	60·0	10·0	50·0	16	12	28·0				
	61·0	10·0	51·0	15	12·5	27·5	Do. do. per atmosphere ...	K	$\frac{H}{P}$	0·957
	60·8	10·0	50·8	7	20	27	Cubical compression, per 10^6 per atmosphere	N	$3K$	2·871
			50·30			27·00				

VIII	72·0	10·0	62·0	7	26	33	Date, 7th June 1880			
	72·8	10·1	62·7	18	16	34	Temperature (Centigrade)	T	—	12·5°
	72·6	10·2	62·4	14	20·5	34·5	Number of observations ..	—	—	7
	71·7	10·3	61·4	16	17	33	Pressure in atmosphere ...	P	$A \times 3{\cdot}13$	195
	73·0	10·4	62·6	9	25	34	Linear compression	F	$0{\cdot}000417D$	0·0140
	73·0	10·7	62·3	10·5	23	33·5	Do. per million	H	$\frac{10^6}{75{\cdot}05}F$	187·2
	73·0	10·9	62·1	10	24	34				
			62·21			33·71	Do. do. per atmosphere ..	K	$\frac{H}{P}$	0·961
							Cubical compression, per 10^6 per atmosphere	N	$3K$	2·882
IX	79·8	11·1	68·7	11	26·5	37·5	Date, 7th June 1880			
	80·6	11·2	69·4	12	25	37	Temperature (Centigrade)	T	—	12·8°
	80·5	11·2	69·3	11	25·5	36·5	Number of observations ..	—	—	10
	80·5	11·4	69·1	12	26	38	Pressure in atmospheres ..	P	$A \times 3{\cdot}13$	216
	80·8	11·4	69·4	11	27·5	38·5	Linear compression	F	$0{\cdot}000417D$	0·01547
	80·5	11·5	69·0	20	15·5	35·5	Do. per million	H	$\frac{10^6}{75{\cdot}05}F$	206·1
	80·5	11·6	68·9	13	24	37				
	80·5	11·8	68·7	20	17	37	Do. do. per atmosphere ...	K	$\frac{H}{P}$	0·957
	80·0	11·8	68·2	11	25·5	36·5				
	80·5	11·8	68·7	12	25·5	37·5	Cubical compression, per 10^6 per atmosphere	N	$3K$	2·870
			68·94			37·10				
X	88·0	12·0	76·0	11	29	40	Date, 7th June 1880			
	89·9	12·1	77·8	11	29	40	Temperature (Centigrade)	T	—	12·9°
	88·9	12·2	76·7	11	28	39	Number of observations ..	—	—	6
	88·5	12·2	76·3	14	27·5	41·5	Pressure in atmospheres ..	P	$A \times 3{\cdot}13$	240
	88·5	12·3	76·2	17	22·5	39·5	Linear compression	F	$0{\cdot}000417D$	0·01671
	88·5	12·3	76·2	11·5	29	40·5	Do. per million	H	$\frac{10^6}{75{\cdot}05}F$	222·6
			76·53			40·08	Do. do. per atmosphere ...	K	$\frac{H}{P}$	0·929
							Cubical compression, per 10^6 per atmosphere	N	$3K$	2·786

$$K = \frac{H}{P}$$

is the same per atmosphere or the linear compressibility of the glass.

$$N = 3K$$

is the cubical compressibility of the same glass.

These results are summarised in Table III.

TABLE III. *Summary of Experiments made on the 3rd, 4th, and 7th June 1880, on the Compression produced on a Glass Rod by Pressures up to 240 Atmospheres.*

Series No.	Number of observations	Temperature, Centigrade	Pressure in atmospheres	Linear compression: Inch	Linear compression: Per million	Linear compression: Per million per atmo.	Cubical compression per million per atmosphere	Greatest deviation of F from the mean	Greatest deviation per cent. from mean
		T	P	F	H	K	N	Q	$\frac{100Q}{F} = R$
1	8	13·5°	64	0·0047″	62·6	0·98	3·0	0·00047″	10·0
6	11	12·2	109	0·0075	100	0·92	2·8	44	5·8
2	8	13·7	118	0·0083	110	0·94	2·9	34	4·1
7	10	12·3	157	0·0113	150	0·96	2·9	42	3·7
3	13	12·8	158	0·0112	149	0·95	2·9	56	5·0
4	8	12·5	177	0·0133	178	1·00	3·0	42	3·2
8	7	12·5	195	0·0140	187	0·96	2·9	33	2·3
5	10	12·5	197	0·0148	197	1·00	3·0	46	3·1
9	10	12·8	216	0·0155	206	0·96	2·9	58	3·7
10	6	12·9	240	0·0167	223	0·93	2·8	59	3·5
						0·960	2·92		

Two columns Q and R are added. Q gives the greatest absolute deviation from the mean total expansion in any series, R gives the deviation per cent.

In the observations recorded I made no attempt to subdivide the micrometer divisions further than to estimate a half. As the micrometer readings are not affected directly by the pressure, the deviation per cent. should be, as it is, less the higher the pressure; and there is no doubt that the higher the pressure is the greater is the accuracy of the observation. The only way in which the pressure affects the reading of the micrometer is that when it is sufficiently high it produces a microscopic distortion of the tube, which throws the point of the rod very slightly out of focus. This is remedied by a slight touch of the fine adjustment screw of the microscope.

The general result of these experiments is that the linear compressibility of the glass experimented on is 0·96, and its cubical compressibility 2·92 per million.

Grassi[1] gives as the means of his observations at pressures up to ten atmospheres—Glass, 2·25; crystal, 2·804 and 2·8584. The agreement between the two is very close.

We have then for the apparent compressibility of water in glass at 2·5° C. 48·68 per million per atmosphere. Adding 2·92 for the compressibility of glass, we have for the absolute compressibility of water at 2·5° C. 51·60. Similarly at 12·5° C. we have 45·39 for the apparent, and 48·31 for the real compressibility. Grassi gives the following values for the true compressibility of water at various temperatures: At 1·5° C. – 51·5; at 4·1° – 49·9; mean – 50·7. At 10·8° C. – 48·0; at 13·4° – 47·7; mean – 47·8. My results agree very closely with these.

Before concluding, I would call attention to a very curious phenomenon, which I have never seen noticed, namely, the peculiar *noise* which accompanies the relief of pressure in a mixture of ice and water. In comparing the piezometer *K*, No. 4, in melting ice with the manometer at 12·5° C., I proceeded gradually from lower to higher pressures. When the pressure which was relieved was 100 or 120 atmospheres, I thought I noticed a slight noise. On raising the pressure higher the noise became more and more distinct, until when the pressure relieved was over 200 atmospheres, it was distinctly audible at a distance of five or six feet. It resembles the noise produced by bending a piece of tin backwards and forwards, and is markedly intensified by accelerating the relief, just as the noise

[1] *Ann. Chim. Phys.* (1851) [3], 31, p. 474.

made by blowing off steam is intensified by enlarging the outlet. When the relief valve is opened very carefully it whispers gently but very distinctly, till the pressure is all down. If opened comparatively briskly, but still with great care, the noise is comparatively loud, but more rapidly used up. I forbear making any reflections until I have been able to study this phenomenon more closely.

Pieces of clear ice which had been subjected to high pressure in the receiver were finely laminated in parallel planes. In each plane there was a central patch surrounded near the sides of the block by a ring of spherules. The annexed figure gives an idea of the arrangement. The size of the spherules is greatly exaggerated.

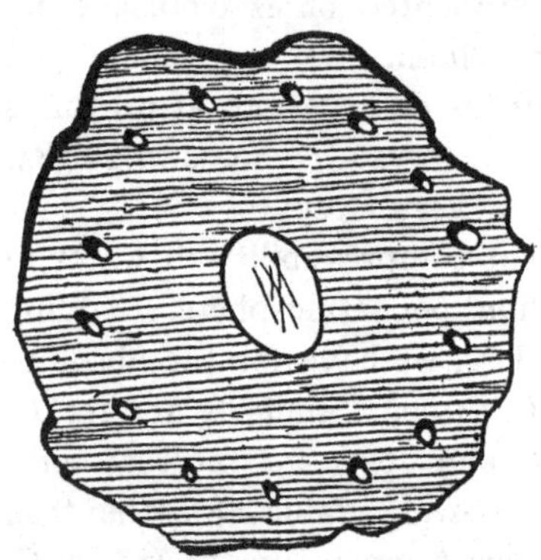

The lamination of ice by pressure in one direction is well known. I am not aware that its production by pressure in all directions has been noticed.

No. 26. [*From Proc. Roy. Soc.* 1904, *Vol.* LXXIII. *pp.* 296—310].

ON THE COMPRESSIBILITY OF SOLIDS

THE solids dealt with in this research are the metals platinum, gold, copper, aluminium, and magnesium. Their absolute linear compressibilities were directly determined at pressures of from 200—300 atmospheres at temperatures between 7° and 11° C. The determinations were made by the same method, and with the same instrument which I used for the determination of the compressibility of glass in 1880[1]. As nearly a quarter of a century has passed since then it will be expedient to recall the principal features of the instrument, and of the method.

The idea of it occurred to me on the evening of March 23, 1875, the day on which the "Challenger" made her deepest sounding, namely, 4475 fathoms (8055 metres), and I was able to put it in practice six days later, on March 29, when, however, the depth was only 2450 fathoms (4410 metres). The observations which I was making during the voyage on the compressibility of water, sea-water, and mercury, were of little value without a knowledge of the compressibility of the envelope which contained them. It was a matter to which I had given much thought. I had studied all the methods which had been used up to that date, but they had all turned out to be faulty.

The idea of utilising the linear compressibility of glass in order to arrive at its cubic compressibility had occurred to me, as it had, no doubt, occurred to many others, before. The difficulty lay in giving the idea experimental expression. It was clear that the instrument would fall to be classed as a piezometer, and would have to be a self-registering one, because what takes place in the depths of the sea is removed from observation. All my piezometers contained a liquid, and this

[1] *Roy. Soc. Edin. Trans.* vol. XXXIX. p. 589, and this volume, paper No. 25, p. 358.

I had recognised to be fatal to absolute measurements. The problem had, therefore, come to be: to design a piezometer which should contain no liquid; and it was the solution of this problem which occurred to me on the evening of March 23, 1875.

The form which the instrument took was very simple. In my laboratory outfit I had included some lengths of tubing suitable for the stems of piezometers, of which I had to make a number during the voyage. In order to be able to use the indices of broken deep-sea thermometers, the tubes had the same internal diameter as the stems of these instruments, about 1 mm. On the outside of the tubes a scale of millimetres was etched. I took the greatest available length of this tube, namely, 60 cms. I then drew out a wire of the same glass and passed it into the tube until it appeared at the other end of the tube. This end of the tube was then sealed up, and the end of the glass wire was fused into it, so that, while free throughout its whole length, longitudinal motion was prevented. The length of the glass wire was 57 cm., so that there was an empty space in the tube of 3 cm. above it. The magnetic index of a broken deep-sea thermometer was re-haired and passed into the tube above the glass wire. The open end of the tube was then sealed up. The result was *a piezometer consisting of nothing but glass.* In principle it was precisely the same as any of the other piezometers. The indioes of these give the difference between the compression produced by the pressure on the contents and on the envelope. In the case of the other piezometers, which contained liquids, the balance was on the side of the contents. In the all-glass piezometers the contents, besides being of the same material as the envelope, were completely protected from pressure, and the whole of the change of length measured fell to the envelope. It has, therefore, a feature which is possessed by no other instrument; *with it the absolute compressibility of a solid is determined by one measurement.*

Before the instrument was attached to the sounding-line, the index was brought down by means of a magnet to rest on the end of the internal glass wire, exactly in the same way as if it had been the mercury column in a *maximum* and *minimum* thermometer. The instrument was then sent to the bottom, or to whatever depth might be decided on.

During the descent the temperature of the glass, both inside and outside, fell with that of the water through which it passed, but as the contraction produced was the same on the wire and on the tube, there was no differential effect to be recorded by the index. On the other hand, the increasing pressure, as the instrument descended, affected only the outside tube, which it shortened. In contracting, it was obliged to pass the index, which was kept in its place by the internal wire. When the instrument was being hove up, the reverse process took place; the tube lengthened, and lifted the index clear of the internal wire by an amount equal to the lengthening of that portion of the tube. As the whole clearance produced by the expansion from the greatest depth did not exceed 1 mm., its amount had to be estimated by the eye with the assistance of a magnifying glass.

The experiment made on March 29, 1875, was quite successful, and it gave 3·74 as the cubic compressibility per million per atmosphere, of the glass of which the tube was made. The exact figure found in 1880 for glass from the same source was 2·92. A number of observations were made with the instrument, both on the sounding-line and in the compression apparatus with which the ship was supplied, and figures from 3—5 per million were found. These were sufficient to give the order of the constant which was sought, but it was impossible with the appliances at hand to measure such small distances with sufficient accuracy to enable a definite value to be determined.

On the return of the ship I embodied the principle in an instrument of precision, which I had constructed in the early part of 1880, and I used it in the month of June of that year for the exact determination of the compressibility of the glass which had been used in the construction of my "Challenger" piezometers.

It is this instrument, and without any alteration, which I have used for the purpose of the present research.

With the assistance of Fig. 1, its features, and the distribution of its parts, will be apparent without any lengthy description. It consists of three parts: the force pump on the left, the receiver for the reception of piezometers or other bodies on the right, and behind these, the block with tubes projecting on either side to receive the rod or wire of the solid, the compressibility of which is to be determined. Every part of the instrument is made of steel.

The characteristic and original part of the instrument is the steel block with its tubular projections. After the experimental

Fig. 1.

rod or wire has been introduced into the steel tube the glass terminals are slipped over the extremities of the rod or wire, and, passing through leather hydraulic collars, close the steel tube. The terminals are then secured in position by open steel caps screwed externally over each end of the steel tube. Each of the glass terminals is commanded by a microscope with micrometer eye-piece.

In 1880, when the instrument was housed in a room with a stone floor, these microscopes stood on three-legged stools, as shown in the figure. As the room with the stone floor was no longer available, I had to instal the instrument close to the windows of the laboratory, which has a wooden floor, and fix metal brackets in the wall to carry the microscopes. *In both cases the micro-*

meters, which measure the expansion or contraction of the body under examination, are independent of the instrument which holds it.

The *manometer*, which indicates the pressure in the instrument, is seen under the steel block which carries the tubes. It is simply a mercurial thermometer with a very thick bulb. The scale on it is an arbitrary one, and its value as a measure of pressure is fixed by observing its reading when the principal piezometer which I used during the voyage of the "Challenger" was in the receiver. This piezometer, known as C. No. 1, contained distilled water, and from very many carefully executed experiments at depths from 800 fathoms (1440 metres) up to 2500 fathoms (4500 metres), made in the South Pacific where the oceanic conditions were most favourable, the apparent compression of distilled water in this particular instrument at the temperature ruling in these depths, which averages in round figures 2° C., and when exposed to measured columns of sea-water, of known quality as regards density, was accurately known. The indications of the manometer are, therefore, equivalent to those of piezometer C. No. 1, the standardisation of which was effected under an open-air water column. The observations made with C. No. 1 on board the "Challenger," which form the basis of the scale of pressures, are collected in Table I. They are expressed in terms of the apparent compressibility of distilled water deduced from them.

In the table the vertical lines represent apparent compressibility in volumes per million per atmosphere, rising by steps of 1 per million from 45—55, so that all the values of the compressibility falling between say 45 and 46, or 49 and 50 are arranged in one column. Above each entry of apparent compressibility will be found the depth in metres to which the instrument was sent, and the temperature (° C.) of the sea-water at that depth. The depth is expressed in metres because it so happens that the average density of the water in this part of the South Pacific, allowance being made for the vertical distribution of temperature, compression, and salinity, is such that a vertical column of it 10 metres high exercises very exactly the same pressure as 760 mm. of mercury. So that the depth in metres, divided by 10, gives the pressure in ordinary atmospheres. At great depths a very slight correction has to be made; the nature of this will be apparent from the following table, in which, for different depths in metres, D, the pressure, P, in atmospheres is given:

D....	1400	2000	3000	4000	5000	6000
P....	139·96	200·14	300·82	401·98	503·62	605·75

Owing to the preponderance of water of low temperature and of very uniform salinity in a vertical column of water in any part of the open ocean, the pressure exercised by it per thousand metres does not differ appreciably from 100 atmospheres.

Inspection of Table I shows at once in which column the true value of the apparent compressibility is most likely to be found. It is the one which includes values between 49 and 50. Outside of this column it is only the adjacent one containing values between 48 and 49 which enters into competition. The mean of all the values in these two columns is 49·16, and this figure forms the basis of the measurement of pressure in this investigation, and it is used in interpreting the pressure-value of the readings of Piezometer C. No. 1, when being compared with those of the manometer used for the ordinary measurements of the pressure in the apparatus.

The change of apparent compressibility per million of water with change of temperature for the small range of temperatures with which we are concerned was found in 1880 to be at the rate of 0·33 per degree (Celsius), and this figure is used in the present research.

Micrometers. The same microscopes and micrometers, which served in 1880, were again used in this research. Their value was determined by reference to a stage micrometer, ruled into hundredths and thousandths of an inch. This was then verified at the National Physical Laboratory. The changes of length measured by the micrometers are therefore given in terms of the standard inch; and, it may be added, the values attached to the readings of the micrometers in 1880 were exactly the same as those now found by reference to the standard of the National Physical Laboratory.

In the microscope which was always placed on the left

hand, one division was equivalent, on the stage, to 0·0004219 inch. In the one on the right hand one division was equivalent to 0·0004167 inch on the stage.

As the contractions or expansions are given directly in terms of the inch, the total length of the rod is given in inches also. In order to bring the ends into a suitable position for observation with the microscopes the length of the rod or wire had to be not less than 75 or greater than 75·5 inches. The actual lengths were measured exactly in each case. The average was 75·32 inches (1·913 metres).

To facilitate the observation of the ends through the thick glass tube a piece of microscopic covering glass was moistened with a drop of water and laid horizontally on the tube, producing the same effect as if a flat surface had been ground and polished on it.

The effect observed and measured is the lengthening of the rod when the pressure is relieved. As the compressibility of solids is very small, the highest pressures have been used which were found to be compatible with the reasonable persistence of the glass terminals; the usual pressure was in the neighbourhood of 200 atmospheres. Very few of the glass terminals stood over 300 atmospheres. The pressures actually chosen were as nearly as possible those at which the manometer had been compared with the "Challenger" piezometer.

The body under observation is in the form either of a rod or a wire. If it is in the form of a rod then it is fitted with wire ends of sufficiently small calibre to enable them to enter the glass terminals.

During an experiment with a rod it contracts while the pressure is being raised, and expands again when the pressure is relieved. The steel tube which holds it, however, acts in the opposite sense, it expands while the pressure rises and contracts while it falls. If the two surfaces were perfectly smooth, one half of the change of length would be measured at the one end and the other half at the other end. As the surfaces are not perfectly smooth, this does not usually occur. Moreover the steel tubes are prolongations of the central steel block which holds them. The block is bored with holes at right angles to each other in the three principal directions. Consequently for a distance of about an inch and a half in passing through the block the rod is not supported at all. With the exception of this small portion, however, the rod is supported throughout the whole of its length by the steel tube. Now, although it is thus nominally supported equally throughout the whole of its length, we know that in reality this is pretty certain not to be the case. At some place, either in the right arm or in the left arm of the apparatus, the rod is sure to bear more heavily than in any other part. The contraction under pressure and the expansion under relief of pressure will then apparently take place as from this point as origin. Supposing this point itself to be motionless, it is evident that the change of length measured at the two ends will be in the same proportion to each other as would be the arcs which they would describe if the rod were a lever oscillating on the point as a fulcrum. As there is no support at all at the centre, this point must lie on one side or on the other of it and the motions of the ends must be unequal. But the fixed point of the tubular receiver is the central block; therefore any point in, let us say, the right-hand tube will, when pressure is being raised, move to the right, and, on relief of pressure, retreat by an equal amount to the left. Consequently when we observe and measure the change of position of, for instance, the right-hand extremity of the rod, when the pressure is relieved, that change of position is composed of two motions, the expansion of the part of the rod which lies between the right-hand extremity and the point in it whose motion with respect to the steel carrying tube is *nil*, along with the proper motion of that point. Similarly, when we measure the change of position of the left-hand end, it also is composed of two parts, the expansion of the part of the rod which lies between the left-hand extremity and the same point in the length of the rod where its motion with respect to the steel tube is *nil*, along with the proper motion of that point. But at the left-hand end the motion of expansion is to the left, and at the right-hand end it is to the right, while the proper motion of the position of the common point on the rod and on the tube is always in one direction, and in this case, to the left. Therefore the distance measured in the right-hand microscope is the expansion of the portion of the rod which lies to the right of the point on it which is motionless relatively to the tube *minus* the proper motion of this point: and the distance measured at the left-hand end is the expansion of the remainder of the rod *plus* the proper motion of the common point. Consequently the algebraic sum of the two motions measured is the expansion of the rod under the relief of pressure.

When the substance is used in the form of a rod, as, for instance, in the case of glass, its ends are drawn out into wires, such that they can enter and be visible in the glass terminals. What we really measure then is the change of length under change of pressure of the axial glass wire in the rod, which may be looked on as a *fascine* of a very large number of similar but somewhat shorter wires. The sole function of these other wires is to maintain the wire that falls under observation in an axial position. It is obvious that this function can be performed with equal efficiency by wires of any other material, and that the conditions are in no way altered if these are fused into a tube of which the wire to be measured may be regarded as the core. Consequently by my method the linear compressibility of a solid can be determined as well on a wire as on a rod; and there is no limit to the thinness of the wire, so long as it can be handled, and be perceived in the microscope.

These two conditions are, in a way, antagonistic, because for the microscope the finest possible point is desirable, while for the handling of the wire a sensible thickness is essential. Only in the case of glass can a good working compromise be effected, because the wire which enters the glass terminal can be drawn out at the end to the finest possible hair, and the end of the hair can be fused into the minutest possible sphere, which can then be observed in the microscope with the sharpness with which a barometer can be read with a good telescope.

When the substance under observation is in the form of a wire, it lies in a glass tube which fits the bore of the steel tube as closely as possible. Its bore is a very little larger than that of the glass terminals, or about 1 mm. This tube acts as a bearer, and its length is as nearly as possible equal to the distance which separates the inside ends of the glass terminals when in position. When the pressure in the apparatus is raised, both the wire and the glass tube which carries it are shortened, while the steel tube which carries both of them is lengthened, and when the pressure is relieved the reverse takes place. The glass tube behaves exactly like the glass rod, that is, it is liable to a slight motion of translation. Similarly, the wire, which is carried by the glass tube, generally expands and contracts under pressure at a less rate than does the glass, producing again a slight apparent motion of translation. But again, as in the case of the rod, the algebraic sum of the observed motions is the expansion or contraction of the wire.

There is an advantage in having a very slight leak in the apparatus. The routine of an observation is then that the observer in charge of the pump and the manometer gets the pressure up somewhat higher than that desired; he then settles himself with the relieving lever in his hand and calls out as the mercury in the manometer in falling passes each division. The observers at the microscopes read their micrometers at the same moment. When the pressure has fallen a little below the desired pressure, the pressure is very carefully relieved, and the readings of the micrometers and of the manometer are taken at atmospheric pressure. The algebraic sum of the movements of the two ends on the micrometers gives the linear expansion of the body which has taken place, and the difference of the two readings of the manometer gives, when interpreted by the help of Piezometer C. No. 1, the difference of pressure which has caused the expansion. The micrometer measurements are then reduced separately to their absolute values in terms of the inch. The algebraic sum then gives the linear expansion in terms of the inch. It is then divided by the length of the rod or wire in inches and by the pressure in atmospheres; the resulting quotient is the linear compressibility of the metal or other substance. Multiplying this by three, we obtain the cubic compressibility of the substance, if truly isotropic.

It will be evident that, to work with this instrument, three observers are necessary, namely, one for each microscope, and one to raise and relieve the pressure and observe the manometer. I was fortunate in being assisted during this investigation by Mr Andrew King, who was formerly my regular assistant, and is now of the Heriot-Watt College, Edinburgh, and by Mr J. Reid, Demonstrator in the chemical laboratory of that institution. These gentlemen gave up their Christmas vacation for this work, and I owe them a deep debt of gratitude for the willingness and the efficiency of their help. The metals experimented with have been used in the form of wire, and the size chosen was No. 22 of the standard wire gauge (S.W.G.). In the case of aluminium, however, the size was No. 20. The dimensions corresponding to those numbers are given in the following table:

No. of wire	Diameter of wire		Sectional area of wire	Length of 1 c.c.
S.W.G.	inch	mm.	sq. mm.	metre
20	0·036	0·914	0·656	1·524
22	0·028	0·711	0·397	2·519

The degree in which the actual wires corresponded with the tabular specification was checked by weighing measured lengths of them. The weight of 1 metre of each wire was as follows: Platinum, 8·156 grammes; gold, 7·320 grammes; copper, 3·375 grammes; aluminium (No. 20), 1·642 grammes; and magnesium, 0·552 gramme. Neglecting the magnesium which, being pressed and not drawn, is very uneven in its calibre, these figures show that the actual wires were very slightly smaller than they should be by the gauge. Thus, in the case of the platinum wire 1 c.c. occupies 2·636 metres (lineal) instead of 2·519 metres as by the table.

The platinum and gold wires were pure specimens obtained from Messrs Johnson and Matthey in the year 1880. The copper was "high conductivity" copper, and it as well as the aluminium and magnesium wires were of the best quality obtainable at the present day. The platinum and gold wires were heated to redness over a Bunsen lamp before use, so that they were thoroughly annealed. The aluminium wire was also heated, though to a much lower temperature, so as to soften it. The other metals were used in the state in which they were supplied. All the wires were straightened, but not stretched, before use.

The temperature of the wires during the operations was always that of the laboratory, and every care was used to keep it as uniform as possible, and it was as nearly as possible that of the air outside. Working in the middle of winter and in a comparatively high latitude, I hoped to be able to do so in conditions which, as regards temperature, would be similar to those which obtain in the depths of the sea, but the extraordinary mildness of the weather this year made it impossible, and the temperatures fell, mostly between 9° and 11° C.

The results of the investigation are set forth in detail in Tables II to VI, and they are summarised in Table VII.

TABLE II. *Platinum. Date, Jan.* 9, 1904. *Temperature,* 7° C. *Wire No.* 22 *S.W.G. Length,* 75·35 *inches.*

No.	Pressure P	Changes of length			Compression per million $10^6 \frac{s}{75 \cdot 35} = S$	Linear compressibility $S/P = \lambda$
		Right arm r	Left arm l	Sum $r+l=s$		
	atm.	in.	in.	in.		
1	204	0·003750	−0·000844	0·002906	38·57	0·188
2	204	0·003750	−0·000970	0·002780	36·88	0·180
3	204	0·003750	−0·000928	0·002822	37·45	0·184
4	300	0·005292	−0·001181	0·004111	54·56	0·182
						0·1835

In the summary, Table VII, the compressibilities of English flint glass and of the glass of which ordinary German tubing is made as well as that of mercury have been included for purposes of comparison. The compressibility of mercury rests upon a large number of observations made in the "Challenger,"[1] by

[1] *Chem. Soc. Jour.* (1878), vol. XXXIII. p. 453.

TABLE III. *Gold. Date, Jan.* 10, 1904. *Temperature,* 10·6° C. *Wire No.* 22 *S.W.G. Length,* 75·4 *inches.*

No.	Pressure P	Changes of length			Compression per million $10^6 \frac{s}{75\cdot4} = S$	Linear compressibility $S/P = \lambda$
		Right arm r	Left arm l	Sum $r+l=s$		
	atm.	in.	in.	in.		
1	231·5	0·005208	−0·000591	0·004617	61·21	0·264
2	230·0	0·005208	−0·000970	0·004238	56·20	0·244
3	230·0	0·005417	−0·000844	0·004573	60·64	0·264
4	204·0	0·005000	−0·000590	0·004410	58·48	0·287
5	247·0	0·005834	−0·001181	0·004653	61·70	0·254
6	230·0	0·005208	−0·001094	0·004114	54·55	0·237
7	238·5	0·005000	−0·000422	0·004578	60·70	0·255
8	238·5	0·005000	−0·000422	0·004578	60·70	0·255
9	238·5	0·005208	−0·000633	0·004575	60·66	0·254
10	273·5	0·006042	−0·000548	0·005494	72·85	0·264
11	273·5	0·005834	−0·000211	0·005623	74·56	0·273
12	273·5	0·006042	−0·000591	0·005451	72·28	0·264
13	273·5	0·006042	−0·000464	0·005578	73·96	0·270
14	273·5	0·006042	−0·000717	0·005325	70·61	0·258
15	273·5	0·006250	−0·000717	0·005533	73·37	0·261
16	273·5	0·006042	−0·000506	0·005536	73·41	0·268
17	269·0	0·005834	−0·000422	0·005412	71·76	0·267
18	273·5	0·006042	−0·000548	0·005494	72·85	0·266
						0·260

TABLE IV. *Copper. Date, Jan.* 9, 1904. *Temperature,* 10° C. *Wire No.* 22 *S.W.G. Length,* 75·3 *inches.*

No.	Pressure P	Changes of length			Compression per million $10^6 \frac{s}{75\cdot3} = S$	Linear compressibility $S/P = \lambda$
		Right arm r	Left arm l	Sum $r+l=s$		
	atm.	in.	in.	in.		
1	195·5	0·005664	−0·001687	0·003980	52·85	0·270
2	230·0	0·006334	−0·001687	0·004647	61·70	0·268
3	195·5	0·005875	−0·001814	0·004061	53·93	0·276
4	195·5	0·006125	−0·001772	0·004353	57·81	0·296
5	247·0	0·007417	−0·001856	0·005561	73·85	0·299
6	230·0	0·006750	−0·001434	0·005316	70·60	0·307
7	282·5	0·007751	−0·001519	0·006232	82·76	0·293
						0·288

TABLE V. *Aluminium. Date, Jan.* 11, 1904. *Temperature,* 9° C.
Wire No. 20 *S.W.G. Length,* 75·35 *inches.*

No.	Pressure P	Changes of length			Compression per million $10^6 \frac{s}{75\cdot35} = S$	Linear compressibility $S/P = \lambda$
		Right arm r	Left arm l	Sum $r+l=s$		
	atm.	in.	in.	in.		
1	195·5	0·005542	0·002616	0·008158	114·22	0·584
2	161·5	0·004900	0·002152	0·007052	93·58	0·579
3	230·0	0·006667	0·002742	0·009409	124·86	0·543
4	178·5	0·005334	0·002109	0·007443	98·77	0·553
5	256·0	0·007251	0·002995	0·010246	135·96	0·531
						0·558

TABLE VI. *Magnesium. Date, Jan.* 17, 1904. *Temperature,* 9° C.
Wire No. 22 *S.W.G. Length,* 75·2 *inches.*

No.	Pressure P	Changes of length			Compression per million $10^6 \frac{s}{75\cdot2} = S$	Linear compressibility $S/P = \lambda$
		Right arm r	Left arm l	Sum $r+l=s$		
	atm.	in.	in.	in.		
1	204	0·009167	0·007120	0·016288	216·61	1·062
2	204	0·010001	0·006202	0·016203	215·48	1·056
3	204	0·009917	0·006329	0·016246	216·07	1·059
4	204	0·010418	0·005991	0·016409	215·57	1·057
5	204	0·010543	0·005442	0·015985	212·60	1·042
6	204	0·011251	0·004852	0·016103	214·16	1·050
						1·054

which its apparent cubic compressibility was found to be 1·5 per million per atmosphere. The piezometers which were used for this purpose were made by myself on board. The divided stems were of lead glass, because I had no other, and the bulbs or reservoirs, which had a capacity of about 20 c.c., were made of German glass, for the same reason. I have, therefore, applied to the values then found for the apparent compressibility of mercury, the value of the absolute compressibility of

German glass found in January of this year, and the result is that the absolute cubic compressibility of mercury at temperatures between 1° and 3° C. is 3·99.

TABLE VII. *Summary.*

Substance	Year	Atomic weight	Density	Compressibility	
				Linear	Cubic
Platinum	1904	194	21·5	0·1835	0·5505
Gold	,,	197	19·3	0·260	0·780
Copper	,,	63	8·9	0·288	0·864
Aluminium	,,	27	2·6	0·558	0·1674
Magnesium	,,	24	1·75	1·054	3·162
Mercury	1875	200	13·6	1·33	3·99
Glass, flint	1880	—	—	0·973	2·92
,, ,, ...	1904	—	2·968	1·02	3·06
,, German ...	,,	—	2·494	0·846	2·54

With regard to the metals quoted in the tables, the figures speak for themselves. The number of different metals is very small and, until the investigation has been extended so as to include at least the greater number of the metals which can be easily procured in the form of rod or wire, it is not likely that any very general features or laws will be apparent. It will, however, be observed that in the case of the five metals used as wire, *their compressibility increases as their density and their atomic weight diminish*, yet there is no reason to suppose that the compressibility is a continuous function of the atomic weight, like the specific heat. Mercury, although in the fused state, shows this clearly. But besides this, it happens that two pairs out of the five metals, namely, platinum-gold and aluminium-magnesium, are contiguous in the atomic weight series, yet the compressibility of magnesium is, roughly, double that of aluminium, and the compressibility of gold is half as much again as that of platinum. If, however, we compare gold and copper, which occupy parallel positions in Mendeléeff's scheme, we see that they are very much alike, and the same holds with regard to magnesium and mercury which occupy a homologous position. If these facts indicate anything more general, we should expect the metals of the palladium and the iron groups to have a low compressibility like platinum, zinc and cadmium to have a very high compressibility like magnesium, and thallium an intermediate but still considerable compressibility like aluminium.

It will be observed that the two kinds of glass mentioned in Table VII are more compressible the greater their density. This may, however, be due to a specific feature of the oxide of lead which enters largely into the composition of the flint glass.

It is obvious that there is here a great field for interesting research, and fortunately the method is capable of great refinement; only, the successful application of it requires considerable manipulative skill, as well as great patience. The necessity to have, as part of the apparatus, two glass tubes which are exposed to the high pressure on the inside only, introduces an element of chance into the work which is sometimes annoying and sometimes exciting. It is impossible to say beforehand whether a particular glass terminal will stand or not. It is necessary to be provided with a large reserve of them before beginning work, and when one fails another is put in its place without loss of time. Hitherto I have taken no particular care of my glass terminals, because I can always depend on finding plenty of them which will stand from 200—300 atmospheres, and there is abundance of work to be done at these pressures. When, however, it is desired to use higher pressures, it will be prudent to take some measures for preventing the points of the wires scratching the internal surfaces of the terminals. When some precaution of this kind has been taken, casualties will be less frequent, and the attainment of higher pressures will be merely a question of how many glass ends the observer is prepared to sacrifice in the service.

In the work connected with this paper, which extended over the greater part of four weeks, fifteen glass terminals gave way; and oddly enough, the failures were as nearly as possible equally distributed between the two ends; eight of them fell to the left arm and seven of them to the right arm. The bursting of a terminal causes no inconvenience beyond the trouble of replacing it, because the construction of the instrument enables air to be completely excluded from it, and the quantity of water in it to be kept within such limits that its resilience is of no account. When a tube bursts it usually splits longitudinally up the middle into two slabs. One of these almost always remains entire, the other is sometimes broken into fragments, but there is never any projection of material unless the instrument has been carelessly put together and air admitted.

Microseismic Effects. In a research like the present where the primary object is the numerical determination of a physical constant, the secondary phenomena which reveal themselves are often of equal and sometimes of greater interest, because they generally affect preferentially the natural history side of physics. To this class belong the phenomena observed in connection with the behaviour of ice under the relief of high pressure in my earlier investigation[1]. In the present case the frequent bursting of the glass terminals afforded the opportunity of observing another and very interesting phenomenon. It is illustrated in Fig. 2. It was first noticed when copper wire was being experimented with. The pressure had been raised to 300 atmospheres, and had begun to fall when the tube gave way. On proceeding to replace the broken tube with another I was astonished to find the copper wire twisted into a regular spiral in the tube. It made three complete turns in the length of an inch, and the undulatory form was visible throughout one-half of the length of the wire. Instead of fixing new glass terminals, I cut off the end of the copper wire, which showed this curious seismic effect, and put another wire in its place. An exactly similar effect was produced on the magnesium wire, when a glass terminal burst; only the effect was even more marked. The spiral produced in the glass end was closer, and, indeed, the wire had been shoved over itself and broken, for magnesium wire is very brittle. The undulations of greater amplitude extended through the whole length of the wire, and there were *maxima* at distances of

[1] *Roy. Soc. Edin. Trans.*, vol. XXIX. p. 598.

about 35 cm. and 85 cm. from the seat of the explosion. The bursting pressure in this case was no more than 150 atmospheres, yet the effect produced was very much greater than it was in the case of copper.

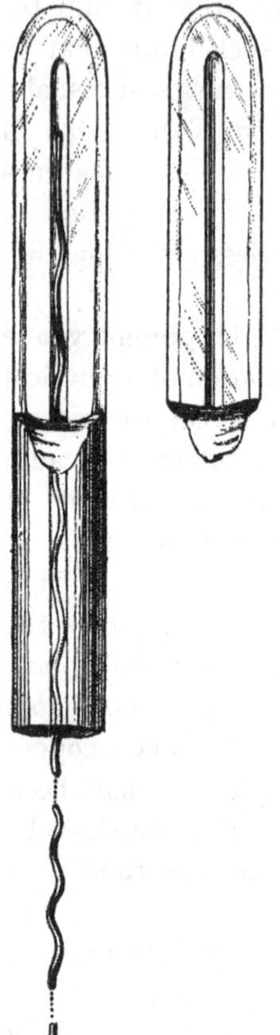

Fig. 2.

The experiments with gold and aluminium were carried out without the loss of a terminal.

In the case of platinum, a terminal burst at about 250 atmospheres, but it produced no apparent seismic effect. On the last day of my experiments I proposed to determine the compressibility of a wire of mild steel, but, owing to hurry in putting the apparatus together, it was impossible to get any satisfactory observations, but one of the terminals burst, and at a pressure over 250 atmospheres. Here again there was no seismic effect. The platinum wire had been thoroughly annealed before being used, and the mild steel wire was as soft and ductile as copper, yet, though copper showed the seismic effect beautifully, it was imperceptible in both platinum and steel. Before the experiment with the steel, I supposed that the high density of platinum caused the shock to be opposed by more inertia than it could overcome, but the density of steel is less than that of copper, therefore its immunity to shock must be due to something other than its density.

The open ends of the glass terminals which are inside of the watertight collars are cut sharply off and the edges are not rounded in the flame. Special directions were given to the glass blower about this, because the effect of it would be the production of considerable tension in that part of the glass. Notwithstanding my directions, some of the tubes were rounded off in the lamp and the effect was as I had foreseen. The only one of these ends which I used burst. In the case of ends which have been cut off and not heated, the fracture is confined to the part of the tube outside the apparatus. In the case of the end with rounded edges the outside part was fractured in the ordinary way, and in addition the rounded portion, which was exposed to no difference of pressure, exploded out of sympathy, much after the fashion of a Prince Rupert's drop.

Postscript.—For the purpose of the report on the results of the "Challenger" Expedition, the general thermometric work was referred to Professor Tait of Edinburgh.

In the course of his examination of the thermometers themselves, he felt the necessity of a knowledge of the absolute compressibility of the glass used in their construction, and he came to me and asked me if I would lend him my apparatus for this purpose.

In making this request he stated to me that, before doing so, he had examined every other instrumental method which claimed to furnish by direct observation the absolute compressibility of solids, and had found them either faulty in design, or so defective mechanically as to shake his confidence in the value of the results which he would obtain by their use. With regard to my instrument, he acknowledged that the principle of it was sound, that the mechanical expression of it was successful, and that the results obtained with it, by competent observers, must be trustworthy.

I was naturally much gratified by receiving such a tribute to the excellence of my apparatus by one of the most distinguished physicists of the day; and it is hardly necessary to add that the instrument was conveyed without delay to the Physical Laboratory of the Edinburgh University, where it remained until the work connected with the "Challenger" thermometers was concluded.

The account of this work is to be found in Professor Tait's report in the second volume of the Chemical and Physical work of the Expedition.

No. 27. [*From Phil. Mag., January,* 1874.]

ON AN APPARATUS FOR GAS-ANALYSIS

A VERY important subject of investigation in the chemistry of the ocean is the nature and quantity of the atmospheric gases dissolved in it. These are extracted by boiling *in vacuo* in an apparatus recently described by Jacobsen[1], and at the end of the operation are obtained hermetically sealed in a glass tube, in which they are capable of indefinite preservation. Although it would be absurd to waste time in making necessarily imperfect analyses of valuable specimens at sea, which are capable of being kept and analyzed with the greatest accuracy on shore, still cases might occur during a three years' voyage where it would be very desirable to make an approximate analysis of a specimen of gas, not necessarily proceeding from sea-water, but, for instance, from hot springs or volcanoes on shore.

The first desideratum for such an apparatus for use on shipboard appeared to be freedom from a mercurial trough. All existing apparatus, besides having this source of inconvenience, were far too large and cumbrous to be at all suited to the modest allowance of laboratory space available on board ship. What seemed to give most promise of success was in some way or other to adapt the original Ure's eudiometer to the purpose. The form of apparatus finally adopted is that represented in Fig. 1. It was constructed, according to my drawings and instructions, by Dr Geissler, of Bonn; and it is needless to add that as a piece of glass workmanship it is a *chef-d'œuvre*.

Before describing it, it may be as well to state the objects sought to be attained, when the motives in the design of the

[1] *Ann. der Chem. und Pharm.*, vol. CLXVII. p. 1.

various parts will be more apparent, and the measure in which they fulfil their end more correctly appreciated.

First of all, the size of the instrument must be reduced to a minimum; nor must its shape be so eccentric as to interfere with its being packed into a reasonable space of symmetrical form. Again, when packed it must be safe against rolling, and be easily unpacked and mounted, and as easily dismounted

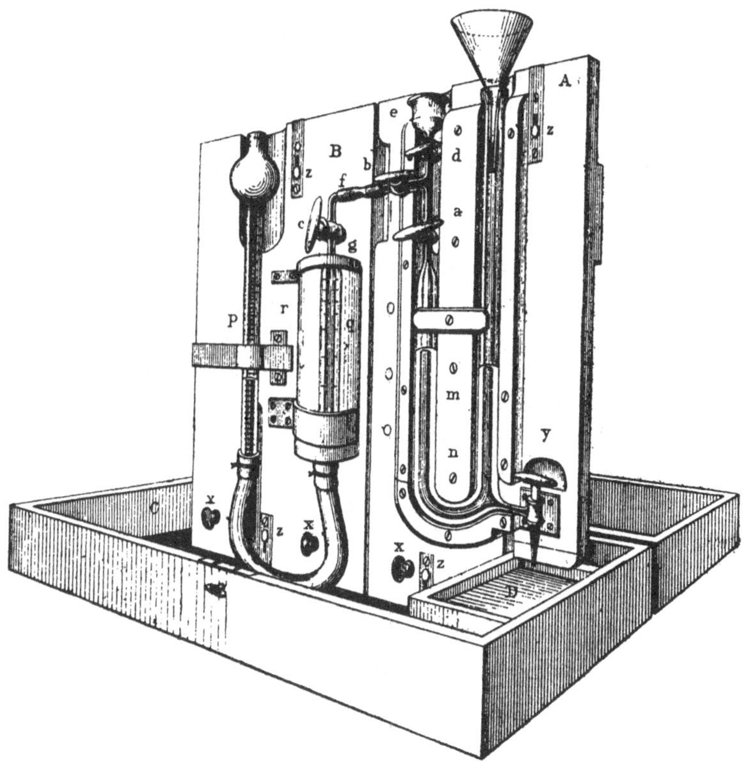

Fig. 1.

and packed away again. I wished further to be able to use eudiometrical as well as absorptiometrical methods, using in the latter liquid reagents. The necessity for a special gas-analysis room had to be dispensed with; and, as before mentioned, the use of a mercurial trough was to be avoided. The advantages secured by the peculiar way of packing, so that the case, in which it is packed when out of use, forms its support, and a working tray for saving spilled mercury, when in use, were not contemplated from the beginning, but suggested themselves only after the glass work was finished, as until then the form and size of the case could not be determined on. It was made for me by Messrs Kemp and Co., of Edinburgh; and the workmanship is in every way satisfactory.

The apparatus consists essentially of two U-tubes. The one, which, according to precedent, we may call the "laboratory tube," is wholly of glass; the other (the eudiometer) has the legs, of glass, united by an india-rubber tube of suitable length. These are affixed to mahogany boards *A* and *B*, which fit into the wings of the box *C*, where they are secured, each by two bolts (*z*, *z*), passing into the sides of the box. *D* is a strong mahogany box for holding mercury.

The glass U-tube, which, for convenience, we shall call *A*, as well as the board to which it is attached, is 0·42 metre high; the shorter leg is 0·18 metre from the bend to where the capillary tube is joined on, and its diameter is 0·02 metre. The diameter of the capillary tubes is 0·003 metre. In the U-tube *B*, the eudiometer (*q*) is 0·17 metre from capillary to india-rubber; the movable leg (*p*) is 0·34 metre long; and the diameter of both is the same as in *A*. The length of the box *C* is 0·45 metre, and the width and depth of each wing 0·195 metre and 0·105 metre respectively. The mercury-box *D* measures 0·12 metre by 0·10 metre by 0·08 metre, and when in use fits into the place where it is represented in the figure. When dismounted it is not packed in the case with the rest of the instrument.

When in use, the parts *A* and *B* are screwed to the back of one wing of the box, *A* with one, and *B* with two screws (*x*, *x*, *x*). The back of the box carries four nuts, so that *A* may be fixed to it close up to the end of the box and with the stopcock *y* projecting over the side, leaving a space between *A* and *B*, the object of which will be explained further on.

The different parts of the apparatus will be best explained by describing the manipulations, occurring during the analysis of a gas, as well as those necessary for introducing the gas into the apparatus. But first we must describe the capillary part of the apparatus between the eudiometer *g* and the laboratory tube *m*. The part belonging to *A* is shown separately in two sections, Figs. 2 and 3. The stopcock *a* has two tubes—the one affording direct communication upwards between the two portions of the capillary tube, shown in Fig. 3, the other communicating through the prolongation of the stopcock with the air, shown in Fig. 2; *b*, *c*, and *d* are ordinary stopcocks pierced to the same bore as the tubes they connect. The capillaries of *A* and *B* are connected by a piece of india-rubber

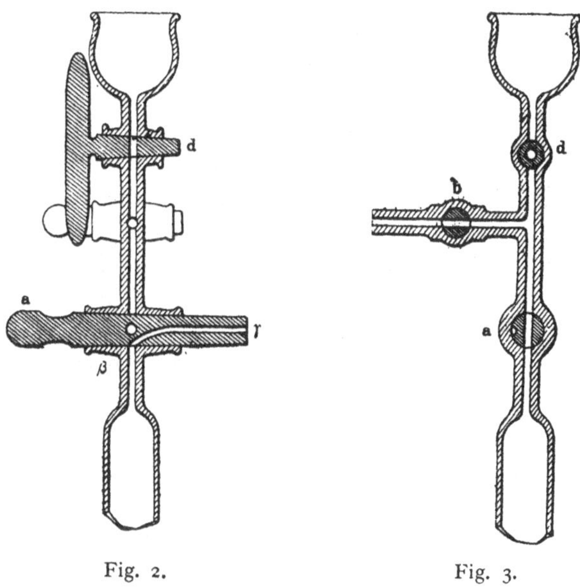

Fig. 2. Fig. 3.

vacuum tubing. This tube is 0·003 metre diameter in the bore; and the thickness of its walls is likewise three millimetres. The stopcock *d* communicates with the air through the cup *e*. When the instrument has been set up as in Fig. 1, it is filled with mercury by pouring in at the open legs of *A* and *B*, all the stopcocks being open, so that finally the mercury rises a little way up in the cup *e*. The apparatus being full of mercury, the gas to be analyzed is introduced into it, either through the opening, γ, of the stopcock *a*, or, if it is contained in a sealed tube, by interposing it at *f* between the two parts *A* and *B*, *A* being for that purpose shifted to the further end of the box, and the two ends of the tube in which the gas is collected connected by vacuum-tubing, filled with mercury, with the ends of the capillaries of *A* and *B*. The pressure in one side of the apparatus is now reduced, either by running mercury from *A* by means of the stopcock *y*, or by lowering the movable leg *p* of *B*. When the points of the tube are broken the gas rushes into the parts of the apparatus where the pressure is least, its place being supplied by mercury from the other side. The stopcocks are then closed, the tube full of mercury removed, the parts *A* and *B* reunited by *f*, and the analysis proceeded with. When the gas is to be admitted through *a*, an india-rubber tube is connected at γ and filled with mercury by opening *a*, as in Fig. 2, and running mercury through it. The point of the collecting-tube, having previously been touched with the file, is pushed into the india-rubber tube; the other end having also received a stroke of the file, is immersed in a cylinder full of mercury and the lower point broken off against the bottom. The mercury immediately rises in the tube; and by supplying its place in the cylinder and reducing the pressure in *A* or *B*, as the case may be, the whole or part of the gas may be transferred to the instrument. Of course in all these cases the india-rubber tubes must be moistened inside with solution of corrosive sublimate.

The gas, having been introduced in one way or the other, is brought all into the eudiometer *q*, mercury being allowed to run over from *A* into *q* until the whole capillary part, down to where it joins the eudiometer at *g*, is full of mercury. *q* is a short eudiometer with a scale of only six divisions, like the measuring-tube in Frankland's apparatus[1]. It is immersed in the cylinder *r*, which is filled with water. The leg *p* is divided into millimetres, and is lowered or raised until the meniscus of the mercury stands exactly at one of the lines. The difference in height of the two columns is then measured by means of the

[1] The tube *q* in the figure is shown divided as stated in the text. I have found, however, that it is much more convenient to have it divided into millimetres; and the labour of calibration is no greater.

millimetre-scale on *p*, and the temperature of the water in *r* ascertained. These, with the height of the barometer, give the volume of the gas. On the tube *r* marks are made (not shown in the drawing) corresponding to and on the same level with those on *q*. When the difference of level of the mercury in the two tubes is to be measured, *p* is applied to *r* and the level of the mercury in *q* read off, the marks on *r* giving the direction for the eye; that in *p* is then read off in the ordinary way, the difference giving the column of mercury to be added to or subtracted from the barometer-reading.

Let us suppose that we have a sample of air extracted from water. It consists of oxygen, nitrogen, and carbonic acid, the last of which is determined first, by absorption with caustic potash. For this purpose mercury is run out of *n* by means of *y*, *p* is raised, and the stopcocks *c*, *b*, and *a* opened. The air is thus drawn over out of *q* into *m*, mercury being allowed to fill the capillary, when the cocks are again shut. The cup *e* is now filled with strong solution of caustic potash, and, the level in *n* being still kept low, the stopcock *a* is opened full, and then *d* very carefully, caustic potash being allowed to run down through *a* into *m*, where it meets the gas in the most advantageous way for quick absorption. When enough caustic potash has been allowed to enter, *d* is closed, some mercury poured into *e*, *d* again opened and the solution in the capillary replaced by mercury. When the absorption is finished, the level of the mercury in *n* is again raised and the stopcocks *b* and *c* opened. The stopcock *a* is now very carefully opened, the flow of the gas being further regulated by raising or depressing *p*, and the gas allowed to pass over into *q* until the potash solution just touches the lower surface of the stopcock *a*, which, being open, has the position shown in Fig. 3. The position of *a* is now shifted to that shown in Fig. 2, when the potash solution is eliminated. The position of *a* is now brought back to that represented in Fig. 3, and the gas remaining in the capillary swept out by mercury and the volume measured as above described.

If the oxygen is to be determined by absorption, the manipulations are exactly the same as in the case of carbonic acid, alkaline pyrogallic acid being used instead of caustic potash. If it is to be determined eudiometrically, then, after the carbonic acid has been absorbed, the gas remains in the eudiometer, the stopcocks *c* and *b* being shut. The stopcock *d* is now opened and *a* turned the reverse way to what it is in Fig. 2—that is, with the side communication $\beta\gamma$ communicating with the capillary above *a*. The capillary is then emptied of mercury and the hydrogen-evolving apparatus connected with γ by means of an india-rubber tube, and the hydrogen allowed to stream through *a, d, e* until all air is swept out; the stopcock *d* is then closed and *a* brought back to its position in Fig. 2, when the gas enters *m*. When enough hydrogen has passed in, *a* is brought to its position in Fig. 3 and the hydrogen-apparatus dispensed with. Mercury is then poured into *e*, and *d* opened and the hydrogen in the capillary driven into *m*. The hydrogen is now passed over into the eudiometer and exploded, the measurements being made as described.

When the analysis is finished the mercury is emptied out of the tubes, the parts *A* and *B* bolted each into its own side of the box, the screws *x, x, x* being lodged in the nuts *w, w, w* let into the board *A*. The box when closed measures 0·48 metre by 0·22 metre by 0·22 metre, outside, and is easily portable. Of course the apparatus is not meant exclusively for work on board ship, where indeed attempts at gas-analysis should be avoided, as at the very best the results must be very uncertain; but it makes a compact laboratory apparatus and is economical in mercury.

For the preparation of the electrolytic gases, either separate or mixed, in a way convenient for gas-analysis, I have had the apparatus represented in Fig. 4 constructed. It is of the well-known lecture-apparatus form, and consists of two tubes, *A* and *B*, united at their lower extremities by the short tube *C*, which connects them at the same time by means of the tube *D* with the reservoir *E*. At their upper extremities *A* and *B* terminate in capillary tubes furnished with stopcocks, *F, F*. Communication with the reservoir can be made or interrupted by the stopcock *G*. One of the tubes, *A*, is furnished with two platinum electrodes; the other, *B*, has but one. Delivery-tubes (not shown in the drawing) of the usual form fit upon the tubes above the stopcocks *F*, *F*. When about to be used, all

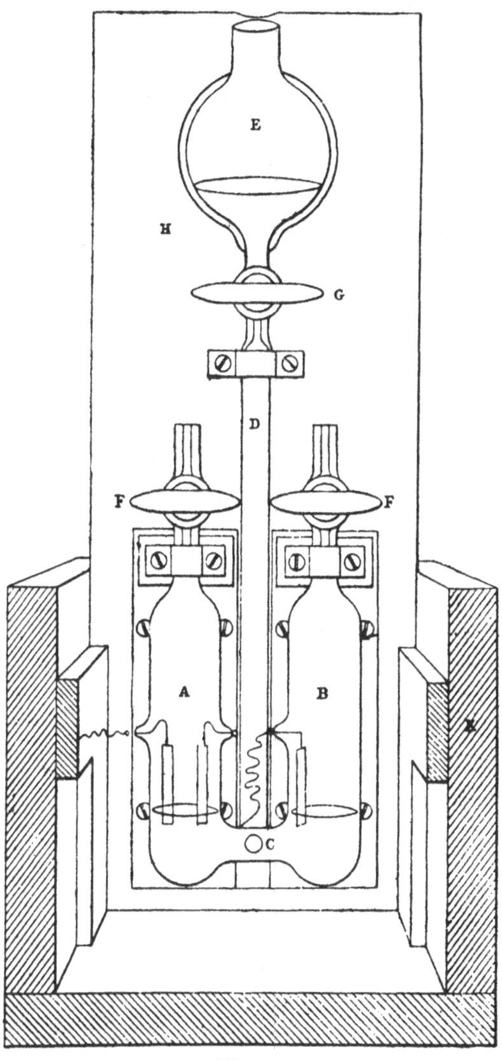

Fig. 4.

the stopcocks are opened and diluted sulphuric acid poured in through the reservoir until it has eliminated all air and is running out itself at the delivery-tubes. The stopcock *G* is now shut and the battery connected, as circumstances may require, either with the two electrodes in *A*, or with one in *A* and one in *B*. Gas is allowed to escape freely until all dissolved air is eliminated. The stopcocks *F*, *F* are then closed, *G* opened, and the liquid in *A* and *B* allowed to sink until it just covers the electrodes. *G* is then closed, and *F*, *F* opened, when the gases may be introduced into the eudiometer in the ordinary way, or into the above-described apparatus in the way there indicated. Connection between the gas-generator, the tubes *A* and *B*, and the reservoir *E* being shut off by the stopcock *G*, there is never any difficulty in forcing the gas through the mercury into the eudiometer.

The whole apparatus is attached to a mahogany board, *H*, which fits into the box *K*, shown cut through the middle in the drawing, either as represented, when in use, or, when not in use, as a lid, with the apparatus attached to its inner side. The box thus fulfils the double purpose of a convenient stand and a safe packing-case. The only alteration which I should be inclined to make, would be to have the lid to which the apparatus is attached made of vulcanite, to avoid the risk of the wood's warping in damp weather.

No. 28. [*From The Times, March* 30, 1889.]

HER MAJESTY'S SHIP "SULTAN"

So little is known or has been allowed to be published about the circumstances under which the "Sultan" went ashore and the character and extent of the damage which she sustained that it may be useful to summarize what has appeared in *The Times*.

In *The Times* of March 8 a telegram, dated March 7, 10.30 A.M., reports the "Sultan" ashore, and another at 4 P.M. on the same day reports her abandoned with water in the hold fore and aft, but most forward. Telegrams of the 8th report the work of floating going on, the sea calm. In the morning the Commander-in-Chief expresses himself as hopeful of getting the ship off, but in the evening he fears there is no hope of saving her. Telegrams of the 10th describe the ship as lying between two rocks abreast the foremast, and state that the water had been got so far under control that the boiler fires had been relighted, and that the pumps of the salvage steamer and the "Sultan" were able to keep the engine room and after hold clear. Attempts were made to drag her off, and on the 11th divers reported that she had been moved 2 ft. On the same day the "Sultan's" pumps broke down, from what cause is not stated, and the position was barely maintained, "another unknown leak" having appeared. This view is somewhat modified by the reports on the 12th to the effect that the pumps had gained full control over the water and that the work of lightening the ship was progressing steadily. It was added that much was expected from air bags placed round the ship at the part where she grounded. On the 13th the "Sultan" was still in the same position, but a heavy swell had begun to run, causing much anxiety, not on its own account only, but also on that of the wind coming up behind it.

The gale of wind thus heralded was not long in putting in its appearance, and at about noon on the 14th, just seven days after she struck, the "Sultan" was forced off the rock by a heavy north-easterly gale with high sea and sank, her upper works, however, still showing above water. The other ships now returned into harbour, and partly owing to the continued bad weather, partly also owing to the altered position of the ship, further efforts to save or raise her appear to have been discontinued. On the 18th Lord George Hamilton stated that the "Sultan" had grounded on a patch of rock not marked on the chart, but he could not say whether this was due to defective survey or to recent volcanic action. In *The Times* of the 20th the Portsmouth correspondent states that experts consider the prospect of raising her hopeless, especially remembering the difficulties which attended the raising of the "Eurydice," a very much smaller ship.

Last night Lord George Hamilton stated that the Commander-in-Chief reported discouragingly of the prospect of raising the ship, and that they had asked a salvage company to examine the wreck and report. He added that the "Sultan" has 460 tons of iron and cement ballast, equal to one foot of draught. This was put in to enable her to carry an upper battery deck.

From the information thus summarized, we may draw the following conclusions: On the morning of the 7th of March the "Sultan" ran upon a rock, of the existence of which there was no suspicion. She was probably steaming at an easy rate, but even if she were going at the slowest speed at which her engines would move, a mass of close on 10,000 tons is not brought to rest in a couple of seconds without much destructive work being done on her hull. Accordingly, we may assume that, from close to the stem and as far as the fore-rigging the bottom of the ship has been ripped open. From the statement made last night about the 460 tons of iron and cement ballast added to balance an extra battery deck, it is probable that this cement and iron filled up a large portion of the double bottom of the ship, thus converting her from a ship with two iron bottoms into one with a single bottom made of a composite substance, having more resemblance to rock than to anything else. If the ballast was placed as supposed, then the important insurance provided by a double cellular bottom was entirely thrown away, and the rent in the inner skin of the ship must be as great as that in the outer one, and the damage to the outer one cannot fail to be of the most extensive character. That this is so is shown by the fact that with all the pump power which was brought to bear on it no impression was made on the water in the forehold.

Immediately after the ship struck she seems to have filled fore and aft with water, from which we may conclude that at the time of the accident there was free intercommunication between the compartments of the ship. On the other hand, from the fact that two days afterwards it was possible to relight the fires, it is evident that there was very little damage done to the ship abaft the forehold, and that as she remained fast on the rock it was possible to shut off the forehold and deal separately with the water in the midship and after sections of the ship. Up to the 13th, when the swell began to set in, had adequate appliances been at hand for dealing with the water in the fore part of the ship she could undoubtedly have been towed off on any one of the five days that she was on the rock and fine weather prevailed. Up to the present no ships, not even salvage ships, so far as I know, have been supplied with means, which, even in theory, could be expected to deal with cases like the "Sultan's." The universal method of dealing with ships which, from one cause or another, have got holes knocked in their bottoms is, by means of suction pumps, to endeavour to remove the water at a greater speed than it can enter through the leak. Where the damage is in any way extensive and it is difficult to get at it to close it this attempt is usually unsuccessful. In the "Sultan's" case it proved hopeless from the beginning.

The main object of all salvage operations, when stated in simplest terms, is to replace the water which has forced its way into the ship's hold by the air which ought by rights to occupy the position. Now this may be done, or attempted, either by pumping out water or by pumping in air. The former, or hydraulic, method is successful if the ship's bottom can be made sufficiently water-tight; the latter, or pneumatic method, is successful if her deck can be made sufficiently air-tight. In the great majority of cases it is much easier to get at the ship's deck and close her hatchways than to get at her bottom and repair a leak. Consequently in most cases the adoption from the first or the pneumatic method would afford the greatest prospect of success. It is necessary, however, that the deck should originally, or by strengthening, be capable of resisting the upward pressure of the air, which by its reaction forces the water out through the leak. In the case of men-of-war and most modern first-class merchant steamers, which are built with iron decks, this condition is fulfilled by their original construction, which makes their hold space comparable to a boiler, with this difference—that the hatchways are not fitted to resist pressure from within. It is therefore necessary to close the hatchways by solid beams pressed up against the under surface of the deck or lower edge of the combing. In a man-of-war this is comparatively easy, as the hatchways are comparatively small. When all the openings of the deck have been made as tight as possible and all communications with the damaged hold have been closed, air-forcing machinery of very moderate power will speedily force the water out and keep it out. All its work is effective, while in the case of water pumping only the residue, after the pumps are throwing more than can enter by the leak, is available. One steamer with air-forcing machinery of the power of the pumps of one of the salvage steamers would, with the resources of the "Sultan" herself and the other ships of the fleet, have cleared her fore-hold of water and enabled her to be towed off.

It is impossible to say what damage or straining the "Sultan" may have experienced in the process of being lifted off the rock by a tempestuous sea, but the pneumatic method remains as available as before. It is unlikely that her iron deck has been seriously damaged; she is lying with her upper works above water; it will therefore be necessary to employ divers for closing the hatchways, where formerly carpenters would have been sufficient. The difference in the conditions is that when she was on the rock the air would have had to work against a column of about 25 ft. of water; now the column is probably about 40 ft. Were the ship otherwise treated as the "Austral" when she sank in Sydney Harbour—that is, were the sides of the ship built up so as to form a dam, or raised bulwark, then by pumping water out of the portion of the ship where the hull is intact and pumping air into the damaged compartments the ship would surely float herself. It is, of course, quite hopeless to think of raising a ship of her weight by means of pontoons or the devices applicable to small craft, but fill her with air and she will float in spite of us, providing that she is not so seriously crippled as to break under the pressure of her own weight.

It may be said that suitable air-forcing machinery is not available; but the use of air bags, which has been mentioned in several telegrams, shows that machinery of some sort is on the spot and might be applied to the useful purpose of inflating the ship instead of air bags; moreover, adequate machinery can be bought, even if it cannot be improvised. Whether, however, the "Sultan" can or cannot be raised is comparatively a matter of little importance. New ships are always being added to the Navy, and in the course of the next few years they are to be added at a greatly increased rate. If by the adoption of very simple mechanical appliances these and the existing ships can be rendered more secure against the ordinary accidents to which all ships are exposed, it is only right that it should be brought as prominently before the public as possible.

In this letter, already much too long, it is impossible to go into details, but the method is easily intelligible. Let the hatchways be constructed so that they can be closed in such a way as to resist pressure outwards in the same degree as the deck itself; then, if the deck were perfectly air-tight, the bottom of the ship might be damaged to any extent and the water would only rise until the pressure of the air had been raised to that exercised by the column of water outside. As the column of water is unable to compress the air further, and as we have assumed that the air cannot get out through the deck, it is impossible for any more water to get in. If, now, a supply of air can be forced into the damaged compartments, then the water which has entered can be easily forced out again through the leak. For this purpose it is only necessary to have adequate air-forcing machinery in the engine-room and to have it connected by pipes with the compartments of the ship. All the most modern ships of war could be very easily fitted so as to deal with leaks by means of air pressure. Moreover, the uses of air as a medium for the conveyance of power are daily increasing, and it cannot but be an advantage if it can be made to contribute to the safety of the ship.

I will close this letter by one remark which is worthy of consideration. A ship is built weighing 10,000 tons; she is fitted with engines capable of indicating 10,000 horse power. These are capable of driving her through the water at a high rate of speed. When she sticks upon a rock they are of no use. Yet engines of 10,000 horse power ought to be able to lift 10,000 tons at a rate of nearly 15 ft. per minute. How can a ship's steam power, or even a small portion of it, be made effective for lifting her as dead weight—in other words, for keeping her afloat? The only suitable "mechanical power" with which I am acquainted is the pneumatic lever acting on an air-tight deck as a fulcrum.

No. 29. [*From Nature, April* 25, 1889, *Vol.* XXXIX. *pp.* 608, 609.]

AIR-TIGHT SUBDIVISIONS IN SHIPS

THE last two months have been unfortunate ones for shipping generally, and more particularly for the navies of at least four of the great Powers. France has lost two torpedo boats under such circumstances as to involve the condemnation of a whole class of vessels. Germany and the United States of America have each lost a small fleet in a hurricane of unusual violence. Besides the material loss of ships these three nations have to bemoan the loss of a considerable number of men. Only little more than a month ago one of the largest ships of the British Navy stranded in waters rightly assumed to be perfectly safe, and has become a total wreck. Fortunately in this case there was no loss of life. Another of H.M. ships only just escaped the disaster which overwhelmed the German and American fleets at Samoa, and the circumstances attending her escape are worthy of a moment's attention.

The storm approached not without warning, and it is evident that the captains of all the ships set about making preparations for meeting it as best they might. They appear all to have got up steam, so as to ease their cables by steaming to their anchors, in case it should be impossible to get out. The only ship that did get out was H.M.S. "Calliope," and without in any way detracting from the merits of her captain and those under his orders, it is evident, from the brief accounts to hand, that all would probably have been unavailing had she not been provided with very powerful machinery. In the Navy List her tonnage is given as 2770, and her horse power as 4020, or one-and-a-half indicated horse power per ton of displacement. The most powerful of the other ships was the German corvette "Olga," which apparently had considerably less than one horse power per ton of displacement.

The other ships, especially the American ones, were so deficient in power that they were unable to make any front to the storm at all. Even with her great power the "Calliope" was only able to attain an effective speed of half a knot per hour in the teeth of the storm. All praise is due to the men who were able to make such good use of this very meagre margin as to save a costly ship and many valuable lives for the further service of their country.

The Samoan disaster has thus, in a dramatic and even tragic way, shown the uses of steam power in saving a vessel by propelling her against a storm. Reflections on the loss of the "Sultan" lead us to ask if steam power cannot be made more useful in succouring and saving a ship after she has struck a rock, or in any other way received such damage to her hull as to render her loss by foundering imminent.

According to convention an engine is working at the rate of one horse power when it is lifting a weight of one ton against gravity at a velocity of 14·74 feet per minute. If, then, a ship is fitted with engines indicating one horse power per ton of displacement, these engines would, if their whole power could be usefully applied and directed against gravity, be able to keep the ship afloat so long as she did not sink at a greater rate than 14·74 feet per minute. The "Vanguard" took 72 minutes to sink. The practical question comes to be, How can the ship's power, of engines or men, be best applied so that the greatest proportion of it may be made available for keeping her from sinking?

Hitherto it has been usual to fit all ships with suction pumps, capable of being worked, some by steam and some by hand power. To use such pumps with effect it is necessary that they be worked at such a rate as to throw overboard more water than can enter the ship in a given interval of time. The lower they bring the water in the hold of the damaged ship, the greater is the facility offered for the water to enter, and the harder becomes the work of lifting it. If the damage to the ship's hull is in any way serious, dealing in this way with its effect is almost always hopeless, unless it is possible to get at the leak and reduce its dimensions or close it altogether.

The bottom of a ship at sea is very inaccessible. If she remains fast on the rock it is usually impossible to get at the leak either from the outside or from the inside. If she is afloat, and will keep afloat long enough, the leak can often be efficiently dealt with by passing a tarpaulin or sail under her bottom. But this is by no means a simple or easy operation, even when performed as a matter of drill with plenty of time, and in the absence of excitement or danger.

When a ship is sinking, she does so because water has got into her either from above or below, and has displaced the air with which she was charged. In order to stop her sinking and to raise her to her original level, it is necessary to reverse the operation and replace the water again by air. If the water has come in from above, by shipping seas, this can be effected by suction pumps, which throw it overboard again. If it has entered and is entering through a hole in the bottom of the vessel, it is necessary not only to remove the water which has entered, but to stop any further entry, and this is achieved by any means which enable us to thrust the water out again by the same way as that by which it entered.

If we consider a ship's hold, and assume that the deck covering it above, and the bulkheads shutting it off fore and aft, are all sufficiently strong and air-tight, then if the whole bottom were allowed to drop out, her stability being otherwise assured, she would be very little the worse; the water would rise in her hold only until it had so far compressed the air that its tension exactly balanced the pressure of the column of water outside, and matters might safely remain in this condition of equilibrium almost indefinitely. Thus, by making the main deck of a modern ship, to which the water-tight bulkheads are carried up, air-tight, she would be practically proof against all risk of sinking from damage to her bottom.

I do not think that there would be any difficulty in making the compartments of a ship perfectly air-tight or more properly, in fitting them so that the rise of tension quickly produced by the entry of water through a serious leak, would at once close any joints or small openings, in the same way as the door of the air lock giving entrance to a submarine caisson is kept closed and air-tight by the pressure of the air within. But inasmuch as the smallest leak of air, whether through the deck or through the bulkheads, would represent an equivalent of water entered and of buoyancy lost, it is necessary to be able to make good the loss by mechanical means. The more carefully the decks and bulkheads have been fitted in the first instance, the less will be the amount of air which will be required to be supplied by engine or man power in order to keep the water out in the event of serious damage to the ship's bottom.

Dealing with leaks in this way is equivalent to transferring the leak from the ship's bottom to her deck, and dealing with it there in the shape of an escape of air in place of an entrance of water.

In order to make successful use of this method it is necessary that the ship's deck and bulkheads should be not only air-tight, but also sufficiently strong to resist a pressure which, in the case of even the largest ships, would not exceed one atmosphere, or 15 pounds per square inch. Each compartment would have to be about as strong as an old low-pressure marine boiler.

Modern men-of-war are built in such a way that they require nothing but the air-tight hatches, and air-forcing pumps to make them quite secure against the most extensive damage to their bottoms. Indeed, as regards the stoke-holds, they are already fitted with the air-tight hatches in order to be able to use forced draught for the furnaces. Modern merchant ships are built with an iron deck, so that there is no difficulty about providing the strength. Their hatchways are, however, always very large; but, on the other hand, there is little traffic through them, so that they could be treated in a more substantial way than the smaller hatchways of a man-of-war with her large complement of men. The bulkheads which subdivide the hold into compartments always profess to be water-tight, and to be able to resist the pressure exercised by the water filling the compartment. There should therefore be no difficulty about them. Indeed, if ships were built to withstand air pressure, a very simple method would be provided for testing the efficiency of the bulkheads without the disagreeable process of filling the compartment with water. It would be only necessary to close the legitimate openings and get the air in it up to a pressure equal to that of the ship's draught of water, and the result would be unequivocal. It is proper to observe that the construction of an air-tight bulkhead would differ slightly from that of a water-tight bulkhead, inasmuch as it will be exposed to the maximum pressure over its whole surface, whereas the water-tight bulkhead is exposed to a graduated pressure, being greatest at the keelson, and least under the deck.

A further advantage of fitting a ship with air-tight subdivisions is, that it not only gives her greater security against foundering, but it affords a means of largely insuring her against risks of fire. This has more especial reference to merchant ships. If the contents of a ship's hold catch fire, the easiest way of putting it out is to stop the supply of air, and this can be done if the hold is air-tight.

So far the damage to the ship is supposed to be a rent in the bottom. If it is not in the bottom, but somewhere above it, then the air can only expel the water down to the level of the breach, when the air will begin to escape through its uppermost part. It will now depend on the supply of forced air available, how large a hole can be kept continuously filled by a stream of air rushing out. The area so occupied is necessarily closed to the entrance of water, and if the machinery can supply air at a sufficient rate, the whole rent can be filled by a current of air, which, so long as it is kept up, is as efficient a leak stopper as a plate of iron would be, and meantime the bottom of the hold can be cleared by the ordinary bilge pumps.

Rents in a ship's side, such as are produced when she is run down, or rammed by another, are usually so extensive and serious that, unless the ship is protected by an inner skin, immediate destruction ensues before there is time to take any measures for rescuing her. But with an inner skin the damage may be so far reduced as to make it possible to deal with it as above indicated. The higher up on the ship's side is the damage the less suitable is the pneumatic method for dealing with it, if it is of a really extensive character; but, on the other hand, the more easy is it (given the time) to get at it, and deal with it from the outside. In all cases where the ship has been damaged by touching the ground, or by torpedo explosion under the bottom, and not involving the destruction of the ship, the pneumatic method affords the readiest means of combating the results.

It must be remembered that a ship's hold when filled with compressed air will be *habitable*; that is, if an air-lock is provided, men can descend into it and repair the damage, just as they can descend into a caisson and dig out the foundations for the pier of a bridge.

The pneumatic method is, however, not only adapted for keeping damaged vessels afloat, it is also useful for raising sunken or stranded ships. For this purpose the salvage steamer must be provided with air-forcing pumps as well as the suction pumps which she usually carries. Having closed, and if necessary strengthened the deck, by means of divers if below water, she then pumps air into the holds of the ship, and at once restores a large proportion of her original buoyancy to her. If she does not rise, the other methods of salvage can be applied in addition, and with much increased chance of success.

The principle of this method is not new. A very old device in endeavouring to float, or to keep afloat, ships, is to fill as much of their damaged hold as possible with empty casks. A later modification of this method is to use inflatable india-rubber bags. It may be remembered that after the "Vanguard" sank Admiral Popoff, of the Russian Navy, sent a large apparatus of this kind in order to render assistance in trying to float her. Both these appliances are cumbersome. A ship's hold is seldom quite empty when she sinks, and even if it were, it is not easy to fill it under water with casks full of air, or even with inflatable air bags; and in any case it is difficult in this way to fill more than a fraction of the hold with air. The simple and efficient way of dealing with the matter is to treat the ship's hold itself as the vessel to be filled with air.

Compressed air is every day occupying a wider field as a

means of transmitting power. It is already used as a substitute for gunpowder in the guns for firing shells with high explosives. It seems to me that if it can be used for largely increasing the safety of life and property at sea it is right that the fact should be brought as prominently forward as possible, in the hope that it may receive practical application in the hands of the shipbuilder and the engineer.

No. 30. [*From The Engineer*, 1894, *Vol.* LXXVIII. *p.* 382.]

THE NORTHALLERTON ACCIDENT, 1894

I HAVE read the leading article on "Locomotives and Trains," in *The Engineer* of 26th October, with great interest. I was a passenger by the train that met with the accident, and being unhurt, I was able to make some observations on the state of the train, which very much surprised me, and have continued to furnish abundant matter for reflection. Like the other correspondents to whom you allude in your article, I possess no special acquaintance with the conduct of railway traffic. But I think the observations which I was able to make, and which in their general lines were published in a letter to *The Times* of the 5th October, are worth stating with a little more precision in a paper such as yours, where they will come under the notice of professional railway engineers, who may possibly be able to make some use of them.

I was in the third car from the rear. The accident was due to the express running into some trucks that were being shunted. I heard the danger signal blown by the goods driver, and noticed that the brakes were put down instantly. After probably not more than five or six seconds' violent jolting the train stopped. On looking out of the window nothing seemed to be amiss with the train, and I expected every moment that it would start again. I only learned the truth when the guard came along with his head covered with blood, and reported that the whole of the front part of the train was demolished. I then got out to see what had happened, and to render any assistance that I could. Fortunately and miraculously all the passengers were—like myself—unhurt, and the wounded drivers and stokers were being attended to by several medical gentlemen. Bonfires were soon made out of the *débris* of the wagons, and it was possible to see the damage which had been done.

The condition of the train was roughly as follows: The two engines were wrecked, the pilot one being thrown over on to the down line, and partially down the embankment. The guard's van next the engine and two third-class carriages behind it were demolished, the two third-class carriages being squeezed out to the left by the Pullman car, which came immediately behind. The body of the Pullman had been wrenched separate from the bogie truck frame which carried it, and slid along it, driving the third-class carriages out sideways, and penetrating into what remained of the guard's van. Behind the Pullman was an ordinary sleeping car which also had slipped forwards on its bearings, but was otherwise undamaged. The carriage behind this was very little hurt, and the remainder—some seven or eight carriages—were standing on the line, just as if they had been stopped in the ordinary way, and neither they nor their occupants had suffered any damage at all. Shortly, the engines and the great part of the train up to the Pullman car were destroyed. One or two carriages in the middle of the train suffered slightly, and the rear half of the train, consisting of at least seven carriages, was unhurt. The train was running at a high speed when the danger signal was given by the goods engine driver, and the brakes were put down instantly. Had there been no continuous brake, there would have been no sensible reduction of speed before the collision, and the carriages would have been piled one on top of another, and the loss of life would have been appalling. As it was, none of the passengers were hurt, but the guard in the front van got an ugly knock on the head, and the four men on the engines were seriously hurt. Could these men and their engines not have been saved? Looking at the train as it stood after the collision, it is evident that if, at the moment of putting down the brakes, the forward part of the train had been made to disappear, the rear part of the train would have stopped under the influence of its own brakes before reaching the mineral train which was on the line, and no collision would have taken place. If, on the other hand, at the moment of putting down the brakes the rear half of the train had been removed, the forward portion of the train would have run with greater violence into the mineral train, because it would not have had the drag of the after portion, which was efficiently braked. Hence the conclusion is forced, that if the forward half of the train had been as efficiently braked as the after portion no collision would have taken place.

An ideal distribution of brake-power would be such that, on shutting off steam and applying the brakes, the train as a whole should come to a standstill in the shortest possible distance, and each individual vehicle making up the train should have its speed reduced at the same rate as every other one, so that there should be neither pull nor push between neighbouring vehicles. The brake distribution in the ordinary carriages of the Scotch express seems to approach this standard sufficiently for practical purposes. It is when we consider the engines that the great defect is apparent. The tender is always fitted with powerful brakes on its wheels; but so are the ordinary coaches, which do not weigh half as much as the loaded tender. Formerly, the engine was entirely wanting in brakes, but gradually they have come to be fitted with them, but the friction surface supplied is quite insignificant when their great mass is considered. We may take the engine of such an express to weigh at least four times one of the ordinary coaches, yet the brake-blocks applied to its wheels are seldom of greater surface than those of the ordinary coaches. The leading wheels are never supplied with brakes at all.

As the velocity of every portion of the train is the same, the brake friction or retarding power ought to be proportional to the weight of each vehicle. If the pressure per unit of area of rubbing surface is the same in each vehicle, then the area of rubbing surface ought to be in proportion to the weight of the vehicle. This proportionality does not exist in any of our railways, when the ordinary carriages are compared with the locomotives. To the ordinary lay observer there would seem to be no insuperable difficulty in fitting an engine which has four 7 ft. driving wheels and four bogie wheels in front, with four times the brake surface which can be applied to an ordinary six-wheeled carriage. It would be very interesting to know, from the professional side, how it comes that this great difference between the power which the driver has of stopping the train of carriages and that of stopping his engine has been allowed to continue. Had there been no such difference, the East Coast

Railway would to-day be the richer by three or four carriages, two express engines, and the services of four experienced men.

Postscript.—The professional disclaimer on the part of the Railway appeared in due course, and I fully expected that things would just go on as before. Still, after the accident, I never travelled by rail without paying particular attention to the engine; and very soon I perceived signs of movement which increased my interest. The outward sign of this movement was the increase of the existing insignificant brake surface applicable to the wheels of the engine itself. The movement was general, but its effect on existing engines was necessarily incomplete. when sufficient time had elapsed for engines, constructed after the date of the accident, to appear on the road, I had the satisfaction of perceiving that earnest and effective measures were being taken to fit them with adequate brake surface to enable them to utilise the power developed on board of them for the control of their own speed as well as that of the load which they hauled.

After the accident and on arrival in London by the relief train, I ascertained that the luggage of passengers by the wrecked train would come forward by the train due at 2 P.M. Accordingly I returned to King's Cross Station at that hour and, when the train drew up, the porter took my luggage out of the baggage compartment of the coach in which I had deposited it at Edinburgh. In fact, the greater part of this train consisted of coaches which had stood their ground in the accident during the night.

A further, and more personal, incident left an impression on me. When the relief train arrived at the scene of the disaster, and all the passengers had got into it, one or more medical men visited the whole train and interrogated each passenger if he or she had been hurt. As every passenger was at that moment dominated exclusively by the sense of satisfaction, if not of thankfulness, for escape from imminent death, the answer which the doctors received was unanimous that there was nothing wrong about anybody—and the Railway Company stood scatheless.

No. 31. [*From The Times, Feb.* 21, 1902.]

THE WRECK OF "SANTOS DUMONT NO. 6" AT MONACO, 14 FEBRUARY 1902

M. Santos Dumont's ascent and accident of Friday, the 14th inst., have an interest beyond the personal one, which, of itself, is of a high order. Not for the first nor for the second time has he shown that he is a man of rare presence of mind, as well as of indomitable courage, and to tempt the air in travel requires both these qualities in a high degree. It may be said that each ascent of M. Santos Dumont marks a step in advance in the dawning history of aerial locomotion, and is fraught with lessons to the attentive onlooker. To M. Dumont himself each fresh ascent, whether the public term it a success or a failure, is full of lessons on a quantity of matters of detail of which the uninitiated can have no perception. Indeed, the more complete the apparent failure the greater is the value of the experience to the air pilot, provided he escape so as to be able to utilise the experience himself. The apparent failure but real achievement of last Friday did not occupy more than five minutes from start to total wreck. By the luck which always accompanies courage, M. Santos Dumont has got off this time with a wetting, and he may be trusted to give effect to his experience without any unnecessary delay. But while he alone, by previous preparation, is able to learn fully the lesson that the accident has taught, the spectator, by intelligent observation, may pick up a few of the crumbs of learning.

Owing to the kindness of his Highness the Prince of Monaco, I was able to accompany the steam launch of his yacht "Princesse Alice" both on the day of the accident and on that of the previous ascent. I was thus able to observe and to photograph the balloon at pretty close quarters both in success and in failure, and I think that the observations which I have been able to make may be of interest to the readers of *The Times*, all the more so as the accounts which have been published up to the present date are in many particulars inexact. At the same time, complete accuracy is not claimed for this account. It is in the main a log corrected by photography.

Since I have been here M. Dumont has taken his balloon out three times. The first time, on the 10th inst., he took his balloon a little way outside the harbour and returned. I witnessed it from the Rock of Monaco, which has a greater altitude than that which the balloon usually attains. The first impression was one of astonishment at the ease with which it was got out of the *hangar* and started, and at the speed which it attained almost immediately. But the most striking and at the same time unfavourable feature was the heavy pitching of the balloon, which at times attained an amplitude of not far from 45 deg. on each side of the vertical. The period of the pitching was like that of the rolling of a large ship—that is, from five to six complete excursions in the minute. The air was not calm. Outside there was a smart breeze blowing. In the harbour it took the form of a back eddy in shore. In going out the balloon had this rather uncertain wind against it, but it went through it without apparent effort at a speed which entirely outstripped the "Princesse Alice's" steam launch, which is a new and very fast boat. It was estimated, and I think the estimate was within the mark, that the balloon was frequently going against this breeze at the rate of 15 knots over the ground. On the journey home the balloon had the back eddy as a fair wind, and M. Dumont's work in taking it in and berthing it was comparable with that of a pilot taking a steamer down a river with a strong current, turning her round at a given point, and taking her stern first into a dock the entrance to which is little greater than the beam of his ship. This very difficult operation was accomplished with complete success.

On the afternoon of the 11th M. Santos Dumont took his balloon out again. He intimated his intention of proceeding towards Cap Martin, and the Prince sent two of his boats to take up positions on the line from Monaco to Cap Martin, while he himself took the steam launch to follow and, if necessary, assist the balloon. The excursion was quite successful, the balloon proceeding very rapidly towards Cap Martin; it turned, however, at about half-way and came back, being brought to land without accident. The balloon pitched much less than on the previous occasion.

On Friday, the 14th, at half-past one, M. Santos Dumont telephoned to the Palace that he was going out at a quarter to two. The Prince had already put all the resources of his yacht, and especially her boats, at M. Dumont's service for his assistance and in case of accident. On each occasion when the balloon went out the steam launch and two of the other boats of the "Princesse Alice" had attended it. While the yacht is lying at Monaco the steam launch is not in daily use, and, although a new boat of the latest type, something more than a quarter of an hour is required for her to get steam. That this did not naturally occur to M. Santos Dumont is his good fortune, not his fault. He is young enough to have made his first really intimate personal acquaintance with *prime movers* in the form of the modern *moteur à pétrole* of high velocity, which starts from cold by turning a handle. We do not at once realize what this new departure means.

Orders were telephoned to the yacht to get up steam in the launch with all speed, and to send one of the whale-boats to take up a position outside the harbour. Owing to engagements the Prince was not able to follow the manœuvres himself. Captain Carr took charge of the launch, and I was able to accompany him.

Very fortunately, as it turned out, the balloon did not appear outside the *hangar* or shed (Plate I, Fig. 1) until half-past 2. By this time the launch had plenty of steam, and she proceeded at once in the direction of the probable track of the balloon. After a short delay the balloon started (Fig. 2) and proceeded rapidly towards the entrance of the bay, pitching heavily. Arrived abreast of the pigeon-shooting ground, the pitching became more violent and the balloon rose, taking the guide-rope, which usually trails on the surface of the water, entirely out of the water and to a height of 50 yards or more above it.

Plate I

Fig. 1. The Balloon leaving the hangar.

Fig. 2. The Balloon passing overhead, the guide-rope trailing in the sea.

The situation was now evidently becoming critical. In pitching the balloon came to be standing (Plate II, Fig. 3) very nearly vertically, first on the one end and then on the other. The after and, at the time, lower end was now seen to shrink; the valve had been opened and gas was escaping. The mechanical effect of the loss of gas was the diminution of stiffness; the balloon buckled upwards (Fig. 4) about one-third of its length from the rear end. The rudder is stepped into a socket outside the propeller. Its upper end is held by a spar close to the body of the balloon and secured to the netting. The moment that this part of the balloon lost its stiffness and floated upwards it took the rudder out of its lower support, and it was thereby rendered useless. The balloon was now unmanageable and was falling. The launch was steamed full speed to catch the guide-rope and to render assistance as might be required. The situation was one of great anxiety, and for a minute or so it was impossible to give attention to anything except the jeopardy of the aeronaut. The guide-rope was got into the launch before the balloon reached the water, and a course was made as quickly as possible for the landing-place. M. Santos Dumont's balloon carries a light girder made of wooden spars, laced by wire, and it is suspended by a network of wire. This is, properly speaking, the car of the balloon, and it carries everything—the aeronaut in his basket, the motor with propeller and shaft, the ballast, etc. Its length is three-fifths of that of the balloon, and it hangs between four and five yards below it. The length of the balloon is 34 metres, or about 37 yards. In falling it was the after part of the girder with the propeller which touched the water first. It is important to observe that a photograph taken at this moment (Plate III, Fig. 5) shows that deflation had not yet assumed considerable proportions, and that the propeller was uninjured, but the rudder was hanging from the netting. The launch now steamed towards the landing stage. The wind, a light eddy, was blowing towards the Casino. The combined effect of the strain of the tow rope, the pressure of the wind, and the still considerable lifting power of the balloon was to make it rise again as a combined kite and balloon (Fig. 6), and in the launch it was hoped that this condition of things would last long enough to get the balloon ashore. But deflation was proceeding too rapidly, and the second ascent was short-lived. Had M. Dumont been able to stop deflation at this time the balloon could certainly have been got ashore without damage. That he was not able to do so, or, at any rate, did not do so, lends probability to the statement that he had "pulled the emergency cord" and that the escape of gas was beyond his control. It is interesting to note that M. Richard, the director of the Musée Océanographique, who was on board the "Princesse Alice" at this moment, observed the escaping hydrogen rising in the denser air, exactly like the hot air from a chimney. The balloon sank till the girder reached the water. The condition of the balloon as regards inflation had now changed. At the time of its second ascent the after end was deflated. During the short time that it was being towed as a kite and was in a horizontal position the gas seemed to be equally distributed through the balloon. When it sank for the last time to the water the foremost end was hanging empty while the after end was inflated. A photograph (Frontispiece) at this time shows the whole of the girder flush with the water. If the escape of gas could have been arrested the balloon could still have been towed to land. By this time Mr Wedderburn, the Prince's whaling master, had pulled up and put his boat alongside the sinking girder. There was still plenty of gas in the balloon to keep it clear of the wreckage in the water, and the situation ceased to cause anxiety. M. Santos Dumont could be rescued at any moment.

During all this trying time M. Santos Dumont's presence of mind and coolness provoked the admiration of every one who was able to witness it. It was not till the water had reached his chin that he let Wedderburn haul him into his whaler.

The operation now became one of ordinary salvage. The launch tugged away at the tow-rope, but it was soon apparent that the girder had anchored itself as the water shoaled. The bow end to which the tow-rope was secured was above water; the stern end, with wrecked propeller and rudder enveloped in an inextricable entanglement of kinked pianoforte wire, had hooked on the bottom. One of M. Dumont's artificers now arrived, and he detached the balloon and took it ashore. The

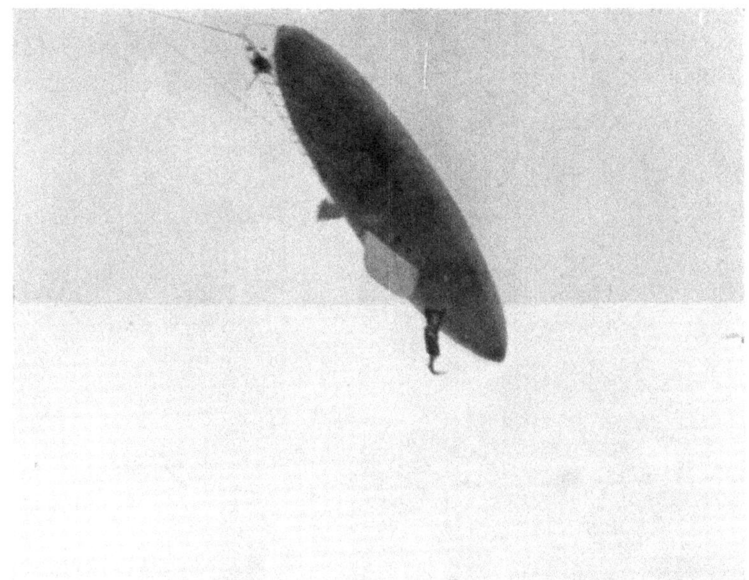

Fig. 3. The Balloon pitching: critical situation.

Fig. 4. The Balloon buckling on the escape of gas.

Fig. 5. The after-end of the girder being immersed, the Balloon is in equilibrium in the air, while the guide-rope is being got into the launch.

Fig. 6. The Balloon being towed as a kite into the bay.

girder, in the efforts made to raise it, had broken. The upper part was detached and taken ashore. The submerged portion, which contained the motor, was buoyed to a boat by one of the loose iron wires. Grappling gear was got from the "Princesse Alice," and the sunk portion was recovered and delivered on shore by 5 o'clock. Thus ended the first wreck of an airship at sea.

The ideas which suggest themselves on witnessing excursions such as those of M. Santos Dumont are manifold. I may be permitted, before concluding, to refer to two. Taking the balloon or airship, when it was in a perfectly airworthy condition and under way, the unfavourable feature was the pitching. It is not possible to say what part the local circumstances may have played in producing or in starting it, but the liability to excessive pitching is there, and it played a very important part in producing the accident. It naturally suggests itself to apply the same remedy to it which is found effectual in reducing the rolling of ships—namely, the application of bilge pieces or keels. In the case of the balloon, these would take the form of horizontal aeroplanes, which would be useful for giving direction as well as for producing steadiness. On the first day's excursion, when the balloon, being at a distance and rather below the level of the eye, was in a position of truer perspective than on the other days, it was impossible not to picture to oneself the aeroplane in conjunction with the motor developing itself more and more, while the balloon successively took smaller and smaller dimensions, until it finally disappeared and left the true flying machine.

While the pitching was, in my opinion, the principal cause which produced the accident, the accident would not have become a disaster if the balloon had behaved as a balloon ought to behave. In an ordinary balloon the aeronaut rises by discharging ballast and sinks by discharging gas. In M. Santos Dumont's balloon the case is the same as regards rising by discharge of ballast, but it is very different as regards sinking. The ordinary balloon when it loses gas shrinks, but it preserves its geometrical form, the sphere, the natural form of equilibrium of an air-bell. M. Santos Dumont's balloon is an ellipsoid of revolution, having six diameters in its length. It is stable only when blown up tight. When it loses gas the tension of the envelope disappears, and it is only the tension of the envelope which forces the enclosed hydrogen to take the unnatural and therefore unstable form of an ellipsoid with its minor axis vertical. So soon as coercion is relaxed, nature resumes her rights, and the gas strives to shape its envelope into one or more spherical shells. No conviction is more firmly rooted in my mind by what I saw on the 14th than that all forms of balloon are false except the sphere.

Postscript.—The importance of M. Santos Dumont's unsuccessful flight, described in the article, was even greater than I estimated. The accidents during flight showed the imperfections of the balloon; the details of the wreck, as they developed from minute to minute, showed the remedy to be applied. It was demonstrated before the representative crowd who witnessed the experiment that the only stable form which can be given to a body of gas is the spherical. The conviction took root, and from that date the dirigible balloon, though retaining externally the "cigar-shape," was composed of spherical units. It is therefore unquestionable that the practical usefulness of the dirigible balloon dates from the *wreck of "Santos Dumont No.* 6" on 14th February, 1902, just as Flying, properly so-called, dates, as an achievement, from M. Blériot's successful flight across the channel on 25th July, 1909.

As a coincidence, let it be recorded that, while the historic flight and fall of M. Santos Dumont's balloon was taking place in the bay, a function of the highest artistic importance was being celebrated on the Mount, namely, the *Répétition Générale* of the first representation of Massenet's *Jougleur de Notre Dame*, under the direction of the Master and in presence of His Highness the Prince.

No. 32. [*From Nature, April* 6, 1911, *Vol.* LXXXVI. *p.* 177.]

THE DAINTINESS OF THE RAT

ONE of the principal sights in Bordighera is the garden of the Villa "Charles Garnier." It is called after its late owner, the famous designer of the Grand Opera at Paris and the Casino at Monte Carlo. It was probably when he was engaged in building the Casino that he acquired the large and valuable property which bears his name.

The garden covers more than five acres, and is a mass of tropical and subtropical vegetation. The banana trees bear fruit which ripens every year. The gardener, a very civil and intelligent man, speaks English, and he told me he had worked for more than a year in Kew Gardens, which he much admires, and endeavours to imitate when his means permit. He was especially proud of his ticketing the plants after the Kew model.

Not being a botanist, I was unable to appreciate adequately his attention in showing me the treasures of his garden, and it was perhaps owing to his perceiving this that he directed my attention to an allied circumstance which interested me much.

He informed me that the greatest trouble which he had was with rats, the *forest rats* he called them. They spend the winter in the garden, and, while causing a good deal of general damage, they evince a remarkable selectiveness in their taste for fruit. It appears that they have a great preference for the lemon and the mandarin, which are abundant in the garden. A number of the trees carried zinc guards on their trunks to prevent the ascent of the rats; but in a terraced garden on a steep slope there are many which cannot be protected in this way, because a rat easily jumps from a higher terrace on to the upper branches of a tree growing on the terrace below. By this means they had boarded and plundered quite a number of trees. They touched only the lemon and the mandarin trees, disregarding the common orange. But what seemed to be most remarkable was the different way in which they treated the früits of these two trees. Of the lemon they eat the rind, removing it completely, and leaving the peeled fruit, clean and without a blemish, still attached to the branch which carried it. On one tree there were eight or ten such freshly peeled lemons still in their places on the tree, and they presented, among the others, a very curious aspect of nakedness. Having boarded a mandarin tree, the rat treats the fruit in the opposite way; he eats the inside and leaves the empty skins hanging on the tree. On one tree that I saw, nearly the whole of the fruit had been treated in this way. Something similar may be witnessed with us on gooseberry bushes in a summer when wasps are abundant. The reason for the different treatment of the two fruits is probably not to be sought further than in the fact that the inside of the mandarin is sweet and that of the lemon sour.

So long as there are mandarins and lemons, the common orange remains untouched, but when there are no more of these two they eat the common orange. By the time these are finished the fields and woods outside are beginning to furnish food, and the rats leave the garden, not to return until the winter begins again.

In answer to my inquiry, the gardener said the rats never attempt to enter the villa; they are forest rats. I asked him if they were a special kind; he said they were brown rats; and I asked him if they were different from the rats he had seen in England, and he said no.

The daintiness of the rat, shown not only in the choice of his fruit, but also in the part of it which he will eat, is not the only feature of rat life which is illuminated by the experience of the gardener of the Villa "Charles Garnier." The annual migration back and forward from the open and natural surroundings of the field and forest, where in summer food is being naturally produced in abundance, to the restricted environment of the highly cultivated garden, where in winter food is produced only by artificial devices, becomes more remarkable the longer it is contemplated. The whole area of semi-tropical garden on the Riviera is an insignificant quantity compared with that of the open ground, so that the proportion of the rat population which is able to enjoy the winter *villegiatura* must be very small, and must be chosen or evolved by a rigorous system of selection, which probably rests on the fundamental principle, the right of might.

In a fertile country like that of Liguria the rats, which are obliged to remain *fore le muri*, are no doubt able to pick up a subsistence during the winter, but they cannot afford to be so dainty as those that are able or privileged to occupy the gardens. In any case, I suppose, it may be taken to be true that a hungry rat will not hesitate to eat a healthy brother rat if he can waylay him or overcome him in combat. It is not improbable that this is the natural winter food of many tribes of rats which inhabit countries where food has its seasons of plenty and scarcity. The shortage thus produced in the winter is quickly made up by the splendid fruitfulness of the mother rats when the food season returns, and the population, over the year, need show no diminution; indeed, there is nothing to prevent it showing an increase. In nature there are accumulators of all kinds.

We have seen, on the evidence of the fruit trees of the garden, that the rats occupying it must live in that state of luxury in which the sensation of real hunger is not felt. How do they keep such a garden of Eden to themselves?

That the common oranges remain as a reserve to the end of the season shows that overcrowding is effectively prevented. We have seen that the lemon and the mandarin are preferred by the rats actually occupying the garden, and apparently indifferently, because the two fruits are consumed *pari passu*. As it is contrary to the animal nature for the strong to give way to the weak, we may feel certain that there is no relative aversion to either the lemon or the mandarin as there is to the common orange, or one of them would be consumed before the other.

All these facts go to show that the occupying force must be a very well-organised body, and must be directed by that degree of intelligence which teaches it, not only to drive and keep out strangers, but also rigorously to keep down its own numbers to the point at which it can, on the basis of experience, expect to pass the winter without being reduced to the necessity of eating common oranges.

No. 33. [*From Nature, November* 23 *and* 30, 1911.]

FISH AND DROUGHT

THE summer of the year 1911 will long be remembered for its excessive heat and dryness. These were especially trying to the inhabitants of streams and shallow lakes or ponds. I had the opportunity of studying a remarkable instance of this, which I think is worth recording.

The Château of Marchais, with its magnificent domain, the property of the Prince of Monaco, lies about 16 kilometres east of Laon, in the department of Aisne, and is well known as one of the best shooting estates in France. The sketch (Fig. 1) represents the park. It occupies a rectangle surrounded by a ditch or moat, *A*, *B*, *C*, *D*, consisting of four canals, each 1250 metres long and 16 metres wide, and carrying usually a depth of 1½ metres of water. These canals form a continuous sheet of water, five kilometres long, and there is a bridge, *a*, *b*, *c*, *d*, over each of them. The country, though well-wooded, is flat and peaty, and the level of the water in the ditch is that of the water in the ground all round it. Like the ground-water, it is subject to rise and fall according to the wetness or dryness of the season.

When I arrived on the morning of September 29, I observed that the ditch was quite dry, with the exception of the small tank or enclosure (*f*) for ducks at the lodge known as the Porte Rouge, where entry to the park is obtained over the bridge marked (*b*). Yet the water of the ditch is always full of fish, principally carp, tench, perch, and pike. Now there was nothing but dry mud. *With the water the fish had entirely disappeared, and without leaving a single dead one to mark where they had before abounded.*

On the evening of September 29 a violent storm of wind and rain broke, and it raged over the whole of northern Europe until October 1. I was curious to see the effect which this first important rain, which marked the breaking of the drought, would have on the ditch.

It is right to say that the full significance of the dryness of the ditch and the absence of dead fish had not sufficiently impressed me. I only felt that I was witnessing a remarkable

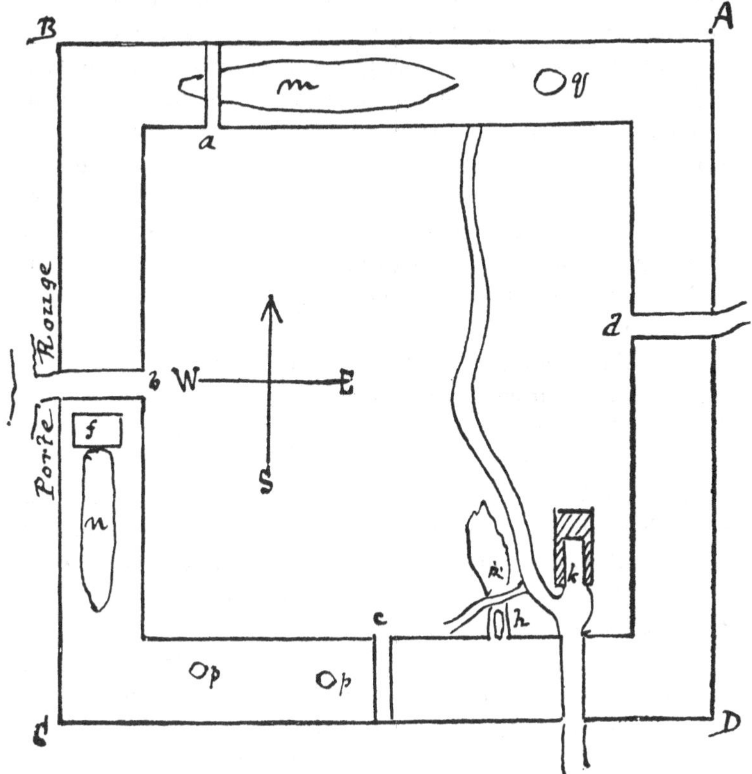

Fig. 1. Sketch Plan of the Park of Marchais.

occurrence in nature and it excited my curiosity, but at first this curiosity went very little beyond considering how long it would take for the water to get back. With only this in my mind, I went round the park on the afternoon of September 30, when the weather had moderated a little, and I found that pools of water had begun to collect in places in the ditch, but

I did not examine them, and I arrived at the conclusion that a little steady wet weather would soon fill the ditch up again.

On October 1 the weather was still very bad, but between the showers I took a walk along the margin of the ornamental water (*h*) on the west side of the château (*k*), which is connected with the south canal of the moat. A pool of water had collected here, and there was quite a quantity of small fish, not more than 10 or 12 centimetres in length, swimming about in the water, which did little more than cover its muddy bottom to a depth of at the most four or five centimetres, in which these small fish were able to swim. I noticed that the water was turbid and that the mud was everywhere being stirred up by the fish. They were darting hither and thither, being disturbed by my presence on the bank, and, whenever they altered their course, they contrived to raise a dense cloud of mud, in which they were able for a short time to conceal themselves from view.

Two days before these fish were invisible, and now they had reappeared in an isolated shallow pool, which also had no existence two days before. *It was evident that all these fish had been covered by the dry mud, and must have released themselves the moment they thought there was enough water for them to swim in.* There was not by any means too much water for the crowd that was moving about in it. I was fascinated by what I saw, especially as it seemed to be in every way likely that the process of release was still going on. But the release of a buried fish would be sure to be accompanied by a cloud of mud, which could not be easily distinguished from that produced by a fish already in freedom and swimming about. Still, considering the shallowness of the water and the very favourable position for following everything that went on in it which I occupied on the bank, I was convinced that there would be some noticeable difference between the two classes of cloud, and I was not mistaken. *After waiting and watching for some time, I saw a mud-cloud rise in the very shallow water, bringing a fish with it to the surface belly upwards.* It lost no time in righting itself and swimming away with the others. A living fish can adopt this attitude only when it has not got full control over its movements, and this is pretty sure to be the case at the moment when it releases itself from burial. By waiting a little longer I witnessed two or three repetitions of this remarkable act.

In a pool such as that which I had under observation, in which the water was not more than two or three centimetres deep, the liberated fish reaches the surface almost as soon as it quits the bottom. In any case it is highly probable that the fish would arrive at the surface before it had fully seized the situation, and the nervous impulse had arrived at the muscles and started them in their righting and locomotive activity.

In water a very little deeper it is probable that the fish would be able to right itself before floating clear of the cloud of mud produced by its struggles in the act of self-exhumation.

In the afternoon I went round the park, and found an extensive pool (*m*) which had collected in the north canal and occupied its western half. There were great shoals of fish, principally perch, darting about, and, in their alarm at being surprised in unusually shallow water, stirring up clouds of mud everywhere. Amongst them was one large pike, quite 40 centimetres in length. The water shoaled off to nothing at both ends of the pool, and in the middle it was perhaps 30 centimetres deep. The pool was far too large to be watched like the smaller one, and I was not able to observe any individual release from the mud. However, I noticed that, with the exception of the large pike, the fish were all small; I estimated them to be not more than 10 to 12 centimetres long.

From the western end of this pool, round the north-west corner and along the western canal to the bridge (*b*), at the Porte Rouge, the canal was quite dry. At the Porte Rouge there was the small enclosed tank for ducks (*f*), and beyond it the canal was again dry. The bottom of this canal does not consist of the fine mud which is found in the north canal; it is hard, sandy, and marly, and there was no sign of life in it anywhere. The same class of bottom with absence of life prevailed in the south canal.

On October 2 the weather had improved, although it was

very cold, and I studied the canals both in the morning and the afternoon. In the large pool in the north canal I found the quantity of fish augmented, and I especially noticed that there were many more large fish present than before, and the average size of the fish was decidedly greater than on the previous day; perch of at least 20 centimetres in length were present. The bigger fish had probably started to bury themselves earlier than the smaller ones, and had buried themselves deeper. I saw the big pike of the day before, and two others of the same species, but very much smaller, had appeared. *One dead perch was floating in the pool, the first dead fish that I had encountered.* In the afternoon it seemed to me that the average size of the fish in the pool was still on the increase. The west and south canals presented the same appearance as before.

On October 3 the weather kept fair but cold. I went round the park and found the big pool in the north canal much the same as on October 2, the number of large perch having apparently increased, but there was no new feature of importance. In the course of my tour, after passing the Porte Rouge, I found that the part of the west canal south of this lodge had begun to collect water, which already covered the bottom, forming a pool (*n*), extending to within a short distance of the south-west corner. Under the water, which was perfectly clear, the light grey bottom was still cracked and apparently unsoftened, and there was not a trace of life of any kind. In the south canal the bottom was generally dry, with, however, every here and there pools of water (*p*) measuring about one metre across, and in these also there was no trace of life. I learned that anglers never fish in these canals, because they know that they will catch nothing.

In the afternoon I revisited the north canal, and instead of following the big pool westwards, which promised nothing new, I turned to the right and followed it eastwards towards the north-east corner. I had not gone very far when I encountered a phenomenon of which I had already perceived the possibility, namely, *a premature resurrection resulting in widespread death.* This part of the canal was apparently dry, in the sense that the bottom was exposed to the air, but nevertheless moist enough to be called wet. Owing, probably, to a slight general rise of the ground-water of the neighbourhood, enough water had been able to filter through the mud of the bottom and to rise to the surface and overflow, producing a very shallow pool (*q*), not more than two or three centimetres deep or more than a metre across. The wetting of the mud below by this infiltration must have aroused the sleepers, who then all started to rise at the same time. But when they released themselves from the mud there was not enough water to float them all, and a formidable struggle for existence was going on at the time of my visit, and the quantity of dead and dying fish lying all round the edges of the pool furnished sufficient evidence of its fierceness. The poor fish would no doubt have willingly re-buried themselves and so saved their lives when they perceived their mistake, but the stronger ones, which were in possession of the only part of the pool which could be called liquid, kept shouldering them outwards on to the mud, where they died in the air. When I left the struggle was still going on, and it looked as if the level of the water was falling, so that it is unlikely that many, if any, would be able to retrieve their mistake by burying themselves again. As I left for England the next morning I was not able to continue my observations.

Although the years vary much as regards humidity, and in dry summers the supply of water in the ditch has often fallen to a pretty low ebb, I was informed that the last time that the ditch became quite dry was in the year 1814, almost a hundred years ago; therefore *the experience of the summer of* 1911 *must have been a new one for all the fish in the ditch, yet the general manœuvre of protective self-burial was carried out without a casualty*. In order to accomplish this a fine instinct was necessary to perceive the impending drying up of the canal, and then to commence the operation betimes so as to finish it before desiccation was complete.

It must be remembered that the area of canal having the muddy bottom, which alone is capable of receiving the fish, is very restricted; and from the number of these that came out of it in the short time that I was observing they must have been packed very closely, and in such an orderly way that, with the return of water in sufficient quantity, they were able to take to it again apparently without having suffered at all.

Although the instinct of the fish seems to have sufficed to make them foresee and provide for the dryness, it does not seem in all cases to have been sufficient to enable them to judge correctly the moment for beginning their release.

Of the different species which inhabit the waters of the ditch, the carp and the tench have the habit of burying themselves in the mud every winter; but the perch and the pike have not this habit; both can be caught at any time in winter, even under a covering of ice; yet both the pike and the perch must have buried themselves with quite as much skill as the carp or the tench.

But in a climate like that of this part of France shallow lakes and ponds may suffer shortage of water by congelation as well as by evaporation; and the Prince informed me that he remembered one winter when in many places the water of the ditch was frozen almost, if not quite, to the bottom, and quantities of pike and perch were frozen into the ice. This form of desiccation did not prompt them to seek refuge in the mud.

It is evident that if the summer of 1911 had marked the beginning of a secular period of dryness, such that the canals were not again to be flooded, the fish which took to the mud in that summer would be kept there. They would die and decay *in situ*, and would be perfectly preserved in well-arranged though crowded masses. Eventually, if the change of climate was final, they would form a rich and interesting bed of fossil fishes. But the interest would depend not only on the abundance of fossils in the muddy matrix of one part of the trough-like formation; it would be intensified by their complete absence in the hardened marly matrix of the other part.

Before serious drying took place the ditch, or trough, was covered by a continuous sheet of water in which the fish and other creatures could circulate freely to all parts. So soon, however, as actual desiccation appeared to be imminent, the fish must have concentrated themselves in a body over the districts of muddy bottom in which they knew they could take refuge as a last resort. *When desiccation was complete every fish in the ditch, without a single exception, had succeeded in burying itself in one or other of these restricted areas of mud. Not one of them appears to have made the mistake of seeking refuge in the marly bottom.* When completely dry the ditch, or trough, consisted of two formations, the more extensive consisting of hard sandy marl and destitute of life, the less extensive consisting of soft mud and teeming with aquatic life. Further, the two formations are contiguous as well as contemporaneous, and together they cover an area of not more than eight hectares.

As illustrating the geological significance of the facts just recorded, the following passage may be quoted from Sir Archibald Geikie's *Text-book of Geology* (1903), p. 1003:

"The water basins of the Old Red Sandstone might be supposed to have been, on the whole, singularly devoid of aquatic life, inasmuch as so large a proportion of the red sandy and marly strata is unfossiliferous. In some of the basins, where the sediment is not red and sandy, it is evident that life was extremely abundant, as is shown, for example, by the vast quantities of fossil fishes entombed in the grey bituminous flagstones of Caithness and Orkney. It may be observed that where grey shales occur intercalated among red sandstones and conglomerates they are often full of plant remains, and may contain also ichthyolites and other fossils which are usually absent from the coarser red sediments. There would appear to have been occasions of sudden and widespread destruction of fish life in the waters of the Old Red Sandstone, for platforms occur in which the remains are thickly crowded together, yet so entire that they could not have been transported from a distance, and must have been covered over with silt before they had time to decay and undergo much separation of their plates."

The last sentence of this passage seems to describe the actual condition of the muddy bed of the moat round the Park of Marchais as it would appear to a geologist after the necessary interval of time had elapsed which is required to separate the date of the death of the crowd of fishes which voluntarily entombed itself in the mud before desiccation was complete, and the date at which the stratum of mud and remains so produced would be entitled to rank as a geological formation. I do not know if there are any adequate data for arriving at a trustworthy estimation of the probable length of this interval. It is quite distinct from what is understood as the age of the geological formation.

The barren districts of sandy and marly matter at the bottom of the ditch would, after the lapse of the same, or perhaps a shorter, interval furnish perfectly unfossiliferous strata, which would suggest to the geologist of later date that the water basin in which it had been laid down had been singularly devoid of aquatic life. Yet, in a sense, it would not be inaccurate to say that the water basin in which the muddy strata holding the crowded fish remains had been "laid down" teemed with life, and that the barren strata had, in the same sense, been "laid down" in water devoid of aquatic life, although the two bodies of water formed one continuous sheet of very restricted dimensions. *It would seem, therefore, that a material barrier is not necessary to separate even a small body of water into two basins and to maintain them distinct, the one of which may be full of life and the other practically barren.*

There is an important point, which should not be missed, in the similarity between what took place this summer at Marchais and what may have taken place in Caithness or Orkney in the Old Red Sandstone period. The fishes which buried themselves in such numbers in the mud this summer, though they were fortunately released, were in the strictest sense contemporaries, and were all buried in the mud within a few days of each other. Moreover, in ordinary circumstances, at least in summer, the mud is untenanted. If the fish were to migrate into the barren waters covering the marly bottom, and their return were barred while the water over the mud evaporated and the secular drought set in, this same mud-bed would be met with in later ages as an unfossiliferous stratum. *So that the fossil fishes which are found in these strata must be held to have gone into occupation only when the signal, intelligible to them, was made that complete desiccation was going to take place. Once in a way this desiccation turns out to be secular, and we have a rich bed of fossils.*

In conclusion, I think that the observations above recorded show that the material of geological formations need not necessarily have been "laid down"; it may have been produced *in situ* like the mud in the ditch round the Park of Marchais, and that the enclosure in it of animal remains may have been in some cases due to a voluntary act of self-inhumation, undertaken, perhaps usually, with a view to self-protection. *They also show that two neighbouring strata, the one carrying abundance of life and the other being destitute of it, may nevertheless be contemporaneous in date and conterminous in locality of formation.*

Printed in the United States
By Bookmasters